Polymer Capsules

Polymer Capsules

Edited by

Ye Liu
Xian Jun Loh

JENNY STANFORD
PUBLISHING

Published by

Jenny Stanford Publishing Pte. Ltd.
Level 34, Centennial Tower
3 Temasek Avenue
Singapore 039190

Email: editorial@jennystanford.com
Web: www.jennystanford.com

British Library Cataloguing-in-Publication Data
A catalogue record for this book is available from the British Library.

Polymer Capsules

ISBN 978-981-4774-60-4 (Hardcover)
ISBN 978-0-429-42873-9 (eBook)

Contents

Preface

To ensure sufficient bioavailability and reduce adverse effects in the formulation of pharmaceutics, cosmetic products, foods, and nutrients, it is indispensable to prepare various types of capsules for encapsulation and the safe and efficient delivery of the active ingredients. Suitable capsules can securely protect, deliver to the targeted sites, and release their active ingredients expeditiously to improve their functions and minimize their adverse effects. Various materials are used in the preparation of capsules with active ingredients, and polymers are among the most versatile and important capsule materials. In this book, we describe capsules using polymers, that is, polymer capsules in terms of polymers developed, formulation processes, and representative applications.

Suitable polymers are important in the preparation of qualified polymer capsules. New polymers are constantly being pursued and explored to develop more efficient capsules. As a demonstration, Chapters 1 and 2 describe two types of polymers developed recently for the preparation of such capsules, that is, redox-responsive polymers and polymer-functionalized carbon nanotubes. Formulation processes play a vital role in the making of these capsules. Chapter 3 describes various formulation processes while Chapter 4 focuses on exploring the coacervation of polymers for encapsulation. Such capsules can be explored for different active ingredients in various products, and several typical active ingredients have been described. Chapters 4 to 7 investigate the polymer capsules of drugs, Chapter 8 explores polymer capsules of functional ingredients for food, and Chapter 9 dwells on the encapsulation of essential oils using such capsules. Chapter 10 discusses the polymer capsules of cells and Chapter 11 is about polymer capsules of bacteriophage.

This book can be used to set the design and devise various formulations of polymer capsules with different active ingredients for pharmaceutics, nutrients, food, cosmetics, and other products in the quest for improved and more efficient performance.

Ye Liu
Xian Jun Loh
Spring 2019

Chapter 1

Redox-Responsive Nanocarriers: A Promising Drug Delivery Platform

Weiren Cheng
*Institute of Materials Research and Engineering, Agency for Science, Technology and Research (A*STAR), 2 Fusionopolis Way, Innovis, #08-03, Singapore 138634*
chengwr@imre.a-star.edu.sg

1.1 Introduction

The redox state of a healthy human is essential for many daily cellular functions [1–6], therefore patients suffering from disorders and diseases like cancer, fibrosis, cardiovascular diseases, diabetes, and neurodegenerative diseases tend to have unregulated redox metabolism [1]. Furthermore, the redox potential of intracellular compartments is significantly greater than the extracellular matrix (100-100 times more reductive) [7], and this disparity in redox potential is even more pronounced in a malignant tumor [8]. As a result, the redox environment in the body is highly favorable for redox-responsive delivery systems which are specifically designed to release drugs only intracellularly. Moreover, with proper design,

Polymer Capsules
Edited by Ye Liu and Xian Jun Loh

ISBN 978-981-4774-60-4 (Hardcover), 978-0-429-42873-9 (eBook)
www.jennystanford.com

polymers can undergo complete redox-induced degradation to facilitate body clearance and avoid undesirable accumulation.

The development and study of polymeric nanocarriers in drug delivery became really popular only in the late 1980s and early 1990s. During that time, different research groups around the world formulated and studied a variety of structurally different polymeric nanocarriers suitable for delivering a range of drugs. One such nanocarrier is the polymeric micelles which K. Kataoka and his colleagues have extensively investigated with diblock copolymers. Through their work they have established many significant principles on polymeric micelles that are still adhered to till this day [9–13]. Another equally well-documented nanocarrier is the liposomes and polymersomes. In the 1970s, liposomes had already demonstrated the potential to deliver enzyme or drugs [14] and chelating agents that could remove intracellular plutonium [15]. However, it is only in the late 1990s and early 2000s that polymersomes were first described [16, 17] and used for controlled release of drugs [18]. At the same time, another novel class of polymeric nanocarriers, the nanogels, was reported by Kabanov *et al.* with their promising capability to deliver genetic material [19, 20]. Nanocapsules and nanospheres which are comparatively less reported nanocarriers were initially prepared from poly(ester)s like poly(D,L-lactide) (PLA) [21, 22] and poly(lactic-co-glycolic acid) (PLGA) [23, 24], respectively. In this chapter of the book, a comprehensive review of redox-responsive polymeric nanocarriers, including polymeric micelles, liposomes and polymersomes, nanogels, nanocapsules, and nanospheres will be presented. Their distinct structures, formulation methods, and application in drug delivery will also be highlighted.

1.2 Redox-Responsive Polymeric Micelles

As a result of more than a century of research on micelles, their behaviors and properties were very well-documented and understood. Furthermore, the application of micelles is very extensive, ranging from food science to cleaning agents to consumer care products. Therefore, these nanoparticles were naturally among the most studied drug nanocarriers. Micelles have a distinct core–shell structure, where typically the core is hydrophobic in nature while the shell is hydrophilic. As a result of this configuration, micelles are an excellent choice for solubilization of hydrophobic drugs. With the exception of polyion complex micelles, most polymeric micelles are derived from amphiphilic polymers, with two distinct portions of opposite affinities towards water. Apart from the most frequently used diblock copolymers [13, 25], polymers of other morphologies like multiblock polymers [13, 25], comb polymers [26–29], branched/hyperbranched polymers [27, 30, 31] are all capable of self-assembling into polymeric micelles with core–shell structures. Though applicable to polymers of all types, hydrophobic amphiphilic polymers tend to yield micelles of greater stability and vice versa.

The self-assembly of amphiphilic polymers into micelles can be generally divided into two groups. The first method is direct dissolution and it works only for moderately hydrophobic polymers [32, 33] and charged polymers [34, 35]. On the other hand, the second method is applicable to polymers which are not readily soluble in water. The polymers are first dissolved in an organic solvent and then added to/with water slowly before the organic solvent is removed [26, 27, 31, 36–38]. Similarly, the loading of drugs in polymeric micelles can be grouped into two different classes, post-micellar formulation or during self-assembly. In the former, drugs are added into micelle solution and left to achieve equilibrium. For example, doxorubicin (DOX)-HCl was added to hydrolyzed core-crosslinked micelles solution and then the solution was adjusted to a neutral condition and left to equilibrate [39]. This method of loading is not very favorable as it is limited to only some drugs like DOX-HCl whose water solubility can be easily adjusted by pH. As for loading of drugs during self-assembly, this process involves an organic solvent and is suitable for more hydrophobic drugs. In brief, both the polymers and drugs are first dissolved in the organic solvent and then either dialysis [33, 38] or nanoprecipitation [26, 27, 31, 36, 37] is performed to form the drug-loaded micelles. Emulsion can also be used to load drugs in micelles. For example, an oil-in-water (O/W) emulsion where DOX was dissolved in the dispersed chloroform phase and the amphiphilic polymers acted as emulsifiers stabilizing the emulsion was left standing in the dark overnight in an open atmosphere to allow evaporation of chloroform resulting in DOX-loaded micelles [40]. The loading efficiency of drugs in polymeric micelles depends hugely on the compatibility of the polymers and drugs. Favorable interaction between the polymers and drugs can significantly improve both the loading efficiency and capacity. For example, π-π interaction between DOX and polymers were reported to significantly improve loading efficiency and capacity of micelles [41, 42].

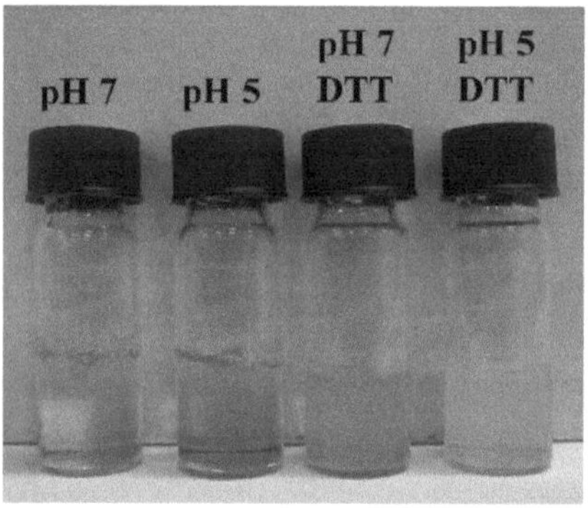

Figure 1.1 Aggregation of micellar hydrophobic cores in the presence of DTT.

To afford redox-responsivity in the polymeric micelles, disulfide bond which is reducible can be integrated in the polymers. These disulfide bonds can be placed in different parts of the polymers in varying concentrations to achieve certain performances. For example, introducing the disulfide bonds between the hydrophobic and hydrophilic portions of the polymers can result in shell-sheddable micelles [33, 36, 37, 43–46]. The shedding of hydrophilic shell improves cellular uptake [47] and facilitates rapid drug release [43] which are both favorable for drug delivery systems. Furthermore, in some polymeric micelles, the shedding of the hydrophilic shells resulted in the aggregation and precipitation of the hydrophobic cores, see Fig. 1.1 [27], thus possibly limiting the efflux of drugs out of the cells. Micelles can only form when the polymer concentration is above the critical micelle concentration (CMC). As a result, micelles tend to face stability issues especially during "sink" conditions. To counter this problem, polymeric micelles were crosslinked using disulfide containing crosslinkers [39, 48–50] or via oxidation of thiols [51, 52] to enhance stability. In an effort to reduce the undesirable consequences of nanocarrier accumulation in tissues and organs, biodegradable polymeric micelles which have disulfide bonds in every mer unit of the polymers to facilitate degradation and clearance are highly valued [26, 27, 30, 31]. The encapsulation of hydrophobic drugs within the micellar core is often very inefficient, resulting in extremely low loading efficiency and drug loading capacity. Thus, to significantly improve the drug loading process, some polymeric micelles have drugs covalently conjugated to the polymers via disulfide bonds [53–57]. Upon exposure to a high reductive environment, the cleavage of disulfide bonds will release the active drugs.

Although there are plenty of studies reporting the *in vitro* redox-responsive drug release profiles of polymeric micelles, it is still extremely vital to demonstrate such sensitivity on the cellular level. Typically, to illustrate this behavior in cells, 3 different approaches are used, each with its pros and cons. In the first approach, non-reducible counterparts of polymers are synthesized [43, 58], for example in these two studies, it was demonstrated in fluorescence microscopy images that micelles from polymers with disulfide bonds could deliver more DOX into the cells as compared to their non-reducible counterparts. However, this approach is not always available as it depends on the feasibility and complexity to synthesize the non-reducible counterparts. The second approach is the pretreatment of cells with glutathione monoester (GSH-oet) which can artificially raise the intracellular GSH level, thus achieving the difference in GSH concentration [30, 59]. However, some do not favor such an approach as the unnatural level of intracellular GSH created may not accurately demonstrate the actual performance and potential of the drug delivery system. The last approach is the pretreatment of cells with buthionine sulfoximine (BSO) which inhibit the synthesis of intracellular GSH, thus lowering the GSH level [31, 35]. The primary limitation of this approach is that the reduction in intracellular GSH could lead to cell death and thus interfere with the experiment conducted. The *in vitro* cytotoxicity performance of reported redox-responsive polymeric micelles varied from system to system but the majority demonstrated a limited and less promising cancerous cells-killing efficiency as compared to free drugs. This is possibly due to the fact that most redox-responsive polymeric micelles were designed to combat the conditions for intravenous delivery and simple incubation with cells does not necessarily highlight the advantages of the systems. Furthermore, encapsulation of drugs within micelles can hinder cell membrane penetration as compared to small molecular weight-free drugs. However, it is worth noting that the shell-sheddable polymeric micelles have frequently demonstrated superior or at least comparable *in vitro* cytotoxicity as compared to free drugs [27, 36, 44]. It is suggested that after redox-induced degradation of the shells, the aggregation and precipitation of hydrophobic cores limits the efflux of the drugs and kills the cancerous cells more effectively [27].

As against the less than ideal *in vitro* performance, some studies have demonstrated promising *in vivo* performance of redox-responsive polymeric micelles. *In vivo* results have shown that encapsulation of drugs like DOX within the micelles can prolong the circulation time [60], and this circulation time can be further lengthened by crosslinking the micelles with disulfide bonds [51, 60]. Upon blood dilution, non-crosslinked micelles suffer destabilization and cause leakage and rapid elimination of drugs, while crosslinked micelles offer more resistance to such destabilization. Together with enhanced permeability and retention effect, long circulating drug-loaded micelles after 24 hours preferentially accumulate in the tumor as compared to other organs like the liver [29, 51, 61, 62]. Furthermore, micelles functionalized with active targeting ligand, dehydroascorbic acid (DHAA), which is specific to GLUT1, an overexpressed glucose transporter on hepatocarcinoma cells, accumulated in atumor in greater concentration than non-targeting counterpart micelles [60]. Interestingly, one of the studies showed that redox-responsive micelles from amphiphilic hyaluronic acid-deoxycholic acid conjugates (HA-ss-DOCA) gathered in a tumor at greater concentration than the insensitive counterpart micelles, HA-DOCA [29]. To improve uptake of the drug-loaded micelles, ligands like hyaluronic acids can be functionalized on the micellar surface to facilitate receptor-mediated endocytosis [29, 63]. As a result of the localized concentration of drugs delivered by the micelles, redox-responsive micelles can effectively suppress tumor growth [60–63] and this performance was even more pronounced with active targeting micelles [60] and crosslinked micelles [51, 60] when compared to their respective counterparts. No significant weight loss in mice treated with redox-responsive drug-loaded micelles was also observed in some studies [62, 63] and in one particular study, the side effects of DOX were limited to the extent that mice treated with redox-responsive active targeting DOX-loaded micelles actually gained weight during the treatment [60].

1.3 Redox-Responsive Liposomes

Liposomes which were first studied by Bangham *et al.* [64], are vesicles with an aqueous core encapsulated by layer(s) of natural or synthetic phospholipid, see Fig. 1.2. Ever since its ability to entrap drugs and enzymes and deliver active ingredients *in vivo* was first demonstrated by Gregoriadis *et al.*, numerous liposomes based drug delivery systems have been reported [14]. Liposomes which have a resemblance to natural cell membranes are divided into multilamellar (MLV) (1–5 μm) which consists of several concentric bilayer membranes around an aqueous core, or unilamellar which is a single bilayer membrane around the aqueous core. Unilamellar liposomes can be further divided into small (SUV, 50–100 nm) or large (LUV, 100–250 nm) [65]. Unlike micelles which are more suitable to load hydrophobic drugs, liposomes can be loaded with both. Hydrophilic drugs are encapsulated within the aqueous core while hydrophobic drugs are entrapped within the lipid membrane. Due to the hydrophobic nature of the lipid bilayer, liposomes have low permeability to hydrophilic drugs but are highly permeable to hydrophobic drugs. Therefore early studies had difficulties in retaining hydrophobic drugs in liposomes. However, the addition of compounds like cholesterol and sphingomyelin transformed the liquid phase lipid bilayer to solid phase, thus reducing leakage [66].

There are numerous reported methods to prepare drug-loaded liposomes. In general, a homogenous lipid solution in organic solvents like chloroform and dichloromethane is first dried completely, sometimes with the help of a vacuum oven. Then the lipid film is hydrated with an aqueous buffer of choice. The resulting liposomes are purified before the final product is collected [67]. The process of loading drugs in liposomes can be divided into either passive or active loading. For passive loading, drugs are encapsulated or entrapped during the formation of liposomes while active loading is done after the liposome is formed. The different techniques of passive loading are summarized in Fig. 1.3 and a comprehensive explanation of each technique can be found in the other review

[67]. Currently there are several successful commercially available liposomal formulations like Ambisome (amphotericin), Doxil® (DOX), DaunoXome (daunorubicin), and Depocyt (cytarabine), and they contain several common compounds like 1,2-distearoyl-sn-glycero-3-phosphocholine (DSPC), 1,2-dioleoyl-sn-glycero-3-phosphocholine (DOPC) and cholesterol. Akin to Doxil which is the first drug-loaded nanocarriers approved by FDA in 1995 [68], many reported liposomes are PEGylated to improve blood circulation time. The polymeric PEG outer layer sterically hinders the interaction and binding of blood components to the liposome surface [69, 70]. There are several ways to modify liposomes with PEG. The first is by physical adsorption of PEG onto the liposome surface. Incorporating PEG-lipid conjugation during liposome preparation is another way of PEGylation. Yet another way is to covalently conjugate PEG onto the reactive groups on the surface of liposomes [69].

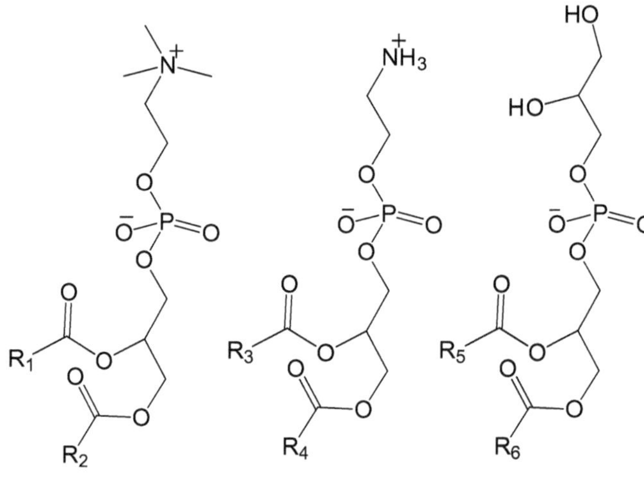

Figure 1.2 Chemical structure of common phospholipid derivatives.

Unlike the previous nanocarrier polymeric micelles which are formulated from a wide variety of precursor polymers, liposomes are made up of a limited selection of phospholipids like DOPC and DSPC which are capable of forming a bilayer structure in an aqueous medium. For example, dioleoylphosphatidylethanolamine (DOPE) which only differs slightly from DOPC (DOPE has primary amine instead of tertiary amine like DOPC), is incapable of turning bilayer in an aqueous medium at neutral pH [71]. Only in the presence of cholesteryl hemisuccinate (CHEMS), can DOPE assemble a bilayer structure [71]. Therefore to afford redox-responsivity in liposomes, the introduction of disulfide linkages is restricted to just two approaches. The first area where disulfide bonds can be integrated is in the linkages between the membrane penetrating ligands/targeting ligands/PEG and the polar head of phospholipids [72–75]. This conjugation via disulfide bonds is typically seen with phosphatidylethanolamines (PE) due to the presence of a reactive primary amine at the polar head, and a common example is PEG-ss-distearoylphosphatidylethanolamine (PEG-ss-DSPE). The other method to yield redox-responsivity in liposomes requires the preparation and introduction of disulfide containing synthetic lipids [76, 77]. These synthetic lipids are integrated in the phospholipid bilayer and upon stimulation the activated-synthetic lipids can disrupt the integrity of the bilayer, causing disintegration of the liposomes.

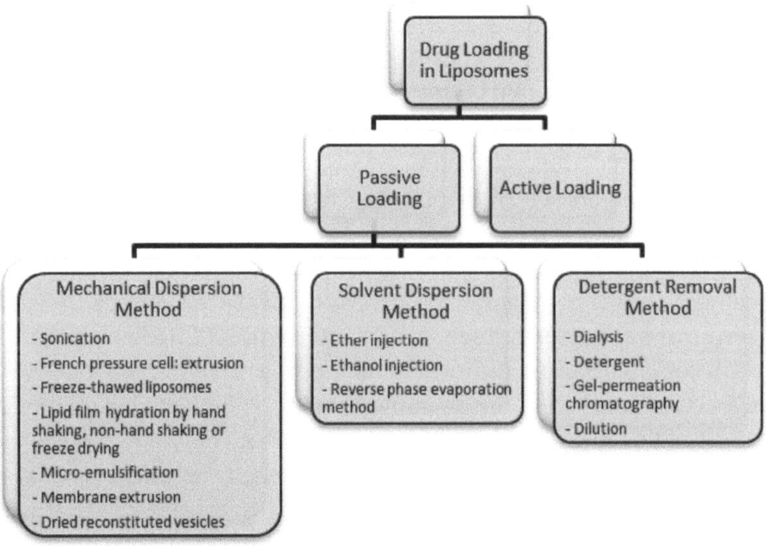

Figure 1.3 Summary of drug loading in liposomes.

As mentioned in the previous paragraph, PEG not only hinders the uptake of liposomes by the reticuloendothelial system (RES) but also the target cells. As a result, liposomes with similar detachable PEG coating to enhance cellular uptake were reported. For these redox-induced detachable PEG, disulfide linkages are incorporated between the PEG chain and the polar head of phospholipid. Upon redox-induced cleavage, the PEG of these liposomes is shed, leaving behind just the typical conventional liposomes for uptake. Studies have repeatedly shown that liposomes with detachable PEG coating have significantly enhanced cellular uptake [72–74]. Furthermore, liposomes functionalized with membrane penetrating ligands like TAT have further improved cellular uptake [72, 74]. Apart from PEG and membrane penetrating ligands, targeting ligands or antibodies can also be functionalized on the liposomal surfaces to improve specificity of the drug delivery systems. However, antibodies are expensive and difficult to handle, therefore a team of scientists reported an active targeting liposomal delivery system which acquires the targeting ligands after intravenous injection [75]. On the liposome surfaces are ortho pyridyl disulfide groups (OPSS) which can form disulfide linkages with exposed sulfhydryl groups on activated C3 in the blood. After incubating the OPSS-liposomes with mouse serum, dot blots analysis showed the strong presence of C3 protein as compared to control liposomes. When the C3 conjugated liposomes were treated with the reducing agent tris(2-carboxyethyl) phosphine (TCEP), the amount of C3 bound significantly reduced. This indicated the association of C3 with liposomes via disulfide linkages. Consequently, the affinity to bind with C3 in serum led to the tremendous improvement in the *in vitro* specificity of liposomes to granulocytic myeloid derived suppressor cells (G-MDSCs). Low cellular uptake of liposomes which were incubated with C3 deficient serum or heat-activated serum or treated with TCEP after incubation with complete serum indicated that binding of OPSS-liposomes with C3 played a key role in the enhancement of the cellular uptake. *In vivo* studies in the 4T1 mammary tumor bearing balb/c mice also revealed similar incidences of high specificity of OPSS-liposomes to G-MDSCs in blood, tumor, spleen, liver, and lungs and reduced uptake by macrophages as compared to controlled liposomes which are predominantly taken up by macrophages.

So far the nanocarriers discussed all depend on disulfide bonds to afford redox-responsivity and endogenous thiols like GSH are the target stimulus to cause this activation. However, non-disulfide based liposomes which are designed to specifically respond to upregulated quinone reductases in numerous cancer types is reported [78]. The reported redox-responsive liposomes were made up of quinone DOPEs (1-Q3) which in the presence of $Na_2S_2O_4$ were reduced to form hydroquinone DOPEs (1-HQ3). Then 1-HQ3 spontaneously underwent cyclization cleavage to cause the dissociation of lactone 3-HQ3 from DOPE. Since DOPE as mentioned earlier cannot adopt a bilayer structure at neutral pH, as the number of 1-Q3 in the liposomes reduced to DOPE rose, the liposomes broke down and released the drugs loaded, see Fig. 1.4.

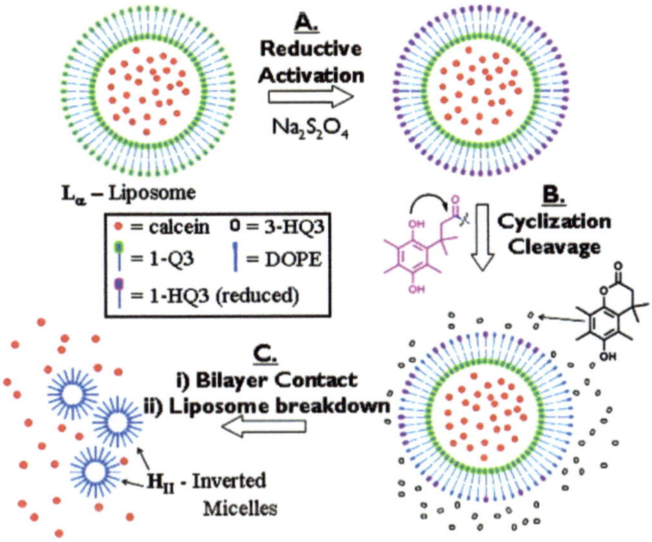

Figure 1.4 Proposed Mechanism for the Conversion of Liposomal **1-Q3** from Lamellar (Lα) Liposomesto Inverted Hexagonal (HII) DOPE Micelles. Step A: $Na_2S_2O_4$ reduces the **1-Q3** lipids thatare located in the outer layer of the liposome to **1-HQ3**. Step B: **1-HQ3** dissociates into lactone **3-HQ3** and DOPE Step C: release of encapsulated calcein dyeoccurs as a result of liposome collapse caused by DOPE layer-DOPE layer contact.

1.4 Redox-Responsive Polymersomes

Akin to liposomes, polymersomes which was first described in the 1990s [16, 17] are nanocarriers with the polymer layer surrounding an aqueous core. However, unlike liposomes, which consist of low molecular weight lipids, polymersomes are made up of amphiphilic macromolecular polymers of various structures like diblock, triblock, graft, and dendritic. A key advantage of polymersomes over liposomes is the highly tunable membrane properties which are closely related to the molecular weight of the polymer. Generally higher molecular weight polymers yield thicker and more mechanically stable membrane, therefore polymersomes are typically more stable and resistant to lysis by classical surfactant than liposomes [79–81]. The formation of polymersomes is mainly determined by the weight fraction of hydrophilic block, molecular weight and effective interaction parameter of hydrophobic block (f_{phil}) with water molecules (χ). Block copolymers with high χ, polymersomes tend to form when f_{PEG} = 20–40%, while at f_{PEG} = 45–55%, cylindrical micelles are favored and at f_{PEG} = 55–70%, spherical micelles dominate [79]. In general, the longer hydrophilic block tends to favor micellar rather than membrane formation [81].

The majority of polymersomes reported were formulated via the self-assembly process, and the simplest method is direct dissolution of the polymers in an aqueous medium, also known as thin film rehydration [17, 82–84]. However, not all polymers can spontaneously self-assemble to form polymersomes upon dissolution. Others require nanoprecipitation [85–89] temperature [90], water-in-oil-in-water (w/o/w) emulsion [91, 92], microfluidic [91, 93] and electroformation [17] to assist in the self-assembly process. An alternative to polymersomes formulation is the layer-by-layer process which involves the coating of sacrificial templates with alternate layers of interacting polymers followed by the removal of the template [80]. This approach can easily control the thickness of the polymer membrane which in turn affects the drug retention, degradation, and drug release profiles. However, the main disadvantage of such a method is the time-consuming sequential polymer coating cycles and purification steps. Like liposomes, hydrophilic and hydrophobic drugs can be encapsulated within the aqueous core and polymer membrane of polymersomes, respectively. Hydrophilic drugs can be loaded either during polymersomes formulation via direct encapsulation [84–86] or post-formulation using pH gradient [86, 89, 94, 95] or w/o/w emulsion [92] while hydrophobic drugs can be entrapped via cooperative incorporation during self-assembly [96], w/o/w emulsion [92] and diffusion [95].

Similar to polymeric micelles, diblock copolymers is one of the most reported polymers used in the formulation of polymersomes and a popular strategy for this category of polymer is to integrate disulfide linkages between the opposite affinity blocks to afford redox-responsivity. One example is PEG-ss-poly(propylene sulfide) (PPS) reported by Cerritelli *et al.* [84]. Ring-opening polymerization (ROP) was used to yield thiolate-terminated PPS which was reacted to thiolate-terminated PEG to form the resulting diblock copolymers, PEG-ss-PPS. The polymersomes were very stable, retaining the loaded calcein for months. However, upon exposure to cysteine, calcein were rapidly released. When the cellular uptake of the polymersomes were investigated, it was found that just after 10 minutes, reduction-sensitive polymersomes were endocytosized and calcein were released intraendosomally. Then after 60 minutes, calcein were able to escape from the endosomes into the cytoplasm by some form of endosomal rupture. Another example is PEG-ss-polyacrylate composed of cholesterol monomers (PAChol) which self-assemble to form ellipsoidal shape polymersomes [85]. Smectic physical crosslinking of the liquid crystal polymer, PAChol, in the hydrophobic layer provided the polymersomes greater stability. Instead of coupling two thiolate-terminated blocks to form the redox-responsive diblock copolymer, in this work PEG with disulfide bond was used as ATPR macroinitiator to polymerize the hydrophobic block. Similarly, these polymersomes exhibited redox-responsive calcein release upon GSH treatment. Polymersomes from diblock copolymers, PEG-ss-PLA, were also used to deliver anticancer drugs, gemcitabine, and DOX to the nucleus via acridine orange active targeting [86]. While diblock copolymers like phospholipid form bilayer structure, triblock copolymers are likely to form "sandwich"-like bilayer membrane polymersomes with the hydrophobic block sandwiched in between two hydrophilic blocks which extend into the aqueous medium inside or outside the core. Folate conjugated DOX-loaded polymersomes from triblock copolymers, pPEGMA-PCL-ss-PCL-pPEGMA which was synthesized via ROP and ATRP could effectively reduce tumor volume by 85% as compared to free DOX of only 42% [88]. This enhanced antitumor activity could be attributed to a few factors, namely longer circulation time, greater accumulation at tumor site due to active targeting and redox-responsive DOX release. For similar factors, another folic acid and trastuzumab-functionalized DOX-loaded polymersomes from PEG-ss-PLA-ss-PLA-ss-PEG, also displayed superior *in vivo* antitumor activity, reducing tumor volume by 90% as compared to free DOX of only 38%[87].

The formation of crosslinks to create a polymer network within nanocarriers like polymersomes is a great approach to enhance stability. In one work, the introduction of dithiolane along the hydrophobic trimethylene carbonate (TMC) chains facilitated the formation of disulfide crosslink within the hydrophobic portion of the polymeric membrane [89]. The disulfide crosslinked polymersomes exhibited superior stability over extensive dilution, against 10% FBS and in DMEM cell culture medium. Furthermore, the crosslinked polymeric membrane has limited drug leakage (<16%) as compared to 66.4% of Lipo-DOX. However, upon exposure to GSH, drug release from polymersomes (PS-DOX) was accelerated, reaching 73.2% in 24 hours. The conjugation of targeting peptide cNGQ on the polymersomes (cNGQ-PS-DOX) greatly improve the cellular uptake via receptor-mediated mechanism, resulting in a greater accumulation of DOX in the cells and thus cNGQ-PS-DOX were more effective in killing A549 lung cancer cells than PS-DOX and Lipo-DOX. Furthermore, *in vivo* studies demonstrated that the cNGQ targeting peptide contributed significantly to the higher accumulation of DOX in a tumor than other major organs. Although both cNGQ-PS-DOX and Lipo-DOX can completely inhibit tumor growth, only mice treated with cNGQ-PS-DOX did not suffer body weight loss and survived the experiment period of 68 days, indicating the low systemic toxicity.

Redox-responsive polymersomes can also self-assemble from polymers with hydrophilic and hydrophobic moieties linked via noncovalent bonds also known as superamphiphiles, see Fig. 1.5 [97]. Compound 1 with disulfide linkage, which is the guest molecules, was caged within the hydrophobic cavity of β-cyclodextrin (β-CD) via hydrogen bonds, hydrophobic and van der Waals interactions. The resulting "Bola" superamphiphiles could self-assemble into "sandwich"-like bilayer polymersomes in an aqueous medium. When treated with dithiothreitol (DTT), the cleavage of the disulfide bonds disassembled the polymersomes and released the hydrophobic drugs entrapped within the membrane. Interestingly, after degradation, the β-CD/3 complex remained stable in water but in pH 7.4 PBS buffer, the complex tends to dissociate.

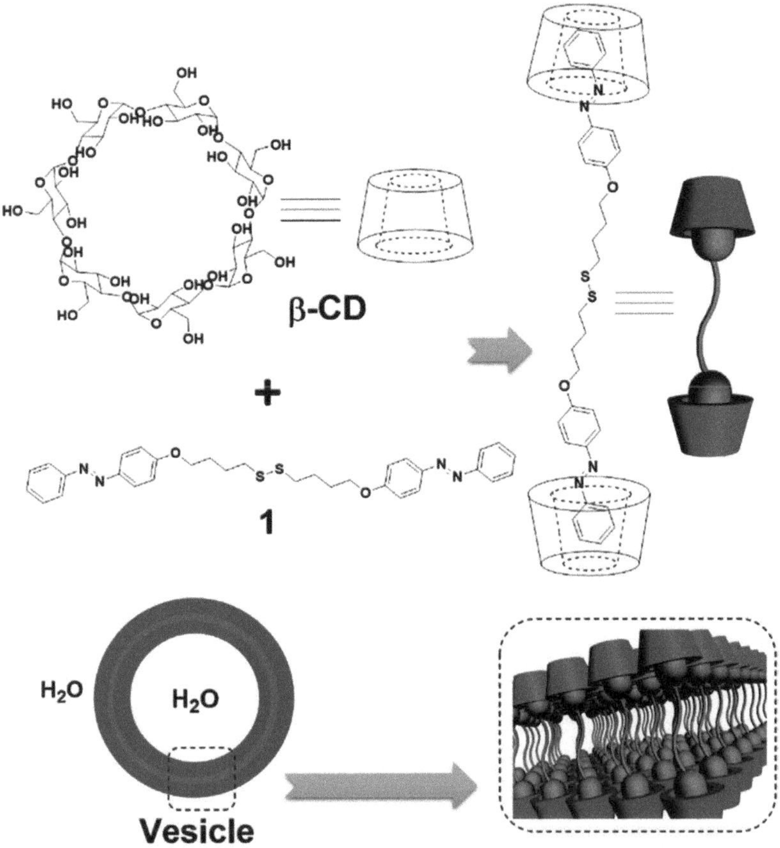

Figure 1.5 Illustration of the formation mechanism for the vesicle and "Bola-type" superamphiphiles.

Besides the commonly reported disulfide approach, ferrocenium cation was also employed to fabricate redox-responsive polymersomes [98]. Conjugating ferrocenium cation head groups to both terminals of pillar[5]arenes formed ferrocenium-capped amphiphilic pillar[5]arenes (FCAP) which self-assembled into cationic polymersomes in aqueous solution. Upon exposure to GSH, ferrocenium cations were reduced to biocompatible ferrocenes and this shift in polarity caused the disassembly of the polymersomes and the release of loaded DOX from the polymersomes. The FCAP polymersomes being cationic in nature are ideal nanocarriers for the delivery of negative charged genetic materials like siRNA. Via electrostatic interaction, cationic FCAP polymersomes complexed with negative charged FAM-siRNA and using flow cytometry and confocal microscopy, the uptake of these polymersomes complexes by cells were quantified (54%) and visualized, respectively. The codelivery of DOX and MRP1siRNA by the polymersomes complexes was evaluated in SKOV-3 cells and the results showed that the cancer cells were re-sensitized by the MRP1 siRNA and the cell death by DOX in cells incubated with the polymersomes complexes improved from 37.9% to 52.8%. With more studies, other non-disulfide approaches like tetrethiafulvalene [99] can also be used to design redox-responsive polymersomes.

1.5 Redox-Responsive Nanogels

In comparison to micelles, nanogels are relatively new nanocarriers, with related studies conducted only in the late 1990s [19, 100]. Nanogels are actually hydrogels in nanoscale, therefore similar to hydrogels. They are made up of three-dimensional polymer network swollen with water molecules. As a result of their hydrophilic nature, nanogels are ideal delivery systems especially for water-soluble drugs. Akin to hydrogel, nanogels can be either chemically or physically [101] crosslinked. However, due to the relatively weak noncovalent interaction, it is often difficult to obtain stable physical crosslinked nanogels. Therefore, predominantly, reported nanogels, especially redox-responsive ones, are chemically crosslinked and exhibit far superior stability.

The preparation of redox-responsive nanogels is generally similar to all other nanogels except for the use of disulfide bond containing precursor polymers or crosslinkers, and the approaches to nanogels are categorized into several methods. Firstly, nanogels can be obtained from the polymerization of monomers in homogenous [102, 103] / heterogeneous [104, 105] phase environment. In heterogeneous phase environment, surfactants are used to form a water-in-oil (W/O) emulsion where the polymerization occurs in the dispersed water phase. Therefore, the size of the nanogels obtained in this manner depends hugely on the dimensions of the water droplets which can be easily adjusted by varying the water to surfactant ratio to make them highly uniform in size. On the other hand, polymerization done in heterogeneous phase environments tends to yield polydispersed nanogels with less control on their size due to the lack of stabilizers. In this precipitation polymerization, monomers, initiators, and crosslinkers are soluble in the initial continuous phase, but upon initiation, the short polymer chains formed are insoluble and begin to precipitate. These precipitates then act as seeds, allowing the polymerization to continue and grow into the resulting nanogels. The second method to fabricate nanogels is to crosslink precursor polymers. Non-crosslinked nanoparticles are first formed via self-assembly and then undergo intermolecular crosslinking to form nanogels. Polymers with intrinsic disulfide bonds can be crosslinked either by adding a catalytic amount of reducing agents [106] or heating [107] to facilitate thiol-disulfide exchange reaction to form redox-responsive nanogels. As for the rest, crosslinkers containing disulfide linkages like cystamine or *N,N*-cystaminebis(acrylamide) (BAC) can be added [108, 109]. Precursor polymers are also deposited on sacrificial templates, such as silica microspheres [110] or Mn_2O_3 cubic particles [109], layer-by-layer and then crosslinked and followed by the removal of the templates to produce nanogels. For example, via hydrogen bonding, alternate layers of thiol-functionalized poly(methacrylic acid) (PMA_{SH}) and poly(vinylpyrrolidone) (PVPON) are deposited on the sacrificial templates. Then under oxidative condition, the PMA_{SH} are crosslinked and PVPON are removed. Lastly, the silica templates are also removed to form the hydrogel capsules. In another example, cubic hydrogels are also fabricated. Using Mn_2O_3 cubic particles, PMA and PVPON are deposited layer bylayer inside the porous template. After crosslinking the polymer layers with cystamine and dissolving the template, nanoporous and biodegradable PMA cubic hydrogel particles are formed. The last method, a "top-down" approach, is an entirely opposite technique as compared to the previous two which are "bottom-up" and it involves photolithography known as particle replication in nonwetting templates (print) [111, 112]. With nonwetting molds, the formation of residual interconnecting films between molded objects are eliminated, and isolated and monodispersed nanogels are produced. Particle size, shape, composition, and surface functionality are strictly controlled in this technique. In one of their studies, cylindrical shape siRNA-conjugated hydrogel nanoparticles obtained via PRINT, successfully release gene materials under reductive environment *in vivo* [113, 114].

The loading of drugs in nanogels can be done either during or post-formulation and the choice largely depends on the intensity of the technique and nature of the drugs. To load drugs during formulation, it is more favorable for techniques that involve a mild reaction or polymerization to avoid damaging fragile drugs like proteins [115, 116]. As a result of this constraint, a vast majority of the reported nanogel studies load their drugs post-formulation. This loading process is highly inefficient and works only with limited drugs which can generate attractive forces like electrostatic attraction, hydrophobic interaction, etc. to the nanogels. Another key problem of post-formulation loading is the encapsulation of macromolecular drugs like proteins and large peptides within the nanogels. These macromolecules are too bulky to penetrate the dense polymer networks and therefore are adsorbed on the nanogel surfaces. However, on the nanogel surfaces, these macromolecular drugs are still exposed to the environment and subjected to degradation.

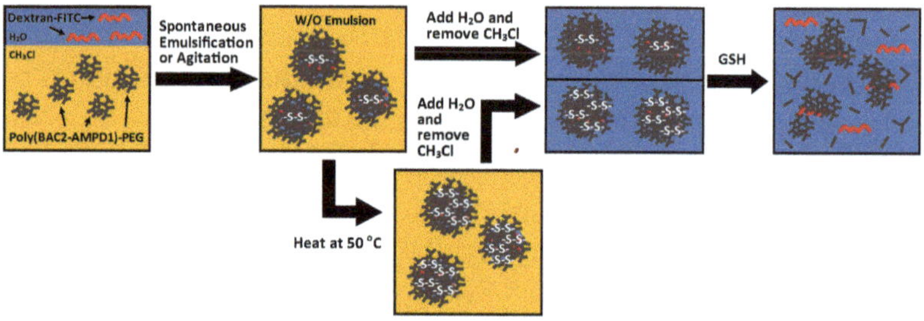

Figure 1.6 Schematic illustration of preparation of FITC-dextran loaded poly(BAC2-AMPD1)-PEG nanogels.

Recently, a surfactant-free emulsion–based approach to redox-responsive nanogels has been reported, see Fig. 1.6 [117]. Unlike the many nanogel preparation techniques mentioned above, this method is simple and no purification is required due to it being devoid of surfactants and other cytotoxic compounds like initiators or crosslinkers. Furthermore, this technique allows the efficient and feasible loading of neutral, macromolecular drugs. In brief, a mixture of water and chloroform solution containing PEGylated hyperbranched poly(amido amine)s, poly(BAC2-AMPD1)-PEG, could feasibly generate W/O emulsion spontaneously or via mechanical agitation. The hyperbranched polymers stabilized and filled the dispersed water droplets in the continuous chloroform phase. The nanogels were then obtained via an intermolecular disulfide exchange reaction. Model neutral macromolecular drugs, fluorescein isothiocyanate-dextran (FITC-dextran) were loaded within the nanogels by simply adding it in water before emulsification. The loading efficiency of FITC-dextran in the nanogels is as high as 100% and upon GSH addition, controlled and sustained release of FITC-dextran lasted for 6 weeks. With a similar hyperbranched poly(amido amine)s [118], it was previously demonstrated that hydrogels with tunable crosslinking density could be easily obtained [119] and to shrink this hydrogel to nanoscale, a emulsion was employed [120]. Under basic conditions, hyperbranched poly(BAC2-AEPZ1)-PEG in the dispersed water droplets of a span80/tween80 stabilized W/O emulsion, undergo *in situ* gelation to form loose hydrogel particles with the size close to the diameter of the water droplets. If the gelling was allowed to continue, further dissociation of the PEG shells would lead to the shrinkage of the loose polymer network to produce compact hydrogel particles. Interestingly, this process of gelation could be terminated and restarted at any moment by adjusting the pH, allowing precise control over the crosslinking density. Multilayer nanogels which have a unique shell/core structure were also obtained using the same polymer and technique. To prepare them, suitable hydrogel particles which are utilized as the seeds are first fabricated with a predetermined gelling time. Then, fresh polymer solution is added to the W/O emulsion with the pre-prepared hydrogel particles serving as the core for the subsequent polymer deposition and crosslinking of the shell. By repeating this step, a succeeding layer can be added, see Fig. 1.7. Another similar hyperbranched poly(amido amine)s which is thermal sensitive undergo temperature induced self-assembly and crosslinking to form nanogels with the size easily controlled by varying the polymer concentration used [121].

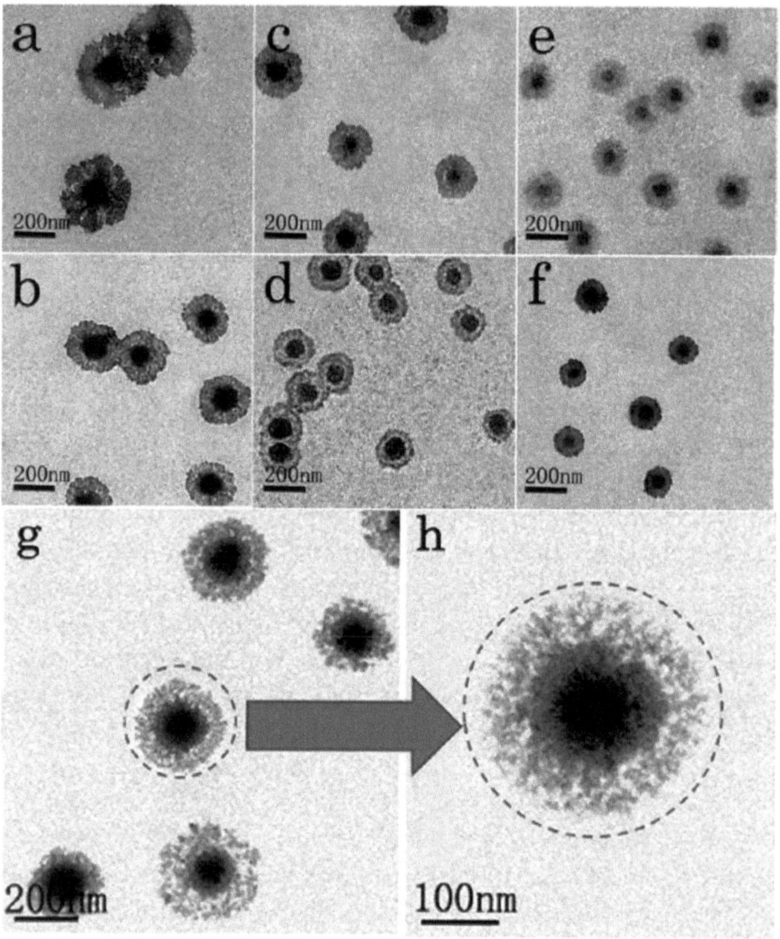

Figure 1.7 (a–f) TEM images of core/shell double-layered nanogels after different crosslinking time: (a) 6 h/1 h, (b) 6 h/2 h, (c) 10 h/2 h, (d) 10 h/4 h, (e) 24 h/4 h, and (f) 24 h/8 h. (g,h) TEM images of (24h/8h/2h) core/double shells triple-layered nanogels.

Being highly hydrophilic in nature, encapsulation of hydrophobic drugs in nanogels can be challenging but Thayumanavan *et al.* have managed to prepare nanogels loaded with hydrophobic drugs like DOX with minimum leakage [122]. Random copolymers that contain oligoethyleneglycol (OEG) and pyridyldisulfide (PDS) units together with hydrophobic drugs first self-assembled to form nanoaggregates were then added with a catalytic amount of DTT to induce disulfide exchange reaction. As a result, the nanogels formed securely trapped DOX within the hydrophobic core. When these nanogels were evaluated in cytotoxicity assay, the nanogels exhibited minimal cytotoxicity but the DOX-loaded nanogels were only slightly less cytotoxic than free DOX.

In another work, using the polymer, poly[(2-(pyridin-2-yl-disulfanyl)-co-[poly(ethylene glycol)]] (PDSEG), multicompartment nanogels are obtained [123]. It was suggested that the intramolecular disulfide bonds formed triggered the assembly of the subcompartment while the intermolecular disulfide bonds formed hold the subcompartments together to form spherical nanogels. Then, upon addition of the reducing agent, tris(2-carboxyethyl) phosphine (TCEP), the multicompartment structure was lost, and uniform and larger nanogels were obtained. Flow cytometry results showed that DOX-loaded nanogels of PDSEG conjugated with targeting ligand, RGD, could enter HCT-116 colon cancer cells more efficiently than free DOX. Furthermore, nanogels loaded with both DOX and paclitaxel, displayed stronger anticancer efficacy and synergism than the free drug cocktail. In another study, results suggested clathrin-mediated endocytosis and macropinocytosis, lipid raft/caveolae-mediated endocytosis play important roles in the HepG2 cellular uptake of poly(*N*-isopropylacrylamide)-ss-acrylic acid (P(NIPAM-ss-AA)) nanogels prepared by precipitation polymerization and then crosslinked with BAC [124]. Furthermore, the increase in hydrophobicity due to its volume phase transition temperature (VPTT) below the body temperature at lysosomal pH 4.5 might have facilitated the lysosomal escape of the (P(NIPAM-ss-AA)) nanogels via fusion of the lysosome membrane and then membrane destabilization. The nanogels, due to the enhanced EPR effect, preferentially accumulate in the tumor and showed deeper tumor penetration than free DOX. Although the *in vitro* cytotoxicity of DOX-loaded (P(NIPAM-ss-AA)) nanogels was only comparable to free DOX, they were much more efficient in inhibiting tumor growth than free DOX and this is likely due to preferential tumor accumulation and superior tumor penetration.

1.6 Redox-Responsive Nanospheres

Polymeric nanospheres are generally defined as matrix-type, insoluble solid-colloidal nanocarriers in which drugs are dispersed, entrapped, encapsulated, chemically bound or adsorbed to the constituent polymer matrix [125]. If the previous nanocarriers, namely micelles, liposomes, polymersomes, and nanogels are considered as "soft" nanoparticles, then nanospheresare "hard" nanoparticles due to the lack of the "soft" hydrophilic outer layer on the unmodified nanospheres. In the early years, poly(ester)s like PLA and PLGA were frequently used to prepare the nanosphere, however, later, the development of the nanosphere shifted to biopolymers like chitosan and gelatin [126]. Nanospheres which have hydrophobic surfaces are highly susceptible to bodily clearance by RES, therefore to prolong blood circulation, the surfaces have to be modified and a common method is PEGylation. In general, the preparation of polymeric nanospheres can be divided into two major approaches, namely dispersion of the preformed polymers [127] and polymerization of monomers [128]. For the former method, there are several different techniques like solvent evaporation [24] which involves emulsifier stabilized O/W emulsion for loading of hydrophobic drugs and W/O/W emulsion for water soluble drugs, spontaneous emulsification/solvent diffusion [23] which involves the diffusion of water-miscible solvent from the oil phase into the water phase, the interfacial turbulence created leading to the nanosphere formation, etc. As for the latter method, in the presence of surfactants, the polymerization of the monomers occurs inside the micelles, followed by the purification to remove the surfactants. The loading of drugs can be done during nanosphere formation [128] or post-formulation [127], and typically more drugs can be entrapped within the nanospheres if the drugs are incorporated during formation rather than adsorbed after formation.

The study on redox-responsive polymeric nanospheres is not as extensive as on the previous redox-responsive nanocarriers, but in recent years, several interesting redox-responsive polymeric nanospheres have been reported. One example is the nanospheres based on viologen cavitand which were prepared from microemulsion polymerization [128]. In the "O/W" emulsion, methylviologen cavitands (MVCAs) acted as emulsifiers stabilizing the diallyl disulfide (SS), which is the monomer, droplets where the polymerization occurred. During polymerization, due to the presence of vinyl terminals, MVCA were conjugated to the nanosphere forming a cavity and shell and hydrophobic drugs can be loaded within the hydrophobic core during polymerization. The resulting nanospheres, p(MVCA-co-SS), were redox-responsive but reacted differently with different reducing agents. In the presence of GSH, the SS cores were degraded and the fluorescence dyes were released, whereas treated with DTT, the nanosphere increased in hydrodynamic size from 130 to 210 nm without releasing the dye. The next example involves multi-stimuli-responsive (redox, pH, and enzyme) nanospheres which were prepared via the electrostatic assembly and disulfide crosslinking of protein, human serum albumin (HSA), polypeptide, and poly-L-lysine (PLL), followed by PEGylation [127]. In the presence of GSH to cleave the disulfide crosslinks within the nanospheres, the photosensitizers (PS) which were adsorbed onto the nanosphere were released. Under irradiation, nanospheres loaded with PS showed superior *in vitro* antitumor activity as compared to free PS which tends to aggregate in aqueous medium, limiting cellular uptake. In animal studies, as compared to free PS, the PS-loaded nanospheres selectively accumulated in the tumor and inhibited tumor growth under irradiation. Besides proteins and peptides, RNAi could also be employed to form nanospheres [129]. S. An and colleagues successfully prepared naked RNAi nanospheres which complexed with disulfide containing cationic polymer (pOEI) to form redox-responsive nanoparticles. Via active targeting and penetrating cyclopeptide iNGR, the RNAi nanoparticles accumulated in greater concentration in glioma than in peripheric organs. Once the RNAi nanoparticles were taken up by the cancerous cells, the high level of GSH present cleaved the disulfide bonds within the cationic polymer coat, thus exposing the naked RNAi nanospheres. The naked RNAi nanospheres were then disintegrated by enzymes to release siRNA to perform effective gene silencing which were supported by the down-regulated luciferase expression *in vitro* and *in vivo*.

1.7 Redox-Responsive Nanocapsules

Polymeric nanocapsules which are also "hard nanoparticles" may be defined as nano-vesicular carriers that have typical core–shell structures in which the drugs are confined to the reservoir or within the cavity surrounded by the polymer membrane or coating [130]. The cavity can be filled with liquid or polymer matrix. Although both polymeric nanocapsules and polymersomes share the same structure, the cavity of nanocapsules typically contain oily liquid while polymersome enclose aqueous media [125]. Similar to the nanosphere, poly(ester)s like PLA and PLGA are popular polymers used in the preparation of nanocapsules and furthermore, due to their hydrophobic surfaces, PEGylation is also done to prolong blood circulation and evade the RES [131]. Polymeric nanocapsules can be prepared via several methods like emulsion–diffusion and emulsion–coacervation. [130]. In general, these approaches employ surfactant-stabilized O/W emulsion which serves as a template for the formation of the polymer membrane or coating around the oily core where hydrophobic drugs are solubilized and very frequently, amphiphilic polymers are the choice of materials [125, 132, 133]. Furthermore, double emulsion, W/O/W emulsion has also been reported to aid in the formation of redox-responsive nanocapsules [132]. In this study, amphiphilic diblock copolymer, hyaluronic acid-cystamine-polylactic-co-glycolic acid (HA-SS-PLGA) was used to form the first W/O emulsion where hydrophobic cyclopamine (CYC), a primary inhibitor was encapsulated in the oil phase while DOX was loaded in the water phase. Then with emulsifiers, polyvinyl alcohol (PVA) and F68, the second emulsion was generated to form the resulting W/O/W emulsion. Then, organic solvent was removed and the nanocapsules were formed. These nanocapsules loaded with dual drugs showed a redox-responsive release profile and superior antitumor activity, inhibiting tumor growth.

As suggested by the definition of polymeric nanocapsules, the cavity of the nanocapsules can also be filled with polymer matrix and a group of scientists led by Y. Tang have made use of protein as the template in which the polymeric shells were built upon [131, 134, 135]. In brief, monomers like acrylamide, *N*-(3-aminopropyl) methacrylamide, *N*-azidodeca(ethylene glycol)ethylacrylamide and crosslinkers like BAC adsorbed on the protein surface via van der Waals or electrostatic interaction were polymerized *in situ* around the protein to form noncovalent shells around the protein, thus yielding the protein nanocapsules. Using this method, the team have successfully encapsulated p53 [134], mature caspase 3 (CP-3) [135], MyoD [131], etc. Furthermore, selecting appropriate monomers can impart properties like surface charge [135], PEGylation [131, 134], and reactive surface group for targeting ligand conjugation [134] to the protein nanocapsules. Enzyme-linked immunosorbent assay (ELISA) analysis has shown that upon treatment with reducing agents, proteins were released from the nanocapsules and remained functional even after encapsulation. In one of their works, LHRH-functionalized p53 encapsulated nanocapsules could accumulate in the nuclei and displayed selective cytotoxicity to MDA-MB-231 cells which overexpressed the LHRH receptors as compared to SK-OV-2 and HFF cells.

However, there are also techniques that generate nanocapsules without the use of templates [133, 136, 137]. An example of such a method is using nanoprecipitation to yield redox-responsive prodrug nanocapsules [133]. Anticancer drug, SN38, was conjugated to oligo(ethylene glycol) (OEG) via thioether linkage which is sensitive to GSH to form the prodrug. Then via nanoprecipitation which involved the slow addition of THF containing OEG-2S-SN38 into water followed by THF evaporation, the prodrug nanocapsules were obtained. These nanocapsules in the presence of H_2O_2, reactive oxygen species (ROS), and GSH were able to release the covalently bonded SN38. Another work on redox-responsive nanocapsules via template-free approach involves the crosslinking of cucurbit[6]uril (CB[6]) with disulfide containing crosslinkers to form very thin shell polymeric nanocapsules with two-dimensional polymer networks [137]. Due to the presence of cleavable disulfide bonds, with DTT treatment, the nanocapsule lost its spherical shape and formed aggregates, and carboxyfluorescein (CF)-loaded nanocapsules displayed a redox-responsive release profile. When the CF-loaded nanocapsules were incubated with HepG2 cells, a sharp increase in fluorescence intensity indicating nanocapsule internalization and CF release, was observed in the cells which was in stark contrast to the non-reducible nanocapsule counterpart.

1.8 Conclusions

This book chapter presented the various polymeric nanocarriers and different approaches to afford redox responsivity in them. The difference among the nanocarrier's structure and characteristic were also highlighted. Polymeric micelles have a distinct core–shell structure and a variety of disulfide containing amphiphilic copolymers have been used to formulate redox-responsive polymeric micelles. Although both liposomes and polymersomes have a very similar bilayer configuration, redox-responsive polymersomes which are made up of a wide range of amphiphilic macromolecular polymers are more chemically diverse than redox-responsive liposomes. Redox-responsive nanogels which are hydrogel nanoparticulates usually consist of three-dimensional polymer networks with disulfide crosslinks. A small section of a redox-responsive nanosphere and nanocapsules were also discussed.

References

1. Sarsour E. H., Kumar M. G., Chaudhuri L., Kalen A. L., Goswami P. C., Redox control of the cell cycle in health and disease, *Antioxidants & Redox Signaling*, 11 (2009) 2985–3011.
2. Oktyabrsky O. N., Smirnova G. V., Redox regulation of cellular functions, *Biochemistry. Biokhimiia*, 72 (2007) 132–145.
3. Wagner S., Rokita A. G., Anderson M. E., Maier L. S., Redox regulation of sodium and calcium handling, *Antioxidants & Redox Signaling*, 18 (2013) 1063–1077.
4. Cadenas E., Davies K. J., Mitochondrial free radical generation, oxidative stress, and aging, *Free Radical Biology and Medicine*, 29 (2000) 222–230.
5. Pani G., Colavitti R., Borrello S., Galeotti T., Redox regulation of lymphocyte signaling, *International Union of Biochemistry and Molecular Biology, 49* (2000) 381–389.
6. Ostrakhovitch E. A., Semenikhin O. A., The role of redox environment in neurogenic development, *Archives of Biochemistry and Biophysics*, 534 (2013) 44–54.
7. Schafer F. Q., Buettner G. R., Redox environment of the cell as viewed through the redox state of the glutathione disulfide/glutathione couple, *Free Radical Biology and Medicine*, 30 (2001) 1191–1212.

8. Stavrovskaya A. A., Cellular mechanisms of multidrug resistance of tumor cells, *Biochemistry. Biokhimiia*, 65 (2000) 95–106.
9. Yokoyama M., Miyauchi M., Yamada N., Okano T., Sakurai Y., Kataoka K., Inoue S., Characterization and anticancer activity of the micelle-forming polymeric anticancer drug adriamycin-conjugated poly(ethylene glycol)-poly(aspartic acid) block copolymer, *Cancer Research*, 50 (1990) 1693–1700.
10. Masayuki Y., Mizue M., Noriko Y., Teruo O., Yasuhisa S., Kazunori K., Shohei I., Polymer micelles as novel drug carrier: Adriamycin-conjugated poly(ethylene glycol)-poly(aspartic acid) block copolymer, *Journal of Controlled Release*, 11 (1990) 269–278.
11. Yokoyama M., Okano T., Sakurai Y., Ekimoto H., Shibazaki C., Kataoka K., Toxicity and antitumor activity against solid tumors of micelle-forming polymeric anticancer drug and its extremely long circulation in blood, *Cancer Research*, 51 (1991) 3229–3236.
12. Kazunori K., K. Glenn S, Masayuki Y., Teruo O., Yasuhisa S., Block copolymer micelles as vehicles for drug delivery, *Journal of Controlled Release*, 24 (1993) 119–132.
13. Kwon G. S., Kataoka K., Block copolymer micelles as long-circulating drug vehicles, *Advanced Drug Delivery Reviews*, 16 (1995) 295–309.
14. Gregoriadis G., Ryman B. E., Liposomes as carriers of enzymes or drugs: a new approach to the treatment of storage diseases, *The Biochemical Journal*, 124 (1971) 58P.
15. Rahman Y. E., Rosenthal M. W., Cerny E. A., Intracellular plutonium: removal by liposome-encapsulated chelating agent, *New York Academy of Sciences*, 180 (1973) 300–302.
16. Zhang L., Eisenberg A., Multiple morphologies of "Crew-Cut" aggregates of polystyrene-b-poly(acrylic acid) block copolymers, *New York Academy of Sciences*, 268 (1995) 1728–1731.
17. Discher B. M., Won Y. Y., Ege D. S., Lee J. C., Bates F. S., Discher D. E., Hammer D. A., Polymersomes: tough vesicles made from diblock copolymers, *New York Academy of Sciences*, 284 (1999) 1143–1146.
18. Ahmed F., Photos P. J., Osmaswa I., Bates F. S., Discher D. E., Controlled release and stealthy circulation of polymersomes, Engineering in Medicine and Biology, 2002. 24th Annual Conference and the Annual Fall Meeting of the Biomedical Engineering Society EMBS/BMES Conference, 2002. *Proceedings of the Second Joint*, 461 (2002), 466–467.

19. Vinogradov S., Batrakova E., Kabanov A., Poly(ethylene glycol)-polyethyleneimine NanoGel™ particles: novel drug delivery systems for antisense oligonucleotides, *Colloids and Surfaces, B: Biointerfaces*, 16 (1999) 291–304.
20. Vinogradov S. V., Bronich T. K., Kabanov A. V., Nanosized cationic hydrogels for drug delivery: preparation, properties and interactions with cells, *Advanced Drug Delivery Reviews*, 54 (2002) 135–147.
21. Fessi H., Puisieux F., Devissaguet J. P., Ammoury N., Benita S., Nanocapsule formation by interfacial polymer deposition following solvent displacement, *International Journal of Pharmaceutics*, 55 (1989) R1–R4.
22. Ammoury N., Fessi H., Devissaguet J. P., Dubrasquet M., Benita S., Jejunal absorption, pharmacological activity, and pharmacokinetic evaluation of indomethacin-loaded poly(D,L-lactide) and poly(isobutyl-cyanoacrylate) nanocapsules in rats, *Pharmaceutical Research*, 8 (1991) 101–105.
23. Niwa T., Takeuchi H., Hino T., Kunou N., Kawashima Y., Preparations of biodegradable nanospheres of water-soluble and insoluble drugs with D,L-lactide/glycolide copolymer by a novel spontaneous emulsification solvent diffusion method, and the drug release behavior, *Journal of Controlled Release*, 25 (1993) 89–98.
24. Sánchez A., Vila-Jato J., Alonso M. J., Development of biodegradable microspheres and nanospheres for the controlled release of cyclosporin A, *International Journal of Pharmaceutics*, 99 (1993) 263–273.
25. Gaucher G., Dufresne M. H., Sant V. P., Kang N., Maysinger D., Leroux J. C., Block copolymer micelles: preparation, characterization and application in drug delivery, *Journal of Controlled Release*, 109 (2005) 169–188.
26. Sun Y., Yan X., Yuan T., Liang J., Fan Y., Gu Z., Zhang X., Disassemblable micelles based on reduction-degradable amphiphilic graft copolymers for intracellular delivery of doxorubicin, *Biomaterials*, 31 (2010) 7124–7131.
27. Cheng W., Kumar J. N., Zhang Y., Liu Y., pH- and redox-responsive poly(ethylene glycol) and cholesterol-conjugated poly(amido amine)s based micelles for controlled drug delivery, *Macromolecular Bioscience*, 14 (2014) 347–358.
28. Ryu J. H., Roy R., Ventura J., Thayumanavan S., Redox-sensitive disassembly of amphiphilic copolymer based micelles, *Langmuir*, 26 (2010) 7086–7092.

29. Li J., Huo M., Wang J., Zhou J., Mohammad J. M., Zhang Y., Zhu Q., Waddad A. Y., Zhang Q., Redox-sensitive micelles self-assembled from amphiphilic hyaluronic acid-deoxycholic acid conjugates for targeted intracellular delivery of paclitaxel, *Biomaterials*, 33 (2012) 2310–2320.

30. Liu J., Pang Y., Huang W., Zhu Z., Zhu X., Zhou Y., Yan D., Redox-responsive polyphosphate nanosized assemblies: a smart drug delivery platform for cancer therapy, *Biomacromolecules*, 12 (2011) 2407–2415.

31. Cheng W., Kumar J. N., Zhang Y., Liu Y., pH- and redox-responsive self-assembly of amphiphilic hyperbranched poly(amido amine)s for controlled doxorubicin delivery, *Biomaterials Science*, 3 (2015) 597–607.

32. Kwon G., Naito M., Yokoyama M., Okano T., Sakurai Y., Kataoka K., Micelles based on AB block copolymers of poly(ethylene oxide) and poly(.beta.-benzyl L-aspartate), *Langmuir*, 9 (1993) 945–949.

33. Chen M., Gao C., Lu S., Chen Y., Liu M., Dual redox-triggered shell-sheddable micelles self-assembled from mPEGylated starch conjugates for rapid drug release, *RSC Advances*, 6 (2016) 9164–9174.

34. Harada A., Kataoka K., Formation of polyion complex micelles in an aqueous milieu from a pair of oppositely-charged block copolymers with poly(ethylene glycol) segments, *Macromolecules*, 28 (1995) 5294–5299.

35. Ping Y., Wu D., Kumar J. N., Cheng W., Lay C. L., Liu Y., Redox-responsive hyperbranched poly(amido amine)s with tertiary amino cores for gene delivery, *Biomacromolecules*, 14 (2013) 2083–2094.

36. Zhong Y., Yang W., Sun H., Cheng R., Meng F., Deng C., Zhong Z., Ligand-directed reduction-sensitive shell-sheddable biodegradable micelles actively deliver doxorubicin into the nuclei of target cancer cells, *Biomacromolecules*, 14 (2013) 3723–3730.

37. Gong C., Shan M., Li B., Wu G., A pH and redox dual stimuli-responsive poly(amino acid) derivative for controlled drug release, *Colloids and Surfaces, B: Biointerfaces*, 146 (2016) 396–405.

38. Yokoyama M., Satoh A., Sakurai Y., Okano T., Matsumura Y., Kakizoe T., Kataoka K., Incorporation of water-insoluble anticancer drug into polymeric micelles and control of their particle size, *Journal of Controlled Release*, 55 (1998) 219–229.

39. Tian K., Jia X., Zhao X., Liu P., pH/Reductant dual-responsive core-cross-linked micelles via facile in situ ATRP for tumor-targeted

delivery of anticancer drug with enhanced anticancer efficiency, *Molecular Pharmaceutics*, 13 (2016) 2683–2690.

40. Kwon G., Naito M., Yokoyama M., Okano T., Sakurai Y., Kataoka K., Block copolymer micelles for drug delivery: loading and release of doxorubicin, *Journal of Controlled Release*, 48 (1997) 195–201.

41. Kataoka K., Matsumoto T., Yokoyama M., Okano T., Sakurai Y., Fukushima S., Okamoto K., Kwon G. S., Doxorubicin-loaded poly(ethylene glycol)-poly(beta-benzyl-L-aspartate) copolymer micelles: their pharmaceutical characteristics and biological significance, *Journal of Controlled Release*, 64 (2000) 143–153.

42. Shi Y., van Steenbergen M. J., Teunissen E. A., Novo L., Gradmann S., Baldus M., van Nostrum C. F., Hennink W. E., π–π Stacking increases the stability and loading capacity of thermosensitive polymeric micelles for chemotherapeutic drugs, *Biomacromolecules*, 14 (2013) 1826–1837.

43. Sun H., Guo B., Cheng R., Meng F., Liu H., Zhong Z., Biodegradable micelles with sheddable poly(ethylene glycol) shells for triggered intracellular release of doxorubicin, *Biomaterials*, 30 (2009) 6358–6366.

44. Sun H., Guo B., Li X., Cheng R., Meng F., Liu H., Zhong Z., Shell-sheddable micelles based on dextran-SS-poly(epsilon-caprolactone) diblock copolymer for efficient intracellular release of doxorubicin, *Biomacromolecules*, 11 (2010) 848–854.

45. Wen H. Y., Dong H. Q., Xie W. J., Li Y. Y., Wang K., Pauletti G. M., Shi D. L., Rapidly disassembling nanomicelles with disulfide-linked PEG shells for glutathione-mediated intracellular drug delivery, *Chemical Communications*, 47 (2011) 3550–3552.

46. Yu C., Gao C., Lü S., Chen C., Huang Y., Liu M., Redox-responsive shell-sheddable micelles self-assembled from amphiphilic chondroitin sulfate-cholesterol conjugates for triggered intracellular drug release, *Chemical Engineering Journal*, 228 (2013) 290–299.

47. Yang X. Z., Du J. Z., Dou S., Mao C. Q., Long H. Y., Wang J., Sheddable ternary nanoparticles for tumor acidity-targeted siRNA delivery, *ACS Nano*, 6 (2012) 771–781.

48. Koo A. N., Lee H. J., Kim S. E., Chang J. H., Park C., Kim C., Park J. H., Lee S. C., Disulfide-cross-linked PEG-poly(amino acid)s copolymer micelles for glutathione-mediated intracellular drug delivery, *Chemical Communications*, (2008) 6570–6572.

49. Kim J. O., Sahay G., Kabanov A. V., Bronich T. K., Polymeric micelles with ionic cores containing biodegradable cross-links for delivery of chemotherapeutic agents, *Biomacromolecules*, 11 (2010) 919–926.

50. Xia Y., He H., Liu X., Hu D., Yin L., Lu Y., Xu W., Redox-responsive, core-crosslinked degradable micelles for controlled drug release, *Polymer Chemistry*, 7 (2016) 6330–6339.

51. Li Y., Xiao K., Luo J., Xiao W., Lee J. S., Gonik A. M., Kato J., Dong T. A., Lam K. S., Well-defined, reversible disulfide cross-linked micelles for on-demand paclitaxel delivery, *Biomaterials*, 32 (2011) 6633–6645.

52. Gyarmati B., Vajna B., Nemethy A., Laszlo K., Szilagyi A., Redox- and pH-responsive cysteamine-modified poly(aspartic acid) showing a reversible sol–gel transition, *Macromolecular Bioscience*, 13 (2013) 633–640.

53. Huang M., Zhao K., Wang L., Lin S., Li J., Chen J., Zhao C., Ge Z., Dual stimuli-responsive polymer prodrugs quantitatively loaded by nanoparticles for enhanced cellular internalization and triggered drug release, *ACS Applied Materials & Interfaces*, 8 (2016) 11226–11236.

54. Tabatabaei Rezaei S. J., Sarbaz L., Niknejad H., Folate-decorated redox/pH dual-responsive degradable prodrug micelles for tumor triggered targeted drug delivery, *RSC Advances*, 6 (2016) 62630–62639.

55. Li X. Q., Wen H. Y., Dong H. Q., Xue W. M., Pauletti G. M., Cai X. J., Xia W. J., Shi D., Li Y. Y., Self-assembling nanomicelles of a novel camptothecin prodrug engineered with a redox-responsive release mechanism, *Chemical Communications*, 47 (2011) 8647–8649.

56. Liu P., Shi B., Yue C., Gao G., Li P., Yi H., Li M., Wang B., Ma Y., Cai L., Dextran-based redox-responsive doxorubicin prodrug micelles for overcoming multidrug resistance, *Polymer Chemistry*, 4 (2013) 5793–5799.

57. McRae Page S., Martorella M., Parelkar S., Kosif I., Emrick T., Disulfide cross-linked phosphorylcholine micelles for triggered release of camptothecin, *Molecular Pharmaceutics*, 10 (2013) 2684–2692.

58. Thambi T., Yoon H. Y., Kim K., Kwon I. C., Yoo C. K., Park J. H., Bioreducible block copolymers based on poly(ethylene glycol) and poly(gamma-benzyl L-glutamate) for intracellular delivery of camptothecin, *Bioconjugate Chemistry*, 22 (2011) 1924–1931.

59. Liu J., Pang Y., Huang W., Huang X., Meng L., Zhu X., Zhou Y., Yan D., Bioreducible micelles self-assembled from amphiphilic

hyperbranched multiarm copolymer for glutathione-mediated intracellular drug delivery, *Biomacromolecules*, 12 (2011) 1567–1577.

60. Guo Y., Zhang Y., Li J., Zhang Y., Lu Y., Jiang X., He X., Ma H., An S., Jiang C., Cell microenvironment-controlled antitumor drug releasing-nanomicelles for GLUT1-targeting hepatocellular carcinoma therapy, *ACS Applied Materials & Interfaces*, 7 (2015) 5444–5453.
61. Wang J., Yang G., Guo X., Tang Z., Zhong Z., Zhou S., Redox-responsive polyanhydride micelles for cancer therapy, *Biomaterials*, 35 (2014) 3080–3090.
62. Huo M., Liu Y., Wang L., Yin T., Qin C., Xiao Y., Yin L., Liu J., Zhou J., Redox-sensitive micelles based on *O,N*-hydroxyethyl chitosan-octylamine conjugates for triggered intracellular delivery of paclitaxel, *Molecular Pharmaceutics*, 13 (2016) 1750–1762.
63. Yin T., Wang L., Yin L., Zhou J., Huo M., Co-delivery of hydrophobic paclitaxel and hydrophilic AURKA specific siRNA by redox-sensitive micelles for effective treatment of breast cancer, *Biomaterials*, 61 (2015) 10–25.
64. Bangham A. D., Standish M. M., Watkins J. C., Diffusion of univalent ions across the lamellae of swollen phospholipids, *Journal of Molecular Biology*, 13 (1965) 238–252.
65. Fan Y., Zhang Q., Development of liposomal formulations: From concept to clinical investigations, *Asian Journal of Pharmaceutical Sciences*, 8 (2013) 81–87.
66. Allen T. M., Cullis P. R., Liposomal drug delivery systems: from concept to clinical applications, *Advanced Drug Delivery Reviews*, 65 (2013) 36–48.
67. Akbarzadeh A., Rezaei-Sadabady R., Davaran S., Joo S. W., Zarghami N., Hanifehpour Y., Samiei M., Kouhi M., Nejati-Koshki K., Liposome: classification, preparation, and applications, *Nanoscale Research Letters*, 8 (2013) 102.
68. Barenholz Y., Doxil(R)--the first FDA-approved nano-drug: lessons learned, *Journal of Controlled Release*, 160 (2012) 117–134.
69. Immordino M. L., Dosio F., Cattel L., Stealth liposomes: review of the basic science, rationale, and clinical applications, existing and potential, *International Journal of Nanomedicine*, 1 (2006) 297–315.
70. Sawant R. R., Torchilin V. P., Liposomes as 'smart' pharmaceutical nanocarriers, *Soft Matter*, 6 (2010) 4026–4044.

71. Ellens H., Bentz J., Szoka F. C., pH-Induced destabilization of phosphatidylethanolamine-containing liposomes: role of bilayer contact, *Biochemistry*, 23 (1984) 1532–1538.

72. Maeda T., Fujimoto K., A reduction-triggered delivery by a liposomal carrier possessing membrane-permeable ligands and a detachable coating, *Colloids and Surfaces, B: Biointerfaces*, 49 (2006) 15–21.

73. Oumzil K., Khiati S., Grinstaff M. W., Barthelemy P., Reduction-triggered delivery using nucleoside-lipid based carriers possessing a cleavable PEG coating, *Journal of Controlled Release*, 151 (2011) 123–130.

74. Kuai R., Yuan W., Qin Y., Chen H., Tang J., Yuan M., Zhang Z., He Q., Efficient delivery of payload into tumor cells in a controlled manner by TAT and thiolytic cleavable PEG co-modified liposomes, *Molecular Pharmaceutics*, 7 (2010) 1816–1826.

75. Kullberg M., Martinson H., Mann K., Anchordoquy T. J., Complement C3 mediated targeting of liposomes to granulocytic myeloid derived suppressor cells, *Nanomedicine: Nanotechnology, Biology, and Medicine*, 11 (2015) 1355–1363.

76. Candiani G., Pezzoli D., Ciani L., Chiesa R., Ristori S., Bioreducible liposomes for gene delivery: from the formulation to the mechanism of action, *PLOS One*, 5 (2010) e13430.

77. Xu X., Zhang L., Assanhou A. G., Wang L., Zhang Y., Li W., Xue L., Mo R., Zhang C., Acid/redox dual-activated liposomes for tumor-targeted drug delivery and enhanced therapeutic efficacy, *RSC Advances*, 5 (2015) 67803–67808.

78. Ong W., Yang Y., Cruciano A. C., McCarley R. L., Redox-triggered contents release from liposomes, *Journal of the American Chemical Society*, 130 (2008) 14739–14744.

79. Meng F., Zhong Z., Feijen J., Stimuli-responsive polymersomes for programmed drug delivery, *Biomacromolecules*, 10 (2009) 197–209.

80. Tanner P., Baumann P., Enea R., Onaca O., Palivan C., Meier W., Polymeric vesicles: from drug carriers to nanoreactors and artificial organelles, *Accounts of Chemical Research*, 44 (2011) 1039–1049.

81. Hamidi M., Shahbazi M.-A., Rostamizadeh K., Copolymers: efficient carriers for intelligent nanoparticulate drug targeting and gene therapy, *Macromolecular Bioscience*, 12 (2012) 144–164.

82. Du J., Armes S. P., Preparation of biocompatible zwitterionic block copolymer vesicles by direct dissolution in water and subsequent

silicification within their membranes, *Langmuir*, 25 (2009) 9564–9570.

83. Lee J. C., Bermudez H., Discher B. M., Sheehan M. A., Won Y. Y., Bates F. S., Discher D. E., Preparation, stability, and in vitro performance of vesicles made with diblock copolymers, *Biotechnology and Bioengineering*, 73 (2001) 135–145.
84. Cerritelli S., Velluto D., Hubbell J. A., PEG-SS-PPS: reduction-sensitive disulfide block copolymer vesicles for intracellular drug delivery, *Biomacromolecules*, 8 (2007) 1966–1972.
85. Jia L., Cui D., Bignon J., A. Di Cicco, Wdzieczak-Bakala J., Liu J., Li M. H., Reduction-responsive cholesterol-based block copolymer vesicles for drug delivery, *Biomacromolecules*, 15 (2014) 2206–2217.
86. Anajafi T., Scott M. D., You S., Yang X., Choi Y., Qian S. Y., Mallik S., Acridine orange conjugated polymersomes for simultaneous nuclear delivery of gemcitabine and doxorubicin to pancreatic cancer cells, *Bioconjugate Chemistry*, 27 (2016) 762–771.
87. Lale S. V., Kumar A., Prasad S., Bharti A. C., Koul V., Folic acid and trastuzumab functionalized redox responsive polymersomes for intracellular doxorubicin delivery in breast cancer, *Biomacromolecules*, 16 (2015) 1736–1752.
88. Kumar A., Lale S. V., Mahajan S., Choudhary V., Koul V., ROP and ATRP fabricated dual targeted redox sensitive polymersomes based on pPEGMA-PCL-ss-PCL-pPEGMA triblock copolymers for breast cancer therapeutics, *ACS Applied Materials & Interfaces*, 7 (2015) 9211–9227.
89. Zou Y., Meng F., Deng C., Zhong Z., Robust, tumor-homing and redox-sensitive polymersomal doxorubicin: A superior alternative to Doxil and Caelyx?*Journal of Controlled Release*, 239 (2016) 149–158.
90. Rank A., Hauschild S., Förster S., Schubert R., Preparation of monodisperse block copolymer vesicles via a thermotropic cylinder–vesicle transition, *Langmuir*, 25 (2009) 1337–1344.
91. Lorenceau E., Utada A. S., Link D. R., Cristobal G., Joanicot M., Weitz D. A., Generation of polymerosomes from double-emulsions, *Langmuir*, 21 (2005) 9183–9186.
92. Beaune G., Dubertret B., Clément O., Vayssettes C., Cabuil V., Ménager C., Giant vesicles containing magnetic nanoparticles and quantum dots: feasibility and tracking by fiber confocal fluorescence microscopy, *Angewandte Chemie International Edition*, 46 (2007) 5421–5424.

93. Thiele J., Steinhauser D., Pfohl T., Förster S., Preparation of monodisperse block copolymer vesicles via flow focusing in microfluidics, *Langmuir*, 26 (2010) 6860–6863.

94. Choucair A., Soo P. L., Eisenberg A., Active loading and tunable release of doxorubicin from block copolymer vesicles, *Langmuir*, 21 (2005) 9308–9313.

95. Ahmed F., Pakunlu R. I., Brannan A., Bates F., Minko T., Discher D. E., Biodegradable polymersomes loaded with both paclitaxel and doxorubicin permeate and shrink tumors, inducing apoptosis in proportion to accumulated drug, *Journal of Controlled Release*, 116 (2006) 150–158.

96. Ghoroghchian P. P., Lin J. J., Brannan A. K., Frail P. R., Bates F. S., Therien M. J., Hammer D. A., Quantitative membrane loading of polymer vesicles, *Soft Matter*, 2 (2006) 973–980.

97. Sun T., Shu L., Shen J., Ruan C., Zhao Z., Jiang C., Photo and redox-responsive vesicles assembled from Bola-type superamphiphiles, *RSC Advances*, 6 (2016) 52189–52200.

98. Chang Y., Yang K., Wei P., Huang S., Pei Y., Zhao W., Pei Z., Cationic vesicles based on amphiphilic pillar[5]arene capped with ferrocenium: a redox-responsive system for drug/siRNA co-delivery, *Angewandte Chemie International Edition*, 53 (2014) 13126–13130.

99. Zhang K.-D., Zhou T.-Y., Zhao X., Jiang X.-K., Li Z.-T., Redox-responsive reverse vesicles self-assembled by pseudo[2]rotaxanes for tunable dye release, *Langmuir*, 28 (2012) 14839–14844.

100. Akiyoshi K., Kobayashi S., Shichibe S., Mix D., Baudys M., Wan Kim S., Sunamoto J., Self-assembled hydrogel nanoparticle of cholesterol-bearing pullulan as a carrier of protein drugs: Complexation and stabilization of insulin, *Journal of Controlled Release*, 54 (1998) 313–320.

101. Sasaki Y., Akiyoshi K., Nanogel engineering for new nanobiomaterials: from chaperoning engineering to biomedical applications, *Chemical Record*, 10 (2010) 366–376.

102. Qiao Z. Y., Zhang R., Du F. S., Liang D. H., Li Z. C., Multi-responsive nanogels containing motifs of ortho ester, oligo(ethylene glycol) and disulfide linkage as carriers of hydrophobic anti-cancer drugs, *Journal of Controlled Release*, 152 (2011) 57–66.

103. Perez E., Fernandez A., Olmo R., Teijon J. M., Blanco M. D., pH and glutathion-responsive hydrogel for localized delivery of paclitaxel, *Colloids and Surfaces, B: Biointerfaces*, 116 (2014) 247–256.

104. Pan Y. J., Chen Y. Y., Wang D. R., Wei C., Guo J., Lu D. R., Chu C. C., Wang C. C., Redox/pH dual stimuli-responsive biodegradable nanohydrogels with varying responses to dithiothreitol and glutathione for controlled drug release, *Biomaterials*, 33 (2012) 6570–6579.

105. Gaulding J. C., Smith M. H., Hyatt J. S., Fernandez-Nieves A., Lyon L. A., Reversible inter- and intra-microgel cross-linking using disulfides, *Macromolecules*, 45 (2012) 39–45.

106. Zhang X., Achazi K., Steinhilber D., Kratz F., Dernedde J., Haag R., A facile approach for dual-responsive prodrug nanogels based on dendritic polyglycerols with minimal leaching, *Journal of Controlled Release*, 174 (2014) 209–216.

107. Chen Q.-B., You Y.-Z., Multi-stimuli responsive hydrogel particles prepared via the self-assembly of PEG-based hyperbranched polymers, *Chemistry Letters*, 44 (2015), 677–679.

108. Ding J., Xu W., Zhang Y., Sun D., Xiao C., Liu D., Zhu X., Chen X., Self-reinforced endocytoses of smart polypeptide nanogels for "on-demand" drug delivery, *Journal of Controlled Release*, 172 (2013) 444–455.

109. Xue B., Kozlovskaya V., Liu F., Chen J., Williams J. F., Campos-Gomez J., Saeed M., Kharlampieva E., Intracellular degradable hydrogel cubes and spheres for anti-cancer drug delivery, *ACS Applied Materials & Interfaces*, 7 (2015) 13633–13644.

110. Becker A. L., Zelikin A. N., Johnston A. P., Caruso F., Tuning the formation and degradation of layer-by-layer assembled polymer hydrogel microcapsules, *Langmuir*, 25 (2009) 14079–14085.

111. Rolland J. P., Maynor B. W., Euliss L. E., Exner A. E., Denison G. M., DeSimone J. M., Direct fabrication and harvesting of monodisperse, shape-specific nanobiomaterials, *Journal of the American Chemical Society*, 127 (2005) 10096–10100.

112. Gratton S. E. A., Pohlhaus P. D., Lee J., Guo J., Cho M. J., DeSimone J. M., Nanofabricated particles for engineered drug therapies: A preliminary biodistribution study of PRINT™ nanoparticles, *Journal of Controlled Release*, 121 (2007) 10–18.

113. Ma D., Tian S., Baryza J., Luft J. C., DeSimone J. M., Reductively responsive hydrogel nanoparticles with uniform size, shape, and tunable composition for systemic siRNA delivery in vivo, *Molecular Pharmaceutics*, 12 (2015) 3518–3526.

114. Dunn S. S., Tian S., Blake S., Wang J., Galloway A. L., Murphy A., Pohlhaus P. D., Rolland J. P., Napier M. E., DeSimone J. M., Reductively responsive

siRNA-conjugated hydrogel nanoparticles for gene silencing, *Journal of the American Chemical Society*, 134 (2012) 7423–7430.

115. Shi L., Khondee S., Linz T. H., Berkland C., Poly(N-vinylformamide) nanogels capable of pH-sensitive protein release, *Macromolecules*, 41 (2008) 6546–6554.

116. Oh J. K., Siegwart D. J., Matyjaszewski K., Synthesis and biodegradation of nanogels as delivery carriers for carbohydrate drugs, *Biomacromolecules*, 8 (2007) 3326–3331.

117. Cheng W., Wang G., Kumar J. N., Liu Y., Surfactant-free emulsion-based preparation of redox-responsive nanogels, *Macromolecular Rapid Communications*, 36 (2015) 2102–2106.

118. Cheng W., Wu D., Liu Y., Michael addition polymerization of trifunctional amine and acrylic monomer: a versatile platform for development of biomaterials, *Biomacromolecules*, 17 (2016) 3115–3126.

119. Wu D. C., Loh X. J., Wu Y. L., Lay C. L., Liu Y., 'Living' controlled in situ gelling systems: thiol-disulfide exchange method toward tailor-made biodegradable hydrogels, *Journal of the American Chemical Society*, 132 (2010) 15140–15143.

120. Zhang J., Yang F., Shen H., Wu D., Controlled formation of microgels/nanogels from a disulfide-linked core/shell hyperbranched polymer, *ACS Macro Letters*, 1 (2012) 1295–1299.

121. Yu Z.-Q., Sun J.-T., Pan C.-Y., Hong C.-Y., Bioreducible nanogels/microgels easily prepared via temperature induced self-assembly and self-crosslinking, *Chemical Communications*, 48 (2012) 5623–5625.

122. Ryu J. H., Chacko R. T., Jiwpanich S., Bickerton S., Babu R. P., Thayumanavan S., Self-cross-linked polymer nanogels: a versatile nanoscopic drug delivery platform, *Journal of the American Chemical Society*, 132 (2010) 17227–17235.

123. Bahadur K. C. R., Xu P., Multicompartment intracellular self-expanding nanogel for targeted delivery of drug cocktail, *Advanced Materials*, 24 (2012) 6479–6483.

124. Yang H., Wang Q., Huang S., Xiao A., Li F., Gan L., Yang X., Smart pH/redox dual-responsive nanogels for on-demand intracellular anticancer drug release, *ACS Applied Materials & Interfaces*, 8 (2016) 7729–7738.

125. Letchford K., Burt H., A review of the formation and classification of amphiphilic block copolymer nanoparticulate structures: micelles,

nanospheres, nanocapsules and polymersomes, *European Journal of Pharmaceutics and Biopharmaceutics:*, 65 (2007) 259–269.

126. Soppimath K. S., Aminabhavi T. M., Kulkarni A. R., Rudzinski W. E., Biodegradable polymeric nanoparticles as drug delivery devices, *Journal of Controlled Release*, 70 (2001) 1–20.
127. Zhang N., Zhao F., Zou Q., Li Y., Ma G., Yan X., Multitriggered tumor-responsive drug delivery vehicles based on protein and polypeptide coassembly for enhanced photodynamic tumor ablation, *Small*, 12 (2016), 5936–5943.
128. Sultanova E. D., Atlanderova A. A., Mukhitova R. D., Salnikov V. V., Osin Y. N., Ziganshina A. Y., Konovalov A. I., Reduction-controlled substrate release from a polymer nanosphere based on a viologen-cavitand, *RSC Advances*, 6 (2016) 70072–70076.
129. An S., Jiang X., Shi J., He X., Li J., Guo Y., Zhang Y., Ma H., Lu Y., Jiang C., Single-component self-assembled RNAi nanoparticles functionalized with tumor-targeting iNGR delivering abundant siRNA for efficient glioma therapy, *Biomaterials*, 53 (2015) 330–340.
130. Mora-Huertas C. E., Fessi H., Elaissari A., Polymer-based nanocapsules for drug delivery, *International Journal of Pharmaceutics*, 385 (2010) 113–142.
131. Biswas A., Liu Y., Liu T., Fan G., Tang Y., Polyethylene glycol-based protein nanocapsules for functional delivery of a differentiation transcription factor, *Biomaterials*, 33 (2012) 5459–5467.
132. Hu K., Zhou H., Liu Y., Liu Z., Liu J., Tang J., Li J., Zhang J., Sheng W., Zhao Y., Wu Y., Chen C., Hyaluronic acid functional amphipathic and redox-responsive polymer particles for the co-delivery of doxorubicin and cyclopamine to eradicate breast cancer cells and cancer stem cells, *Nanoscale*, 7 (2015) 8607–8618.
133. Wang J., Sun X., Mao W., Sun W., Tang J., Sui M., Shen Y., Gu Z., Tumor redox heterogeneity-responsive prodrug nanocapsules for cancer chemotherapy, *Advanced Materials*, 25 (2013) 3670–3676.
134. Zhao M., Liu Y., Hsieh R. S., Wang N., Tai W., Joo K. I., Wang P., Gu Z., Tang Y., Clickable protein nanocapsules for targeted delivery of recombinant p53 protein, *Journal of the American Chemical Society*, 136 (2014) 15319–15325.
135. Zhao M., Biswas A., Hu B., Joo K.-I., Wang P., Gu Z., Tang Y., Redox-responsive nanocapsules for intracellular protein delivery, *Biomaterials*, 32 (2011) 5223–5230.

136. Harada A., Matsuki R., Ichimura S., Yuba E., Kono K., Intracellular environment-responsive stabilization of polymer vesicles formed from head-tail type polycations composed of a polyamidoamine dendron and poly(L-lysine), *Molecules*, 18 (2013) 12168–12179.

137. Kim E., Kim D., Jung H., Lee J., Paul S., Selvapalam N., Yang Y., Lim N., Park C. G., Kim K., Facile, template-free synthesis of stimuli-responsive polymer nanocapsules for targeted drug delivery, *Angewandte Chemie International Edition*, 49 (2010) 4405–4408.

Chapter 2

Smart Polymers–Functionalized Carbon Nanotubes Delivery Systems

Guan Wang and Ye Liu
*Institute of Materials Research and Engineering, Agency for Science, Technology and Research (A*STAR), 2 Fusionopolis Way, Innovis, #08-03, Singapore 138634*
wangg@imre.a-star.edu.sg

Carbon nanotubes (CNTs) with favorable sizes and unique structures such as high aspect ratio, larger surface area, and stable tubular conformation, have become ideal nanocarriers to load and deliver cargo species *in vitro* and *in vivo*. In this chapter, we will present a select group of functionalized carbon nanotubes for the application as delivery systems for various payloads, including small molecular anticancer drugs, genes, and proteins. The relationship between the functionalization and specific application will be discussed.

2.1 Introduction

CNTs are one dimensional nanomaterial of one to multiple rolls of carbon sheets, which usually have diameters of nanometers

Polymer Capsules
Edited by Ye Liu and Xian Jun Loh

ISBN 978-981-4774-60-4 (Hardcover), 978-0-429-42873-9 (eBook)
www.jennystanford.com

sizes and with the length in the range of tens of nanometer sizes [1]. Their exceptional properties of high aspect ratio, large surface area, and stable tubular structures strongly enhance their ability to load various payloads, including small molecular chemicals, plasmids DNA, SiRNA, and proteins [2]. Besides, the nanosizes of CNTs could improve the passive targeting ability to tumors *in vivo* through an enhanced permeability and retention (EPR) effect, which minimizes the side effect caused by indiscriminate delivery of drugs to tumor and regular cells [3]. It is still under debate regarding how CNTs enter mammalian cells. In some reports, CNTs have been demonstrated as uptaken presumably through endocytosis because their sizes are similar to those of the biomolecules such as enzymes, proteins, plasmids DNA, ion channels, and the glomerular filtration barrier [4]. There are also scenarios where CNTs enter mammalian cells through an endocytosis independent route, a unique "needle penetration" mechanism [5].

Water dispersity is a prerequisite for CNTs to be used as delivery systems in a biological environment. Unfortunately, pristine CNTs are insoluble in almost all solvents. Hence, it is vital to functionalize CNTs to bring in water dispersity to facilitate the loading of payloads and to improve biocompatibility and targeting ability. The acid treatment method is well established to cut pristine CNTs into shorter tubes, to remove the impurities, and to surface functionalize CNTs with active carboxyl groups [6]. And further functionalization of CNTs with polymers, either noncovalently or covalently, is widely reported. After functionalization, various payloads could be loaded into the internal-functionalized CNTs physically or attached to the surface or to the polymers on the surface of CNTs through π-stacking or van der Waals interaction. Payloads could also be covalently attached to organic groups on CNTs or the polymer chains on CNTs. In some work, ligands such as folate moiety, arginine-glycine-aspartic acid (RGD), epidermal growth factor (EGF), and biotin have been further attached to CNTs to provide targeting ability.

Significant scientific research on CNTs has been reported towards the understanding and development of their applications as delivery systems for various species *in vitro* and *in vivo*. A few excellent reviews by other authors on the topic have been published and are included here as references for further reading [7–10]. In this chapter, we will focus on the development of selected range CNTs for anticancer drugs, plasmids DNA, SiRNA, and proteins delivery. In particular, we will review the polymeric approaches used for functionalization of CNTs and correlations between a specific payload and the respective functionalization.

2.2 Polymers-Functionalized Carbon Nanotubes for Drugs Delivery

2.2.1 Paclitaxel

The majority of the US Food and Drug Administration approved anticancer drugs for chemotherapy have limited water solubility, which limits their clinical applications. Paclitaxel (PTX) is a potent anticancer drug, which has very poor water solubility. The commercial paclitaxel formulation (Taxol) consisted of Polyoxyl 35 Castor Oil (Cremophor EL) and ethanol. However, Cremophor EL may cause hypersensitivity and other side effects in some patients [11]. To circumvent the drawbacks, use of nanomaterials such as CNTs as drug delivery systems for PTX is being widely investigated.

PEGylation of CNTs is the most commonly used strategy to increase their water dispersity and biocompatibility. Dai *et al.* reported the first successful demonstration on the use of CNTs for *in vivo* tumor treatment with efficacy in mice [12]. This was done by a conjugation of PTX onto PEGylated hipco single wall carbon nanotubes (SWCNTs) through a 2′-OH position on PTX to the terminal amine group of the branched poly(ethylene glycol) (PEG) on SWCNTs via a cleavable ester bond. The resultant SWCNT-PTX conjugate performs better then commercial Taxol, PEGylated paclitaxel (PEG-PTX), and DSPE-PEG conjugated paclitaxel (DSPE-PEG-PTX) in terms of tumor uptake of PTX and tumor suppression. Using a different strategy, the physical loading of PTX onto PEG-graft-single walled CNTs (PEG-g-SWCNTs) or PEG-graft-multi-walled CNTs (PEG-g-MWCNTs) in a saturated solution of PTX in methanol was also reported by Liu *et al.* [13]. *In vitro* cytotoxicity evaluation in HeLa cells and MCF-7 cells indicated that PTX loaded in the PEG-g-SWCNTs or PEG-g-MWCNTs could kill cancer cells more effectively than free PTX.

There are also examples of other polymers used for modification of CNTs. Atyabi *et al.* reported the use of hyperbranched polycitric acid (PCA)–functionalized MWCNTs to conjugate with PTX, forming a cleavable ester bond [14]. The resultant MWCNT-g-PCA-PTX had a more significant cytotoxic effect on both A549 and SKOV3 cell lines, as compared to free PTX. The possible reason is the increased uptake of MWCNT-g-PCA-PTX than free PTX. Geckeler *et al.* reported a PTX-loaded water-soluble supramolecular complex based on SWCNTs by using a synthetic polyampholyte, PDM [15]. They suggested that the PDM copolymer containing the tertiary amines, which may be released from the endosome through the proton sponge mechanism, can bring in increased cellular accumulation. The PDM copolymer could envelope SWCNTs and hydrophobic anticancer drugs non-covalently and the resulting complexes are promising drug delivery systems.

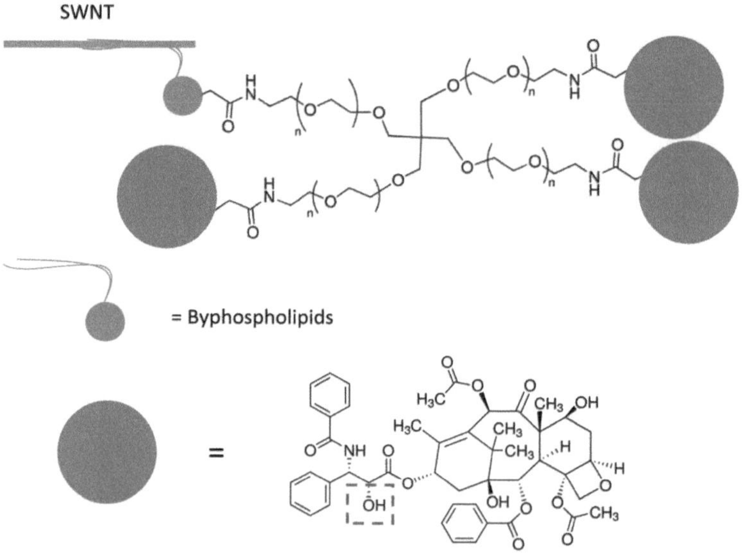

Figure 2.1 Schematic illustration of paclitaxel conjugation to SWNT-functionalized byphospholipids with branched-PEG chains. The PTX molecules are reacted with succinic anhydride (at the OH site, dashed square) to form cleavable ester bonds and linked to the termini of branched PEG, via amide bonds. This allows for releasing of PTX from nanotubes by ester cleavage *in vivo*.

PTX delivery systems based on multifunctional CNTs or CNTs complexes are also reported. Xu and Marshall *et al.* reported a drug delivery system based on CNTs and controlled the release of compounds from within the nanotubes inside cells using inductive heating with an external alternating current or by using a pulsed magnetic field [16]. A 100-fold lower concentration could be used as compared to exogenous treatment, yet through such an encapsulation strategy achieve a comparable ~70% cancer kill rate. Jain *et al.* explored how the surface chemistry of CNTs could affect the trafficking, subcellular translocation, and the therapeutic efficiency of drugs loaded in the respective CNTs [17]. Their findings sparked off the inspiration to embark on the surface chemistry of CNTs for subcellular organelle targeting and to potentially enhance the therapeutic effect. Shin *et al.* reported polymeric microspheres prepared from PTX, polycaprolactone (PCL), 100 nm water soluble carbon nanotubes (w-CNTs) as a drug carrier and trioctylmethylammonium chloride as a room temperature ionic liquid (RTIL) [18]. The PTX-loaded PCL microspheres effectively eliminate the initial burst release and give sustained release of PTX over 60 days.

2.2.2 Doxorubicin

Doxorubicin (DOX) is another potent anticancer drug used in the treatment of a wide range of cancers, including blood cancers like leukaemia and lymphoma [19]. Compared to PTX, DOX has better water solubility and was more widely investigated. Oxidized CNTs have a large adsorption capacity for DOX. The loading and release of DOX to oxidized CNTs are dependent on pH values and the incubation time [20]. Fontana *et al.* reported the use of biocompatible amphiphilic diblock copolymer poly(ethylene glycol-b-propylene sulfide) ($PEG_{44}PPS_{20}$) to disperse MWCNTs in water and achieved good DOX loading as well as efficient cell internalization [21]. Doxorubicin prodrug with a carbamate linker, which cleaves enzymatically to cause temporal release of DOX was also loaded onto the surface of PEGylated CNTs as reported by Sengupta and coworkers [22]. The *in vivo* melanoma model study revealed that the nanotube–doxorubicin conjugate treatment could abrogate tumor growth causing no systemic side-effects associated with free DOX. Khashab *et al.* reported polymer-gated CNTs that are loaded with the anticancer drug DOX with thermo-sensitivity [23]. Polyethylenimine (PEI) was grafted to carbon nanotubes and formed hydrogen bonding with polyvinyl alcohol to control the release of DOX. It was found that there was no release at room temperature of 25 °C and slow release at body temperature of 37 °C. Cremophor EL noncovalently modified SWCNTs were also investigated for the delivery of DOX in a sarcoma tumor model to evaluate the *in vivo* efficacy [24].

CNTs were also modified with ligands for targeted cancer cells treatment. Misra and coworkers reported the use of folate decorated chitosan to encapsulate DOX-loaded CNTs [25]. Due to the hydrogen bonding interaction between CNTs, DOX, chitosan, and folic acid, the release of DOX is pH controlled. Such a folate decorated and CNTs mediated delivery system has the potential for targeted and controlled release to the tumor. Using similar components, Dong and coworkers also reported chitosan and folic acid encapsulated CNTs for DOX delivery [26]. A post modification of folic acid onto chitosan after encapsulating CNTs was carried out instead of pre-modification of chitosan with folic acid as described in Misra's work and DOX was loaded finally into the encapsulated CNTs. Their work focused on both *in vitro* and *in vivo* studies, using the drug delivery system and found that their drug delivery system has higher therapeutic efficiency and lower systemic toxicity as compared to free DOX. Mcfadden *et al.* also reported the application of the carbon nanotube-mediated drug delivery system for DOX [27]. A pH-dependent drug loading and release was optimized to achieve maximum anticancer efficacy. The folic acid target agent–modified analogue of the CNTs delivery system was also demonstrated with enhanced efficacy and selectivity. Lu and coworkers described a system based on polysaccharide [sodium alginate (ALG) and chitosan (CHI)]–modified SWCNTs for controlled release of DOX, which also included folic acid as a targeting group [28]. This doubly modified system with both ALG and CHI appears to provide better control over the release rate of the anticancer drug DOX than the individually modified system. Targeted cell killing was also demonstrated in Hela cells through the folic acid mediated endocytosis. Lu and Meng *et al.* reported the use of folic acid conjugated PEG (FA-PEG) to wrap SWCNTs noncovalently for targeted delivery of DOX to Hela cells [29]. Obvious cytotoxicity to folate receptor expressing Hela cells was observed for FA-PEG wrapped SWCNTs as compared to PEG wrapped SWCNTs and free DOX. The cellular uptake mechanism was also investigated to be a clathrin-mediated endocytosis. Hyaluronate attached MWCNTs (HA-MWCNT) were also prepared for hyaluronan receptors (HR)-mediated tumor-targeted delivery of DOX [30]. The *in vivo* study results indicated that the HA-MWCNT-DOX could inhibit tumor-growth 5 times more efficiently than free DOX in equivalent concentration. Jain *et al.* reported the conjugation of dexamethasone mesylate, a potent and selective glucocorticoid steroid having a steroid-specific binding affinity to glucocorticoid receptors, onto CNTs for the delivery of DOX [31]. Besides ligand–receptor interaction, antibody could also be used for targeting. Zou and coworkers reported the functionalization of water-soluble SWCNTs with the antibody of P-gp (anti-P-gp). The functionalized CNTs were then loaded with DOX, and the complex (DOX/anti-P-gp-SWCNTs) was used to challenge the multidrug resistance (MDR) of K562 human leukemia cells [32].

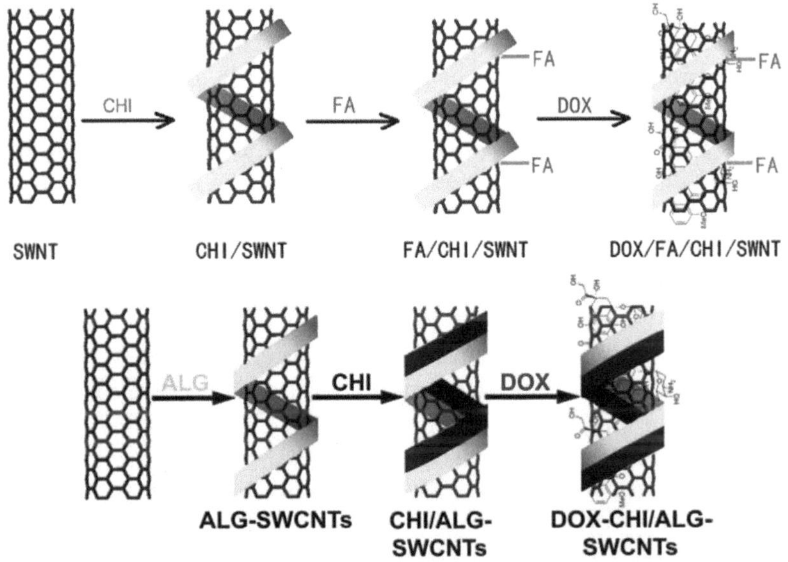

Figure 2.2 (Above) Scheme for preparation of DOX/FA/CHI/SWNTs. Reprinted from Ref. [25], Copyright 2011, with permission from Elsevier. (Below) Modification of SWCNTs (derivatized with–CO_2H groups) with ALG, CHI, and DOX. Reprinted from Ref. [28], Copyright 2009, with permission from Elsevier.

CNTs with multimodal functionalities are also investigated for the delivery of DOX. Mcfadden's group reported triply functionalized

SWCNTs with DOX, the CEA antibody and fluorescein [33]. The fluorescent complex allowed the visualization of the uptake of the CNTs by confocal microscopy and showed enhanced anticancer efficiency to WiDr colon cancer cells. Jiang *et al.* reported a dual-targeting CNTs system cooperated by PEGylated oxidized MWCNTs (O-MWCNTs-PEG) as carriers for passive targeting and an angiopep-2 ligand modification for active targeting to both brain and glioma [34]. Systematical evaluation both *in vitro* and *in vivo* of the DOX-loaded drug delivery system (DOX-O-MWCNTs-PEG-ANG) indicates it as a prospective dual-targeting drug delivery system for the treatment of brain glioma. Based on a different mechanism, a dual-targeted CNTs drug delivery system was also developed by Chen and coworkers [35]. Their method includes the use of magnetic nanoparticles for the magnetic field–guided targeting and a folic acid ligand for the ligand-receptor–mediated targeting. The authors successfully demonstrated the selective killing of U87 cells at the magnetic targeting site, causing little adverse effect to cells at the control site. Using a similar strategy, Zou and coworkers reported the conjugation of folate and iron nanoparticles with oxidized MWCNTs, and then applied them as dual-targeted drug nanocarriers to deliver DOX into cancer cells under an external magnetic field [36]. About 6-fold higher delivery efficiency of DOX than free DOX was achieved thanks to both active (biologically) and passive (magnetically) targeting functions.

In another example, a magnetically guided drug delivery system for DOX was developed by conjugating CdTe quantum dots with Fe_3O_4-filled CNTs for simultaneously targeted optical imaging of cancer cells [37]. The active targeting was realized by attaching transferrin on the surface of the modified CNTs, which facilitate the transport of the anticancer drug DOX into Hela cells. Chen and Dai *et al.* utilized a dual dye (FITC and DOX) labeling approach to functionalize CNTs for investigation of the subcellular release of DOX in living cells using confocal microscopy [38]. Their results indicated that DOX was released from CNTs in the lysosome and moved into the nucleus, while CNTs are trapped in the lysosome and remained in the cytoplasm. Their work again illustrated the uptake drug-loaded CNTs through endocytosis. A dendrimer-, ligand-, and fluorescent dye–modified CNT with pH sensitive and targeting capability was developed by Shi and coworkers [39]. An amine-terminated generation 5 poly(amidoamine) dendrimers were modified with fluorescein isothiocyanate and folic acid (FA) and then were covalently attached to the wall of MWCNTs. After loading with DOX into the multifunctional drug delivery system, the anticancer cytotoxicity turned out to be highly effective. Gu and Cheng *et al.* also reported a pH responsive drug-release behavior of DOX using CNTs [40]. CNTs were firstly wrapped with polyethylene glycol diamine (NH_2-PEG-NH_2) and linked with hydrazinobenzoic acid, which was further covalently conjugate with DOX using hydrazine as the linker. The formed hydrazone bonds are acid cleavable, providing a strong pH-responsive drug release. Hassan *et al.* reported the Luciola cruciate Luciferase-functionalized CNTs with chemiluminescence for *in vivo* imaging of the bio distribution of CNTs and CNTs loaded with DOX [41]. The modified CNTs retain the chemiluminescence activity, decay rate, and half-life through proper chemical conjugation as compared to the physical adsorption of luciferase. Branched gold nanoparticles were also employed to modify CNTs for the delivery of DOX into cancer cells [42]. Such CNTs/gold hybrids have a broad absorption band in the red and near-infrared (NIR) regions, making them useful for imaging applications. Zhang and coworkers reported SWCNTs drug delivery system with targeting and hyperthermia functionalities [43]. AS1411 aptamer was used as the targeting ligand for endocytosis and near-infrared laser excitation on SWCNTs triggers the endosome rupture and releases the drugs from the nanotubes. Fujigaya and Nakashima *et al.* reported the use of gelation from acrylamide derivatives (*N*-isopropylacrylamideacrylamide and *N,N*-dimethylacrylamide) to encapsulate SWCNTs for delivery of DOX [44]. The gel system displays pH- and photo-responsiveness due to the pH sensitive polymers and the NIR driven release of drugs from SWCNTs.

Supermolecular chemistry of water soluble CNTs are well investigated for the loading and delivery of DOX through π-stacking. Kostarelos *et al.* reported the use of the MWCNTs-DOX supermolecular complex for enhanced cytotoxicity to cancer cells as compared to free DOX and DOX in amphiphilic pluronic polymer [45]. Dai *et al.* thoroughly studied the supermolecular interaction between DOX and SWCNTs in the context of prefunctionalization of CNTs with phospholid PEG. Besides the well-known pH dependency, they also found that the drugs loadings are also dependent on diameters of CNTs favoring larger nanotubes [46]. The same group further extended the work by using phospholid-branched PEG-functionalized SWCNTs for the delivery of DOX and studied the *in vivo* biodistribution and therapeutic efficacy [47].

The preparation of nanofibrous scaffolds based on the composite of polylactic acid (PLA)/PEG/MWCNTs via the electrospinning process was reported by Irani *et al.* [48]. Loading of DOX was before the electrospinning by making a bath solution of DOX/polymers. With careful modifying of the electrospinning parameters, DOX-loaded nanofibrous were fabricated without forming crystals.

2.2.3 Platinum Metallodrugs

Platinum metallodrugs are also widely used for cancer treatment with its most famous agent being cisplatin. Tripisciano *et al.* reported the successful embedding of cisplatin into SWCNTs using organic solvent DMF [49]. Through careful characterization, they concluded that the cisplatin interact weakly with the CNTs and was coated on the wall of nanotubes. The anticancer drugs could be successful deployed and render efficient cytotoxicity in cancer cells *in vitro*. Wilson *et al.* reported the use of ultra-short SWCNTs for encapsulation of cisplatin and a controlled release of cisplatin in phosphate-buffered saline (PBS) at 37 °C can be realized by wrapping the CNTs with a pluronic surfactant [50]. Enhanced cytotoxicity against two different breast cancer cell lines, MCF-7 and MDA-MB-231 was achieved over free cisplatin. Curley and coworkers demonstrated the release of cisplatin in pluronic wrapped ultra-short SWCNTs via a radio frequency trigger [51]. The trigger is based on the heat generated by radio frequency irradiation which disrupts the polymer pluronic, as confirmed by a fluorescence quenching experiment of a thermal stable protein (APC). Pastorin *et al.* reported the encapsulation of cisplatin into MWCNTs through a nano-extraction method followed by end capping of the CNTs with gold nanoparticles [52]. A high loading of cisplatin was obtained with the capping and efficient delivery to living cells was also demonstrated.

Another strategy for delivery of platinum metallo drugs is to prepare their prodrugs analogues, which readily undergo reaction to provide anticancer drugs. Dai and Lippard *et al.* reported a CNT prodrug system by attaching a platinum(IV) complex c,c,t-

$[Pt(NH_3)_2Cl_2(OEt)\text{-}(O_2CCH_2CH_2CO_2H)]$ to the surface of amine-functionalized SWCNTs [53]. This prodrug was released into the core compound of cisplatin at low pH conditions via reduction and concomitant loss of the axial ligands which tethered the prodrug to the SWCNT's surface. Such mechanism of release of platinum(IV) prodrugs in endosomes could be applied to others with similar axial ligands tethered to the surface of SWCNTs. Moreover, the toxic character of the SWCNT-Pt(IV) complex surpasses the free cisplatin on a per platinum basis. The same team further extended the system by attaching a targeting moiety, folic acid, onto one axial ligand of the platinum prodrug tethered to SWCNT to form a complex [54]. Significantly improved cell killing properties was obtained for the complex, largely due to the enhanced uptake of the complex through the targeted endocytosis. It was also found that major cisplatin 1,2-d(GpG) intrastrand is formed upon reduction in the cellular environment, which crosslinks with nuclear DNA. Direct imaging the distribution and quantification of CNTs bioconjugated with cisplatin and a targeting ligand (EGF) was reported by Leapman and coworkers using Z-contrast scanning transmission electron microscopy [55]. The same team then demonstrated for the first time the use of a multifunctional CNTs complex by bioconjugation of cisplatin, quantum dots, EGF, onto SWCNT for targeted cancer chemotherapy in living animals [56]. With the help of two photon intravital videos, they obtained primary data to directly prove the rapid uptake of the drug delivery system by cancer tumor cells. Further investigation on the distribution and clearance of PEGylated SWCNTs bioconjugated with cisplatin in mice was carried out [57]. The successful inhibition of growth of head and neck tumor xenografts in animals was achieved due to the targeting effect from ligand EGF on the PEG-SWCNT–cisplatin. Their data suggested that non-PEG-SWCNTs were either excreted rapidly after 1 or 2 days, or got lodged in various vital organs of the animal. Raman signal of SWCNTs in the fecal material indicated a biliary excretion pathway for the CNTs, and SWCNTs Raman bands observed in the kidney also suggested that CNTs are very likely excreted via the renal pathway. More forms of the platinum prodrug have been discovered to provide efficient cytotoxicity after releasing the active platinum cancer drugs. Pastorin and Ang *et al.* reported an effective hydrophobic platinum (IV) complex, which could be loaded into CNTs via π-π stacking

[58]. Similar to other prodrugs, this prodrug undergoes chemical reduction to its cytotoxic and hydrophilic cisplatin form, which is released from the carrier and binds with DNA for cell killing.

The development of second and third generations of cisplatin analogues such as carboplatin, oxaliplatin, heptaplatin, nedaplatin, and lobaplatin has attracted considerable attention, because these second generation or third generation drugs possess higher water solubility and display fewer side-effects than the parent agent cisplatin [59]. The use of CNTs for the loading of carboplatin serves as a continuous depot for the release of this [60]. In contrast, carbon nanofibers (CNFs) could not emit sufficient release of carboplatin despite the fact that the loading of carboplatin is still comparable to CNTs. As a result, the CNTs second-generation drug-loaded carboplatin showed greater anticancer activity as compared to the CNFs-loaded carboplatin and the free carboplatin, which lost its activity after 3 days.

2.3 Polymers-Functionalized Carbon Nanotubes for Gene Delivery

CNTs have also been applied for gene delivery in a similar way to drug delivery. The principle of gene therapy is to introduce nucleic acids that could repair, regulate, replace, and add or delete a certain genetic target causing a particular disease. Both non-viral and viral vectors have been researched for gene delivery; however, the use of viral vectors has brought in concerns on immunogenicity and viral vectors generally have limited genetic payloads and poor targeting to an interested cell population. Hence, the development of biocompatible nanomaterials as non-viral vectors becomes a desirable option for gene delivery. Moreover, nanomaterials could protect the nucleic acid payloads from degradation or cleavage [61].

Genes are negatively charged and have generally been loaded onto CNTs containing positive charges after noncovalent or covalent functionalization. The first example of gene delivery using CNTs involved the functionalization of CNTs with amine terminated functional groups through covalent 1,3-dipolar cyclo addition of azomethineylides [62]. The loading of plasmids DNA was facilitated by the electrostatic binding between the amine groups on the CNTs and the phosphate groups of DNA. A marker gene in CHO cells was used to demonstrate the delivery of genes into nucleus and gene transfections. It was also found that the highest level of gene transfections was achieved when the ratio of charges on CNTs to the phosphate groups in DNA is 2:1 and 6:1.

Liu's group then reported the polyethylenimine-grafted multiwall carbon naotubes (PEI-g-MWCNTs) for DNA delivery [63]. The synthesis of the PEI-g-MWCNTs involved a cationic polymerization of aziridine in the presence of amine-functionalized MWCNTs (NH_2–MWNTs) and PEI was grafted onto the surface of MWCNTs. Due to the high contents of primary, secondary, and tertiary amines in the grafted PEI, DNA are securely noncovalently immobilized onto MWCNTs. PEI-g-MWCNTs exhibited a high transporting efficiency for the DNA delivery, while pristine or amine-functionalized MWCNTs have little effect on DNA migration.

The delivery of short SiRNA into live cells was pioneered by Dai's group [64, 65]. SiRNA was covalently incorporated onto CNTs through an enzymatically cleavable disulfide bond. Once inside lysosomal orendosomal compartments, the CNTs delivery system release SiRNA directly into the cytosol. In a control experiment, SiRNA was covalently linked to the nanotube, which showed internalization but no nuclear translocation or silencing capabilities. An additional advantage of this CNTs system is its ability of transfection human T cells and primary cells, where it is not possible for conventional nonviral cationic liposome-based agents to have transfection. In another example, Kostarelos *et al.* reported the antitumor activity and prolonged survival of the animal model using the above-mentioned amine-functionalized CNTs for the delivery of SiRNA by intratumoral administration [66].

Cai *et al.* reported the development of the "CNT spearing" technique for teffective molecular delivery by the use of nickel (Ni) embedded MWCNTs for transportation into the cell membranes [67]. Driven by an external magnetic field, plasmids DNA, and a green fluorescent protein (GFP) sequence bound to MWCNTs were transported by the spearing of DNA–MWCNTs into the targeted cells. This spearing technique results in higher transduction efficiency and higher viability as indicated by the transduction results in Bal17 Blymphoma, *ex vivo* B cells and primary neurons.

Prato *et al.* prepared dendron-functionalized MWCNTs derivatives, with numerous positively charged tetraalkyl ammonium salts at the periphery of the dendrons [68]. Interestingly, they demonstrated that cytoplasmic delivery of the nucleic acid was increased remarkably throughout the later dendron generations. Bottini and coworkers reported the use of PEGylated CNTs for conjugation of antisense oligonucleotides for the knockdown of PTPN22 in T Lymphocytes [69]. In their experiment, antisense oligonucleotides were tethered to PEGylated CNTs through a cleavable disulfide bond, which enabled efficient knockdown of PTPN22 expression both at the mRNA and protein levels.

2.4 Polymers-Functionalized Carbon Nanotubes for Protein Delivery

The use of CNTs for biological applications has attracted much attention. Besides genes, CNTs could also be loaded with another type of biomolecular cargos, proteins. The functionalization of CNTs with proteins can on one hand improve the biocompatibility of carbon nanotubes and on the other hand work as an intracellular proteins transporter. The proteins immobilized on the CNTs could also be used as potential biosensors. Dai *et al.* reported a noncovalent method to incorporate functionalized SWCNTs with both ferritin, streptavidin, and biotinyl-3,6-dioxaoctanediamine [70]. CNTs were modified with 1-pyrenebutanoic acid succinimidyl ester to give to succinimidyl groups. The pyrenyl groups from 1-pyrenebutanoic acid succinimidyl ester attach to CNTs by strong π–π interaction and the succinimidyl ester groups anchors the proteins to the amine groups in proteins. The same group also reported the nonspecific binding of various types of proteins (molecular weight ≤80 kD) to the sidewalls of SWCNTs in a noncovalent way [71]. The proteins are reported to be readily transported inside different mammalian cell lines via the endocytosis pathway using nanotubes as the transporter. Moreover, the protein-CNT conjugate can perform biological functions after been uptaken by cells by inducing apoptosis.

In a study of complement activation by CNTs, Green *et al.* found that proteins bind to CNTs in a highly selective manner, as very few proteins out of the many different proteins in plasma can bind to the CNTs [72]. Molecular dynamics simulation was also carried out to study the encapsulation of a globular protein into the CNTs [73].

The simulation result indicated that although proteins containing groups with high hydrophobicity and/or aromatic rings have great affinity to CNTs, as a whole, the attraction of protein molecules is nonspecific. This conclusion to some extent also agrees with Green's report.

Based on a specific biomolecular binding between the biotin and streptavidin (SA) pair, SWCNT-SA complex was developed by firstly functionalizing SWCNT with biotin molecules based on EDC-amine reactions followed by incubating in SA proteins for complexes formation [74]. Although SA cannot enter cells by itself, it easily enters nonadherent and adherent cell lines (HL60, CHO, and 3T3) after complex to the biotin-SWCNT transporter. In a covalent way, bovine serum albumin (BSA) was bound to SWCNTs/MWCNTs by diimide-activated amidation [75]. The formed CNT–BSA conjugates possess high water solubility with about 90% of BSA molecules retaining their activity determined by the total protein micro-determination assay.

2.5 Summary and Future Perspectives

In this chapter, we have introduced the application of CNTs based delivery systems for various payloads, including small molecular anticancer drugs, genes, and proteins. Functionalized CNTs are characterized by unique structures of high aspect ratios, large surface areas, and stable nanosized tubular structures, which result in an advantageous performance as delivery systems including the EPR effect, versatile surface functionalization possibilities and high loading of payloads.

The general four ways of loading cargo species into CNTs are as follows: firstly, payloads can be absorbed via π-π interaction or hydrophobic interaction between the aromatic rings of the payloads and the surface of CNTs. Secondly, the payloads can be covalently attached permanently or through cleavable linkers onto the surface of CNTs. Thirdly, the payloads could be encapsulated into the internal cavity of CNTs. Lastly, payloads can be attached to excipient components and form complexes, hydrogel, and nanofibers with CNTs.

Functionalization of CNTs with polymers is used to improve water dispersity and biocompatibility. PEGylation of CNTs is the mostly used strategy. PEG functionalization could be realized using noncovalent wrapping (e.g., phospholid PEG) and covalent grafting PEG onto CNTs. Other biocompatible polymers including CHI, ALG, and PCA are also developed for desirable properties. Ligands such as folate, EGF, RGD, biotin, and antibodies are also functionalized onto CNTs to further bring in active targeting capability. Multimodal functionalities including dual fluorescent imaging, magnetic field responsiveness, NIR laser responsiveness, and pH responsiveness have been harnessed onto CNTs, which further empowers the use of CNTs as delivery systems.

Nevertheless, there are still concerns over the toxicity of CNTs in clinical use. Although there are *in vivo* toxicity studies on CNTs, these studies are of a general, short term nature. For the relatively large nanoparticles of CNTs (a few hundred nanometers length), it does not easily degrade or decompose into small biocompatible components. Unlike nanoparticles of less than 5 nm sizes, which could be cleared by the body through the renal clearance pathway [76], the only metabolic pathway for bigger sized CNTs is to enter the liver and get excreted as part of the bile and later the feces. This biliary process is very slow and could cause long-term toxicity after CNTs are trapped inside the organs.

Moving forward, it is a key issue to develop approaches that ensure uniformity, including the length and the diameter of CNTs and to improve the batch-to-batch consistency. It is well known that the sizes and structures of CNTs could significantly affect various biological responses, including the cell uptake mechanism and biocompatibility. New approaches to introduce functionalities, besides polymer modification which could bring multimodal functionalities, induce biodegradation, and improve the biocompatibility are highly desirable. With the well-established understanding on CNTs and innumerable exciting opportunities, the application of CNTs as delivery systems in future will move forward towards a more nature friendly and human friendly environment.

Acknowledgments

The authors acknowledge Institute of Materials Research and Engineering, Agency for Science, Technology and Research (A*STAR) for its financial support.

List of Abbreviation

arginine-glycine-aspartic acid (RGD)
bovine serum albumin (BSA)
carbon nanofibers (CNFs)
carbon nanotubes (CNTs)
chitosan (CHI)
doxorubicin (DOX)
enhanced permeability and retention (EPR)
epidermal growth factor (EGF)
folic acid (FA)
folic acid conjugated PEG (FA-PEG)
green fluorescent protein (GFP)
hyaluronan receptors (HR)
multidrug resistance (MDR)
multi-walled CNTs (MWCNTs)
near-infrared (NIR)
paclitaxel (PTX)
phosphate-buffered saline (PBS)
polycitric acid (PCA)
poly(ethylene glycol) (PEG)
poly(ethylene glycol-b-propylene sulfide) (PEG44PPS20)
polycaprolactone (PCL)
polyethylene glycol diamine (NH_2-PEG-NH_2)
polyethylenimine (PEI)
polylactic acid (PLA)
polyoxyl 35 Castor Oil (Cremophor EL)
room temperature ionic liquid (RTIL)
single-walled carbon nanotubes (SWCNTs)
sodium alginate (ALG)
streptavidin (SA)
water-soluble carbon nanotubes (w-CNTs)

References

1. Iijima S., Helical microtubules of graphitic carbon, *Nature,* 354 (1991) 56–58.
2. Vashist S. K., Zheng D., Pastorin G., Al-Rubeaan K., Luong J. H. T., Sheu F.-S., Delivery of drugs and biomolecules using carbon nanotubes, *Carbon,* 49 (2011) 4077–4097.
3. Matsumura Y., Maeda H., A new concept for macromolecular therapeutics in cancer chemotherapy: mechanism of tumoritropic accumulation of proteins and the antitumor agent smancs, *Cancer Research,* 46 (1986) 6387.
4. Ruggiero A., Villa C. H., Bander E., Rey D. A., Bergkvist M., Batt C. A., Manova-Todorova K., Deen W. M., Scheinberg D. A., McDevitt M. R., Paradoxical glomerular filtration of carbon nanotubes, *Proceedings of the National Academy of Sciences of the United States of America,* 107 (2010) 12369–12374.
5. Kostarelos K., Lacerda L., Pastorin G., Wu W., Wieckowski S., Luangsivilay J., Godefroy S., Pantarotto D., Briand J. P., Muller S. et al., Cellular uptake of functionalized carbon nanotubes is independent of functional group and cell type. *Nature Nanotechnology,* 2 (2007) 108–113.
6. Datsyuk V., Kalyva M., Papagelis K., Parthenios J., Tasis D., Siokou A., Kallitsis I., Galiotis C., Chemical oxidation of multiwalled carbon nanotubes, *Carbon,* 46 (2008) 833–840.
7. Hong G., Diao S., Antaris L., Dai H., Carbon Nanomaterials for biological imaging and nanomedicinal therapy, *Chemical Reviews,* 115 (2015) 10816–10906.
8. Wong B. S., Yoong S. L., Jagusiak A., Panczyk T., Ho H. K., Ang W. H., Pastorin G., Carbon nanotubes for delivery of small molecule drugs, *Advanced Drug Delivery Reviews,* 65 (2013) 1964–2015.
9. Waseem A., Wani S. P., Sheikh, S., Gómez-Ruiz S., Nanostructured materials functionalized with metal complexes: In search of alternatives for administering anticancer metallodrugs, *Coordination Chemistry Reviews,* 312 (2016) 67–98.
10. Lay C. L., Liu J., Liu Y., Functionalized carbon nanotubes for anticancer drug delivery, *Expert Review of Medical Devices,* 8 (2011) 561–566.
11. Gelderblom H., Verweij J., Nooter K., Sparreboom A., Cremophor EL: the drawbacks and advantages of vehicle selection for drug formulation, *European Journal of Cancer,* 37 (2001) 1590–1598.

12. Liu Z., Chen K., Davis C., Sherlock S., Cao Q., Chen, X., Dai H., Drug delivery with carbon nanotubes for *in vivo* cancer treatment, *Cancer Research*, 68 (2008) 6652–6660.
13. Lay C. L., Liu H. Q., Tan H. R., Liu Y., Delivery of paclitaxel by physically loading onto poly(ethylene glycol) (PEG)-graft-carbon nanotubes for potent cancer therapeutics, *Nanotechnology*, 21 (2010) 065101.
14. Sobhani Z., Dinarvand R., Atyabi F., Ghahremani M., Adeli M., Increased paclitaxel cytotoxicity against cancer cell lines using a novel functionalized carbon nanotube, *International Journal of Nanomedicine*, 6 (2011) 705–719.
15. Lee Y., Geckeler K. E., Cellular interactions of a water-soluble supramolecular polymer complex of carbon nanotubes with human epithelial colorectal adenocarcinoma cells, *Macromolecular Bioscience*, 12 (2012) 1060–1067.
16. Wu C. H., Cao C., Kim J. H., Hsu C. H., Wanebo H. J., Bowen W. D., Xu J., Marshall J., Trojan-horse nanotube on-command intracellular drug delivery, *Nano Letters*. 12 (2012) 5475–5480.
17. Das M., Singh R. P., Datir S. R., Jain S., Surface chemistry dependent "switch" regulates the trafficking and therapeutic performance of drug-loaded carbon nanotubes, *Bioconjugate Chemistry*, 24 (2013) 626–639.
18. Kim S. Y., Hwang J. Y., Seo J. W., Shin U. S., Production of CNT-taxol-embedded PCL microspheres using an ammonium-based room temperature ionic liquid: As a sustained drug delivery system, *Journal of Colloid and Interface Science*, 442 (2015) 147–153.
19. Brayfield A., *Martindale: The Complete Drug Reference*, 38th Ed. (2013), Pharmaceutical Press, London.
20. Wang Y., Yang S. T., Liu Y., Wang H., Adsorption and desorption of doxorubicin on oxidized carbon nanotubes, *Colloids and Surfaces, B: Biointerfaces*, 97 (2012) 62–69.
21. Di Crescenzo A., Velluto D., Hubbell J. A., Fontana A., Biocompatible dispersions of carbon nanotubes: a potential tool for intracellular transport of anticancer drugs, *Nanoscale*, 3 (2011) 925–928.
22. Chaudhuri P., Soni, S., Sengupta S., Single-walled carbon nanotube-conjugated chemotherapy exhibits increased therapeutic index in melanoma, *Nanotechnology*, 21 (2010) 025102.
23. Mashat A., Deng L., Altawashi A., Sougrat R., Wang G. C., Khashab N. M., Zippered release from polymer-gated carbon nanotubes, *Journal of Materials Chemistry*, 22 (2012) 11503–11508.

24. Liu H., Xu H., Wang Y., He, Z., Li S., Effect of intratumoral injection on the biodistribution and therapeutic potential of novel cremophor EL-modified single-walled nanotube loading doxorubicin, *Drug Development and Industrial Pharmacy,* 38 (2012) 1031–1038.

25. Huang H., Yuan Q., Shah J. S., Misra R. D., A new family of folate-decorated and carbon nanotube-mediated drug delivery system: synthesis and drug delivery response, *Advanced Drug Delivery Reviews,* 63 (2011) 1332–1339.

26. Ji Z., Lin G., Lu Q., Meng L., Shen X., Dong L., Fu C., Zhang X., Targeted therapy of SMMC-7721 liver cancer in vitro and in vivo with carbon nanotubes based drug delivery system, *Journal of Colloid and Interface Science,* 365 (2012) 143–149.

27. Heister E., Neves V., Lamprecht C., Silva S. R. P., Coley H. M., McFadden J., Drug loading, dispersion stability, and therapeutic efficacy in targeted drug delivery with carbon nanotubes, *Carbon,* 50 (2012) 622–632.

28. Zhang X. K., Meng L. J., Lu Q. H., Fei Z. F., Dyson P. J., Targeted delivery and controlled release of doxorubicin to cancer cells using modified single wall carbon nanotubes, *Biomaterials,* 30 (2009) 6041–6047.

29. Niu L., Meng L., Lu Q., Folate-conjugated PEG on single walled carbon nanotubes for targeting delivery of doxorubicin to cancer cells, *Macromolecular Bioscience,* 13 (2013) 735–744.

30. Datir S. R., Das M., Singh R. P., Jain S., Hyaluronate tethered, "smart"multi walled carbon nanotubes for tumor-targeted delivery of doxorubicin, *Bioconjugate Chemistry,* 23 (2012) 2201–2213.

31. Lodhi N., Mehra N. K., Jain N. K., Development and characterization of dexamethasonemesylate anchored on multi walled carbon nanotubes, *Journal of Drug Targeting,* 21 (2013) 67–76.

32. Li R., Wu R., Zhao L., Wu M., Yang, L., Zou H., P-glycoprotein antibody functionalized carbon nanotube overcomes the multidrug resistance of human leukemia cells, *ACS Nano,* 4 (2010) 1399–1408.

33. Heister E., Neves V., Tilmaciu C., Lipert K., Beltran V. S., Coley H. M., Silva S. R. P., McFadden J., Triple functionalisation of single-walled carbon nanotubes with doxorubicin, a monoclonal antibody, and a fluorescent marker for targeted cancer therapy, *Carbon,* 47 (2009) 2152–2160.

34. Ren J., Shen S., Wang D., Xi Z., Guo L., Pang Z., Qian Y., Sun X., Jiang X., The targeted delivery of anticancer drugs to brain glioma by PEGylated oxidized multi-walled carbon nanotubes modified with angiopep-2, *Biomaterials,* 33 (2012) 3324–3333.

35. Lu Y. J., Wei K. C., Ma C. C. M., Yang S. Y., Chen J. P., Dual targeted delivery of doxorubic into cancer cells using folate-conjugated magnetic multi-walled carbon nanotubes, *Colloids and Surfaces, B: Biointerfaces*, 89 (2012) 1–9.

36. Li R. B., Wu R. A., Zhao L. A., Hu Z. Y., Guo S. J., Pan X. L., Zou H. F., Folate and iron difunctionalized multiwall carbon nanotubes as dual-targeted drug nanocarrier to cancer cells, *Carbon*, 49 (2011) 1797–1805.

37. Chen M. L., He Y. J., Chen X. W., Wang J. H., Quantum dots conjugated with Fe_3O_4-filled carbon nanotubes for cancer-targeted imaging and magnetically guided drug delivery, *Langmuir*, 28 (2012) 16469–16476.

38. Kang B., Li J., Chang S., Dai M., Ren C., Dai Y., Chen D., Subcellular tracking of drug release from carbon nanotube vehicles in living cells, *Small*, 8 (2012) 777–782.

39. Wen S., Liu H., Cai H., Shen M., Shi X., Targeted and pH-responsive delivery of doxorubicin to cancer cells using multifunctional dendrimer-modified multi-walled carbon nanotubes, Advanced Healthcare Materials, 9 (2013) 1267–1276.

40. Gu Y. J., Cheng J., Jin J., Cheng, S. H., Wong W. T., Development and evaluation of pH-responsive single-walled carbon nanotube–doxorubicin complexes in cancer cells, *International Journal of Nanomedicine*, 6 (2011) 2889–2898.

41. El-Sayed R., Eita M., Barrefelt A., Ye F., Jain H., Fares M., Lundin A., Crona M., Abu-Salah K., Muhammed M., Hassan M., Thermostable luciferase from Luciola cruciate for imaging of carbon nanotubes and carbon nanotubes carrying doxorubicin using *in vivo* imaging system, *Nano Letters*, 13 (2013) 1393–1398.

42. Minati L., Antonini V., Dalla Serra M., Speranza G., Multifunctional branched gold–carbon nanotube hybrid for cell imaging and drug delivery, *Langmuir*, 28 (2012) 15900–15906.

43. Zhang H. J., Chen C. Q., Hou L., Jin N., Shi J. J., Wang Z., Liu Y., Feng, Q. H., Zhang Z. Z., Targeting and hyperthermia of doxorubicin by the delivery of single-walled carbon nanotubes to EC-109 cells, *Journal of Drug Targeting*, 21 (2013) 312–319.

44. Fujigaya T., Morimoto T., Nakashima N., Isolated single-walled carbon nanotubes in a gel as a molecular reservoir and its application to controlled drug release triggered by near-IR laser irradiation, *Soft Matter*, 7 (2011) 2647–2652.

45. Ali-Boucetta H., Al-Jamal K. T., McCarthy D., Prato M., Bianco A., Kostarelos K., Multiwalled carbon nanotube–doxorubicin supramolecular complexes for cancer therapeutics, *Chemical Communications,* (2008) 459–461.

46. Liu Z., Sun X., Nakayama-Ratchford N., Dai H., Supramolecular chemistry on water-soluble carbon nanotubes for drug loading and delivery, *ACS Nano,* 1 (2007) 50–56.

47. Liu Z., Fan A. C., Rakhra K., Sherlock S., Goodwin A., Chen X., Yang Q., Felsher D. W., Dai H., Supramolecular stacking of doxorubicin on carbon nanotubes for *in vivo* cancer therapy, *Angewandte Chemie, International Edition,* 48 (2009) 7668–7672.

48. Aboutalebi Anaraki N., Roshanfekr Rad L., Irani M., Haririan I., Fabrication of PLA/PEG/MWCNT electrospun nanofibrous scaffolds for anticancer drug delivery, *Journal of Applied Polymer Science,* 132 (2015) 41286.

49. Tripisciano C., Kraemer K., Taylor A., Borowiak-Palen E., Single-wall carbon nanotubes based anticancer drug delivery system, *Chemical Physics Letters,* 478 (2009) 200–205.

50. Guven A., Rusakova I. A., Lewis M. T., Wilson L. J., Cisplatin@US-tube carbon nanocapsules for enhanced chemotherapeutic delivery, *Biomaterials,* 33 (2012) 1455–1461.

51. Raoof M., Cisneros B. T., Guven A., Phounsavath S., Corr S. J., Wilson L. J., Curley S. A., Remotely triggered cisplatin release from carbon nanocapsules by radio frequency fields, *Biomaterials,* 34 (2013) 1862–1869.

52. Li J., Yap S. Q., Yoong S. L., Nayak T. R., Chandra G. W., Ang W. H., Panczyk T., Ramaprabhu S., Vashist S. K., Sheu F. S., Tan A., Pastorin G., Carbon nanotube bottles for incorporation, release and enhanced cytotoxic effect of cisplatin, *Carbon* 50 (2012) 1625–1634.

53. Feazell R. P., Nakayama-Ratchford N., Dai H. J., Lippard S. J., Soluble single-walled carbon nanotubes as longboat delivery systems for platinum(IV) anticancer drug design, *Journal of the American Chemical Society,* 129 (2007) 8438–8439.

54. Dhar S., Liu Z., Thomale J., Dai H. J., Lippard S. J., Targeted single-wall carbon nanotube-mediated Pt(IV) prodrug delivery using folate as a homing device, *Journal of the American Chemical Society,* 130 (2008) 11467–11476.

55. Bhirde A. A., Sousa A. A., Patel V., Azari A. A., Gutkind J. S., Leapman R. D., Rusling J. F., Imaging the distribution of individual platinum-based

anticancer drug molecules attached to single-wall carbon nanotubes, *Nanomedicine (Lond.)*, 4 (2009) 763–772.

56. Bhirde A. A., Patel V., Gavard J., Zhang G., Sousa A. A., Masedunskas A., Leapman R. D., Weigert R., Gutkind J. S., Rusling J. F., Targeted killing of cancer cells *in vivo* and *in vitro* with EGF-directed carbon nanotube-based drug delivery, *ACS Nano*, 3 (2009) 307–316.
57. Bhirde A. A., Patel S., Sousa A. A., Patel V., Molinolo A. A., Ji Y., Leapman R. D., Gutkind J. S., Rusling J. F., Distribution and clearance of PEG-single-walled carbon nanotube cancer drug delivery vehicles in mice, *Nanomedicine* , 5 (2010) 1535–1546.
58. Li J., Yap S. Q., Chin C. F., Pastorin G., Ang W. H., Platinum(IV) prodrugs entrapped within multiwalled carbon nanotubes: Selective release by chemical reduction and hydrophobicity reversal, *Chemical Science*, 3 (2012) 2083–2087.
59. Oberoi H. S., Nukolova N. V., Kabanov A. V., Bronich T. K., Nanocarriers for delivery of platinum anticancer drugs, *Advanced Drug Delivery Reviews*, 65 (2013) 1667–1685.
60. Arlt M., Haase D., Hampel S., Oswald S., Bachmatiuk A., Klingeler R., Schulze R., Ritschel M., Leonhardt A., Fuessel S., Büchner B., Kraemer K., Wirth M. P., Delivery of carboplatin by carbon-based nanocontainers mediates increased cancer cell death, *Nanotechnology*, 21 (2010) 335101.
61. Singh R., Pantarotto D., McCarthy D., Chaloin O., Hoebeke J., Partidos C. D., Briand J. P., Prato M., Bianco A., Kostarelos K., Binding and condensation of plasmid DNA onto functionalized carbon nanotubes: Toward the construction of nanotube-based gene delivery vectors, *Journal of the American Chemical Society*, 127 (2005) 4388–4396.
62. Pantarotto D., Singh R., McCarthy D., Erhardt M., Briand J. P., Prato M., Kostarelos K., Bianco A., Functionalized carbon nanotubes for plasmid DNA gene delivery, *Angewandte Chemie, International Edition*, 43 (2004) 5242–5246.
63. Liu Y., Wu D. C., Zhang W. D., Jiang X., He C. B., Chung T. S., Goh S. H., Leong K. W., Polyethylenimine-grafted multiwalled carbon nanotubes for secure noncovalent immobilization and efficient delivery of DNA, *Angewandte Chemie, International Edition*, 44 (2005) 4782–4785.
64. Liu Z., Winters M., Holodniy M., Dai H. J., siRNA delivery into human T cells and primary cells with carbon-nanotube transporters, *Angewandte Chemie, International Edition*, 46 (2007) 2023–2027.

65. Kam N. W. S., Liu Z., Dai H. J., Functionalization of carbon nanotubes via cleavable disulfide bonds for efficient intracellular delivery of siRNA and potent gene silencing, *Journal of the American Chemical Society,* 127 (2005) 12492–12493.

66. Podesta J. E., Al-Jamal K. T., Herrero M. A., Tian B. W., Ali-Boucetta H., Hegde V., Bianco A., Prato M., Kostarelos K., Antitumor activity and prolonged survival by carbon-nanotube-mediated therapeutic siRNA silencing in a human lung xenograft model, *Small,* 5 (2009) 1176–1185.

67. Cai D., Mataraza J. M., Qin Z. H., Huang Z. P., Huang J. Y., Chiles T. C., Carnahan D., Kempa K., Ren Z. F., Highly efficient molecular delivery into mammalian cells using carbon nanotube spearing, *Nature Methods,* 2 (2005) 449–454.

68. Herrero M. A., Toma F. M., Al-Jamal K. T., Kostarelos K., Bianco A., Da Ros T., Bano F., Casalis L., Scoles G., Prato M., Synthesis and characterization of a carbon nanotube-dendron series for efficient siRNA delivery, *Journal of the American Chemical Society,* 131 (2009) 9843–9848.

69. Delogu L. G., Magrini A., Bergamaschi A., Rosato N., Dawson M. I., Bottini N., Bottini M., Conjugation of antisense oligonucleotides to PEGylated carbon nanotubes enables efficient knockdown of PTPN22 in T lymphocytes, *Bioconjugate Chemistry,* 20 (2009) 427–431.

70. Chen R. J., Zhang Y. G., Wang D. W., Dai H. J., Noncovalent sidewall functionalization of single-walled carbon nanotubes for protein immobilization, *Journal of the American Chemical Society,* 123 (2001) 3838–3839.

71. Kam N. W. S., Dai H. J., Carbon nanotubes as intracellular protein transporters: Generality and biological functionality. *Journal of the American Chemical Society,* 127 (2005) 6021–6026.

72. Salvador-Morales C., Flahaut E., Sim E., Sloan J., Green M. L., Sim R. B., Complement activation and protein adsorption by carbon nanotubes, *Molecular Immunology,* 43 (2006) 193–201.

73. Kang Y., Liu Y. C., Wang Q., Shen J. W., Wu T., Guan W. J., On the spontaneous encapsulation of proteins in carbon nanotubes. *Biomaterials,* 30 (2009) 2807–2715.

74. Kam N. W. S., Jessop T. C., Wender P. A., Dai H. J., Nanotube molecular transporters: Internalization of carbon nanotube-protein conjugates into mammalian cells, *Journal of the American Chemical Society,* 126 (2004) 6850–6851.

75. Huang W. J., Taylor S., Fu K. F., et al., Attaching proteins to carbon nanotubes via diimide-activated amidation, *Nano Letters*, 4 (2002) 311–314.

76. Choi C. H. J., Zuckerman J. E., Webster P., Davis M. E., Targeting kidney mesangium by nanoparticles of defined size, *Proceedings of the National Academy of Sciences of the United States of America*, 108 (2011) 6656–6661.

Chapter 3

Smart Polymer Capsules

Ting Wu and Ye-Zi You
Department of Polymer Science and Engineering, University of Science and Technology of China, No. 96 Jinzhai Road, Hefei, Anhui 230026, China
yzyou@ustc.edu.cn

Capsules are usually used for encapsulation and delivery of the ingredients of consumer care products, various pharmaceutics and nutrients and food. Capsules can protect and deliver ingredients to the targeted sites to improve their functions. A diverse range of materials have been used for their preparation, but polymers are one of the most versatile and important capsule materials.

This chapter focuses on the present state of smart polymers capsules. The fundamental knowledge of the formations and formulations, the properties and the performances of smart polymer capsules, together with the applications are to be discussed.

3.1 Introduction

Recently, hollow polymer capsules have attracted the extensive attention of material scientists due to their very promising potential

---

Polymer Capsules
Edited by Ye Liu and Xian Jun Loh

ISBN 978-981-4774-60-4 (Hardcover), 978-0-429-42873-9 (eBook)
www.jennystanford.com

applications in material science, nanomedicine, catalysis and so on [1, 2]. Generally, the functions of polymer capsules are able to encapsulate and protect the sensitive molecules such as drugs, enzymes, DNA fragments, and proteins in their cavities, and further deliver and release them to the targeted sites efficiently. To fully realize the functions, it is significantly important for polymer capsules to be responsive to external stimuli such as pH, temperature, oxidation/reduction and so on, which can be used to achieve the controlled encapsulation and triggered release of the payloads. Polymer capsules responding to external stimuli are considered to be "smart". To date, a wide range of smart polymer capsules have been constructed in various methods [3–5] and significant efforts have been directed to regulate their size, morphology, and functionality to augment their applications.

3.2 Preparation

In recent years, a diverse range of approaches have been developed for the preparation of nanocapsules from polymers, which generally can be classified into two categories, template-free and template-based strategies, such as self-assembly of amphiphilic block copolymers, graft copolymer, hyperbranched copolymer, and template-based polymerization strategies. In addition to the significant progress in preparation methods, building blocks to fabricate polymer capsules have also developed rapidly. The building blocks have been extended from synthetic block copolymers to homopolymers, hyperbanched polymers, graft copolymers, proteins, and peptides [6–11]. Moreover, dendrimers can also be considered as a type of capsule due to its distinct core–shell structure at some stage.

There are both advantages and disadvantages in the preparation of polymer capsules for all the reported preparation methods. According to different building blocks, various self-assembly strategies will be elucidated on in the subsequent sections. As for template-based strategies, the layer-by-layer assembly technique, single-step adsorption assembly technique, template-based polymerization techniques and the self-assembly approaches in nanodroplet are also clarified, respectively.

3.2.1 The Method of the Self-Assembly Approaches of Amphiphilic Block Copolymers

The self-assembly of amphiphilic block copolymers is one of the most widely used approaches in polymer capsule preparation. Amphiphilic block copolymers can self-assemble into multiple nanostructures at different ratios of hydrophobic block and hydrophilic block [12–14]. Similar to surfactants, with the increase of the hydrophobic block, amphiphilic copolymers self-assemble into first micelles, then worms, vesicles, and finally tubular vesicles [15]. The so-called "polymersomes" arc a class of artificial polymer vesicles formed from amphiphilic block polymers at a suitable ratio of hydrophobic to hydrophilic block (Fig. 3.1). The formed polymersomes membrane provides a physical compartmentalization which isolates the encapsulated organic molecules or biomolecules from external materials, such as those biomolecules found in biological systems. In comparison to their lipid counterparts, polymersomes exhibit enhanced stability and reduced permeability [16]. Moreover, polymersomes exhibit a low critical aggregate concentration (CAC), which means that there are only few individual amphiphilic block polymer chains in the solution after reaching CAC [17]. On the other hand, the formed polymersomes have no exchange of block copolymer chains between the separated vesicles, and show a greatly enhanced mechanical stability over liposomes [18].

Three major approaches have been developed in the preparation of polymersomes [20]. One method is a "top down" approach to start from a thin polymer film, also known as the film dispersion technique [21–23]. The opposite of the "top down" approach is the "bottom up" approach starting from a completely dissolved amphiphilic block copolymer. This method is also known as solvent-switching technique [24, 25]. The third way has been developed by Pan [26] and Armes *et al.* [27] and describes formation of polymersomes during polymerization. The authors named it a polymerization induced self-assembly or PISA process. All of these approaches will be covered in detail in the following section.

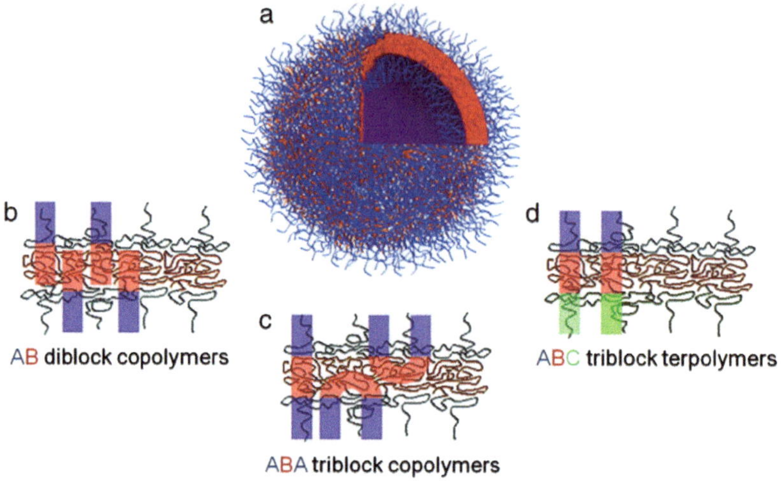

Figure 3.1 (a) Schematic diagram of the structure of a polymersome in water, showing the hydrophobic membrane (red) and hydrophilic corona (blue). Polymersome membrane conformation formed by self-assembly of (b) AB diblock and (c) ABA triblock copolymers, and (d) ABC triblock terpolymers. Reproduced from Ref. [19], Copyright 2009, with permission from The Royal Society of Chemistry.

3.2.1.1 Film dispersion technique

In the film dispersion technique, polymersomes are synthesized from a previously created polymeric thin film which is then rehydrated with a suitable solvent. The polymers are first dissolved in a good solvent for both blocks and then the solvent is removed by vacuum rotary evaporation to form a dry thin film (Fig. 3.2). Subsequently, a good solvent (water or organic solvent systems) for exclusively one of the blocks was added to rehydrate the film. The polymer layers swell and detach from the surface, and eventually close to form vesicles [19]. The sizes of the formed vesicles are related to the concentration of the polymer, thus addition of a different amount of the later solvent (for exclusively one of the blocks) may result in vesicles with different sizes. Moreover, faster rehydration results in nanometer-sized vesicles with the help of sonication and vigorous stirring. Generally, thin film rehydration can yield vesicles of a broad variety in size, from 100 nm till 100 μm, depending on the specific conditions applied and the polymer applied.

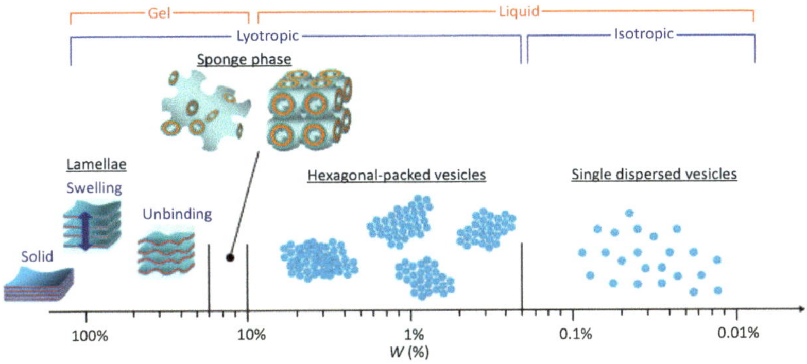

Figure 3.2 Evolution of a polymersome from a thin film to a complete vesicle shown in dependence of the water/polymer ratio and molecular weight. After the first lamellae phases are formed, packed vesicles (sponge phase) are formed initially, and later on free ones appear in solution. Reprinted from Ref. [28], Copyright 2005, with permission from Springer Nature.

3.2.1.2 Solvent-switching technique

The solvent-switching technique is one of the first methods used for the preparation of polymersomes by Discher and Eisenberg [16, 29]. Firstly, the whole polymer is dissolved in a good solvent for both blocks, and then a solvent for exclusively one of the blocks (mostly water) [30, 31] is added. The hydrophobic part of the block copolymer tends to aggregate away from the aqueous phase while the hydrophilic part is exposed to the aqueous phase, hence the self-assembly process is induced [19]. The aqueous phase can be added dropwise or through dialysis or a fast injection [32]. One can see from this approach that only polymers that are soluble in water-miscible organic solvents can be used applying this organic solvent-to-water solvent switch. Polymersomes can also be prepared directly in water. With the help of sonication [19] or detergent [34, 35], amphiphilic block copolymers can directly be dissolved in water. Apart from switching the polarity of the solvent, one can also change the pH [15, 36], temperature [33, 37] (Scheme 3.1), or light [38] of the solvent which yields in the one block of the polymer becoming insoluble to induce self-assembly. Moreover, stimuli-responsiveness is invertible which can result in polymersomes coming apart, further leading to controlled release of the encapsulated materials under external stimuli [39]. For example, temperature-responsive polymers like

PNIPAM with a LCST are soluble at low temperatures, with rising temperatures, the hydrogen bonds keeping the polymer in the solution break and thus inducing the self-assembly process into polymersomes [40].

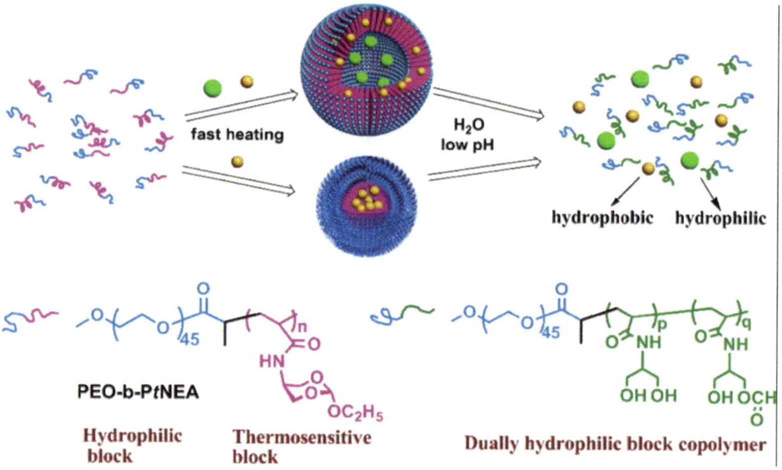

Scheme 3.1 Thermally induced formation and acid-triggered dissociation of spherical nanoparticles and polymersomes. Reproduced with permission from Ref. [33], Copyright 2013, American Chemical Society.

3.2.1.3 Polymerization-induced self-assembly

The above-mentioned self-assembly of polymers is usually performed in dilute solutions and the large-scale preparation of polymeric materials with the same well-defined nanostructure is hard to achieve. Moreover, the procedure involves multi-steps and is time-consuming [26]. Therefore, a new polymerization-induced self-assembly strategy is reported by Pan [26] and Armes *et al.* [27]. In this strategy, the synthesis of block copolymers and their self-assembly could be simultaneously conducted in a high concentration, and the morphology of the formed aggregate could be precisely controlled. Self-assembly is achieved by direct preparation of the block copolymers in a selective solvent using a macro initiator or a macromolecular chain transfer agent (macro-CTA), and during this procedure, the driving force for aggregation is produced mainly due to the change of the block chain length ratio. A typical transition from micelles to vesicles during the PISA process is illustrated in Fig. 3.3.

In the polymerization-induced self-assembly method, the block copolymers are required to have narrow molecular weight distribution [41, 42]. Therefore, anionic polymerization and controlled radical polymerizations (CRP) including nitroxide-mediated polymerization (NMP), atom transfer radical polymerization (ATRP), and reversible addition–fragmentation chain transfer (RAFT) polymerization are suitable for preparation of the block copolymers. For example, Pan and coworkers [43] designed a polymerization system, where the RAFT polymerization of styrene (St) was carried out in methanol using trithiocarbonate-terminated poly(4-vinylpyridine) (P4VP-TC) as the chain transfer agent and corona-forming block. The P4VP used in this study had a M_n = 10 400 g mol/L, M_w/M_n = 1.08. When the PS block increased past a critical chain length, aggregates with PS as the core and P4VP as corona tuning the feed molar ratios and with strict control of polymerization conditions, uniform polymer capsules can be obtained.

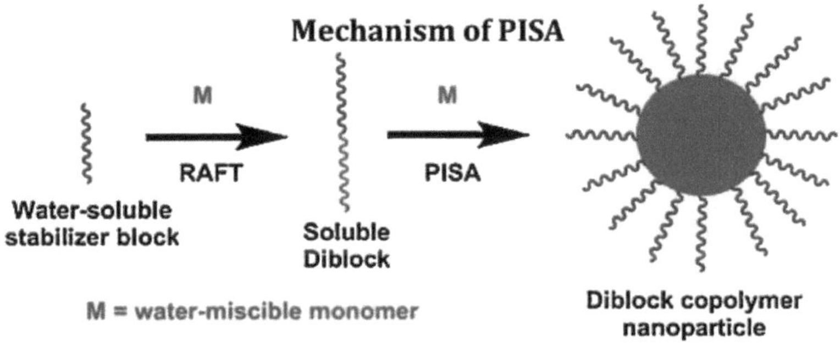

Figure 3.3 Principle of polymerization-induced self-assembly in aqueous media. Reprinted with permission from Ref. [44], Copyright 2006, John Wiley and Sons.

PISA is a powerful technique for the preparation of polymer capsules. In addition, more functional block copolymers responding to external stimuli could be synthesized to lend their assemblies more promising prospects in many applications, like in drug delivery and controlled-release fertilizer [45, 46].

3.2.2 Self-Assembly Approaches of Homopolymers

In recent years, homopolymer self-assembly has attracted the attention of material scientists due to its simplified synthetic route for nanostructures preparation [47, 48]. Amphiphilic block copolymers can self-assemble into multiple nanostructures at a different ratio of hydrophobic block and hydrophilic block due to the repulsion effect between hydrophobic and hydrophilic blocks in water. Similarly, amphiphilic homopolymers incorporated with both hydrophobic and hydrophilic units within the side chain can also self-assemble into multiple nanostructures. Thayumanavan and coworkers [49–51] first reported the self-assembly of amphiphilic homopolymers with a relatively long hydrophobic side chain and a short hydrophilic group on each repeat unit. By changing the hydrophilic/hydrophobic ratio of the monomer, hollow aggregates are formed with high hydrophobic content.

Due to homopolymer's fuzzy hydrophobic and hydrophilic domains, the self-assembly behavior of homopolymers is different from block copolymers which have distinct hydrophilic/hydrophobic blocks. Du group [52] have studies on the morphological transition and the mechanism of homopolymer self-assembly. Amphiphilic homopolymers, poly(2-hydroxy-3-phenoxypropyl acrylate) (PHPPA) with different molecular weights were synthesized by the RAFT process. A wide range of new homopolymer-based nanostructures can be prepared by changing the homopolymer chain length or co-solvent used for self-assembly, such as large compound micelles (LCMs), simple vesicles, large compound vesicles (LCVs), and hydrated large compound micelles (HLCMs), as a result of different degrees of inter-/intra-polymer hydrogen bonding in the homopolymer self-assemblies (Scheme 3.2). Moreover, micro-scale branched cylinders can form from a pre-mixed homopolymers blend (the same homopolymer with different molecular weights).

Recently, You and coworkers [53] have proposed and experimentally verified that anion–dipole interactions can enable homopolymers to self-assemble into multiple nanostructures. When weakly hydrated anion is introduced into the aqueous solutions of common polar homopolymers, such as poly(propylene oxide) (PPO), poly(vinyl methyl ether) (PVME), and poly(*N*-isopropyl acrylamide) (PNIPAM), the anion can interact with some slightly positively charged units in the homopolymers via anion–dipole interactions, which results in many tiny anion-bound-regions and non-anion-bound-regions within the polymer chains. The anion-bound-regions and non-anion-bound-regions can drive the homopolymers to self-assemble into multiple nanostructures in water like an amphiphilic block copolymer. This peculiar self-assembly of homopolymers does not require accurate control of the molecular weight and dispersity; the self-assembly can be easily tuned by the concentration of added anions into a great variety of nanostructures. This method is a significant conceptual advance in self-assembly.

In another example, the pH-responsive homopolymer vesicle was prepared using poly[2-hydroxy-3-(naphthalen-1-yl amino) propylmethacrylate] (PHNA), a well-designed homopolymer with both hydrophilic (pH responsive) and hydrophobic moieties in each repeat unit [6]. Different from ordinary polyelectrolytes, due to the highly hydrophobic naphthalene within the membrane, PHNA vesicles are physically crosslinked. Therefore, instead of a total disassociation in acidic water, these pH responsive homopolymer vesicles display a reversible "breathing" behavior (membrane shrinking/swelling in certain pH range). Also, the "breathing" behavior rather than complete disassociation prevents the outburst of the cargos (drug, siRNA, protein, etc.), which is suitable for controlled release.

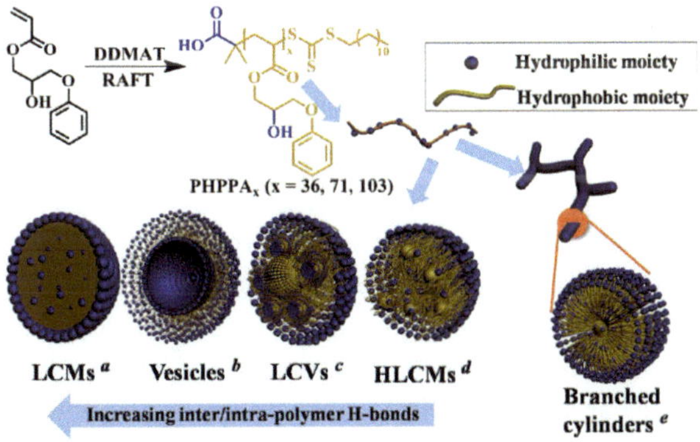

Scheme 3.2 PHPPA homopolymers self-assemblies with different morphologies: (a) large compound micelles (LCMs); (b) vesicles; (c) large compound vesicles (LCVs); (d) hydrated large compound micelles (HLCMs); (e) branched cylinders [52]. (Reproduced with permission from ref. [52]. Copyright (2013) American Chemical Society.)

In summary, various noncovalent interactions, such as hydrogen bond, hydrophobic interactions, cation–dipole interactions and anion–dipole interactions can be used as a driving force for the polymer self-assembly [52, 54–56]. Furthermore, only one step of polymerization is required in the preparation of homopolymer. Thus, homopolymers have great potential to form a large number of nanostructures which can be employed in drug delivery, biosensor, and enzyme-catalyzed reaction.

3.2.3 Self-Assembly Approaches of Hyperbranched Polymers

Hyperbranched polymers are highly branched macromolecules with a three-dimensional dendritic architecture. Over the last 20 years, hyperbranched polymers have attracted extensive attention due to their unique topological structure and unusual physical/chemical properties [11]. Compared with their linear analogues, hyperbranched polymers have a large number of terminal functional groups and better solubility. Furthermore, hyperbranched polymers are synthesized by fast one-step polymerization reactions compared to dendrimers which need multistep synthesis. Up to now, many types of hyperbranched polymers have been developed, such as polyesters, polyamines, and polyethers [57].

Amphiphilic hyperbranched polymers can self-assemble into various supramolecular structures just like amphiphilic linear block copolymers. Similarly, capsules or large compound vesicles (LCVs) can be fabricated using hyperbranched polymers as building blocks (Fig. 3.4). In 2004, Yan and coworkers [58] first reported that amphiphilic hyperbranched polymers could directly self-assemble into vesicles in water. A hyperbranched multi-arm copolymer (HBPO-star-PEO) (HBPO=hyperbranched poly(3-ethyl-3-oxetanemethanol), PEO=poly(ethyleneglycol)) with a small HBPO core can self-assemble into vesicles through a direct hydration method, and the vesicle size can be controlled from 100 nm to 100 μm depending on the hydrophilic fraction (PEO molar content). These vesicles formed from amphiphilic hyperbranched polymers are termed branched polymersomes. Subsequently, several hyperbranched polymers vesicles have been reported [47, 60–63]. For example, Shi and coworkers [61] introduced methacrylate groups into the terminals of commercialized Boltorn™ H_{20} and found that the polymers could self-assemble in water into narrowly distributed hollow polymer capsules with the diameter of around 40 nm. The vesicles were fixed by crosslinking of methacrylate groups in order to obtain a shape-persistent property in various environmental effects. For the biocompatible aliphatic polyester structure of H_{20}, the vesicles are expected to have great application prospects in various biomedicine fields.

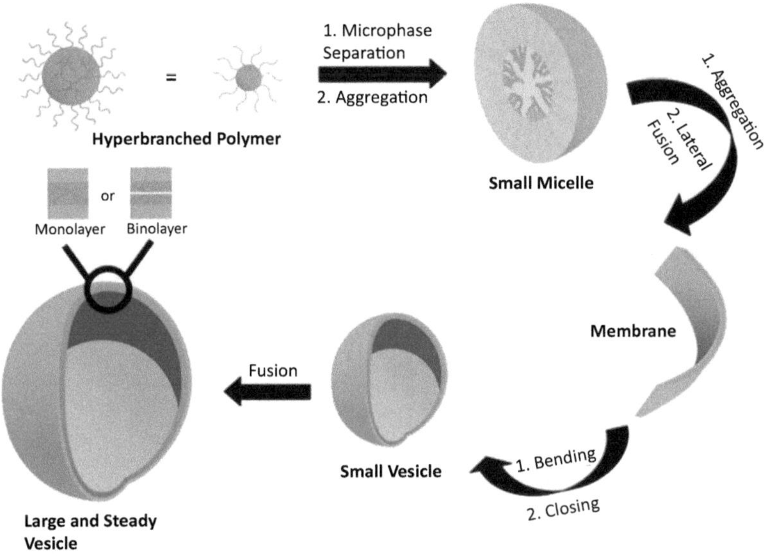

Figure 3.4 Schematic representation of amphiphilic hyperbranched polymers forming micelles, membrane, and vesicle. Reproduced with permission from Ref. [59], Copyright 2014, John Wiley & Sons Ltd.

It is worth noting that hyperbranched polymers can also self-assemble into large compound vesicles (LCVs). For example, Yan and coworkers have synthesized another kind of HBPO-star-PEO with very short PEO arms, and found that the polymers can self-assemble into large compound vesicles (LCVs) ranging from 10 to 100 μm through the solid hydration method [64]. Subsequently, more and more vesicles form and fuse together to become bigger and bigger. These self-assembly processes also extend from the surface to the interior of the polymer layer. Finally, the polymer layer turns into a huge and relatively stable three-dimensional vesicle stack (TDVS) comprising numerous interconnected vesicles.

In another example, Yan and coworkers [65, 66] synthesized hyperbranched polyglycerol grafted from β-cyclodextrin. As shown in Fig. 3.5, Janus hyperbranched polymers were constructed through the specific AZO/CD host–guest interaction from a hydrophobic hyperbranched poly(3-ethyl-3-oxetanemethanol) with an apex of an azobenzene (AZO) group and a hydrophilic hyperbranched polyglycerol with an apex of a β-cyclodextrin (CD) group. The obtained supramolecular amphiphiles further self-assemble into vesicles in water, which can be disassembled under irradiation of UV light (365 nm). Furthermore, the computer simulation of the self-assembly process and the vesicular structure by a dissipative particle dynamics simulation has shown a result consistent with the experimental result.

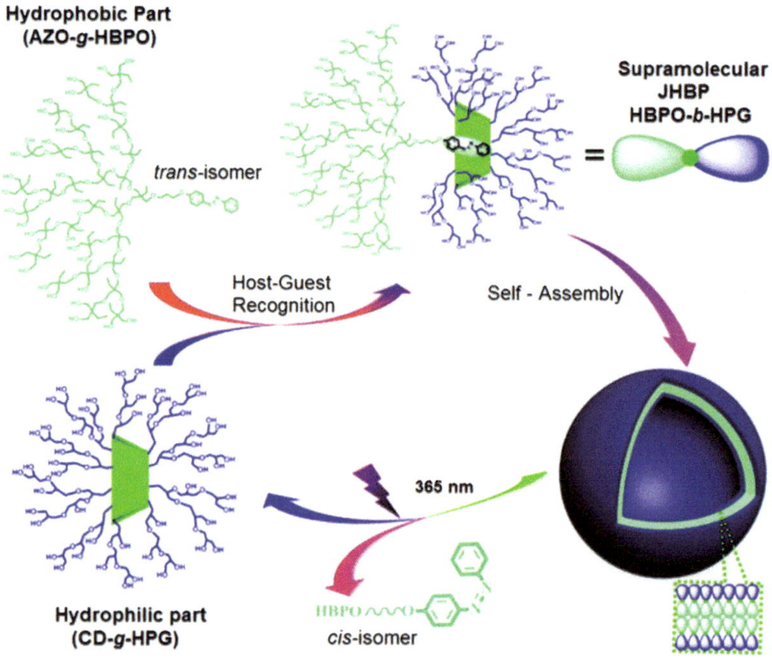

Figure 3.5 Self-assembly and disassembly processes of the supramolecular Janus hyperbranched polymer of HBPO-*b*-HPG. Reprinted with permission from Ref. [64], Copyright 2006, John Wiley and Sons.

3.2.4 Self-Assembly Approaches of Graft Copolymers

Graft copolymers are a special type of branched copolymer in which the side chains are structurally distinct from the main chain. The individual chains of a graft copolymer may be homopolymers or copolymers. Graft copolymers can be categorized into one-component graft copolymers with homopolymer side chains [68, 69] and multicomponent graft copolymers with different pendant side chains [69, 70]. Graft copolymers can also self-assemble in selective solvents into rodlike aggregates [71], spherical micelles [72, 73], and vesicles [74].

For example, Zhao and coworkers [67] synthesized a graft copolymer with a hydrophilic poly(ethylene glycol) (PEO) and poly(styrene-block-*N*-isopropylacrylamide) (PS-*b*-PNIPAM) block copolymer side chains on poly(glycidyl methacrylate) (PGMA) backbone (PGMA-graft-(PEO/PS-*b*-PNIPAM)). The graft copolymers are able to form ordered aggregates in solutions. In aqueous solution, PGMA-graft-(PEO/PS-*b*-PNIPAM) brush molecules are also able to self-assemble into vesicles as shown in Fig. 3.6. Temperature exerts a significant effect on the morphology of the structures. At a temperature above LCST of PNIPAM, the size of the vesicles decreases due to the shrinking of PNIPAM in the corona.

In another example, Liu and coworkers synthesized ternary graft copolymers with poly(3-azido-2-hydroxypropyl methacrylate), P(GMA-N_3) as the backbone polymer and alkynes-functionalized polymers, such as poly(ethylene glycol) methyl ether (MPEG), polystyrene (PS), and poly(2-cinnamoyloxyethyl methacrylate) (PCEMA) as side chains (PGMA-*g*-(PS-*r*-PCEMA-*r*-MPEG)). Evidently, MPEG was water-soluble, PS was soluble in the used oil decahydronaphthalene (DN), and PCEMA was photo-crosslinkable and soluble in neither water nor DN. PGMA-*g*-(PS-*r*-PCEMA-*r*-MPEG) is first dissolved in dichloromethane (DCM), which solubilized all three types of grafts. This solution was then slowly added into a stirred mixture of water and DN (A, Scheme 3.3), where DN solubilized absorbance [75]. Typically, a CEMA double-bond conversion of 60% was used to lock in the capsule structure. The PCEMA chains probably formed a membrane or wall separating the

DN and water phases. Photolysis of this system yielded permanent capsules.

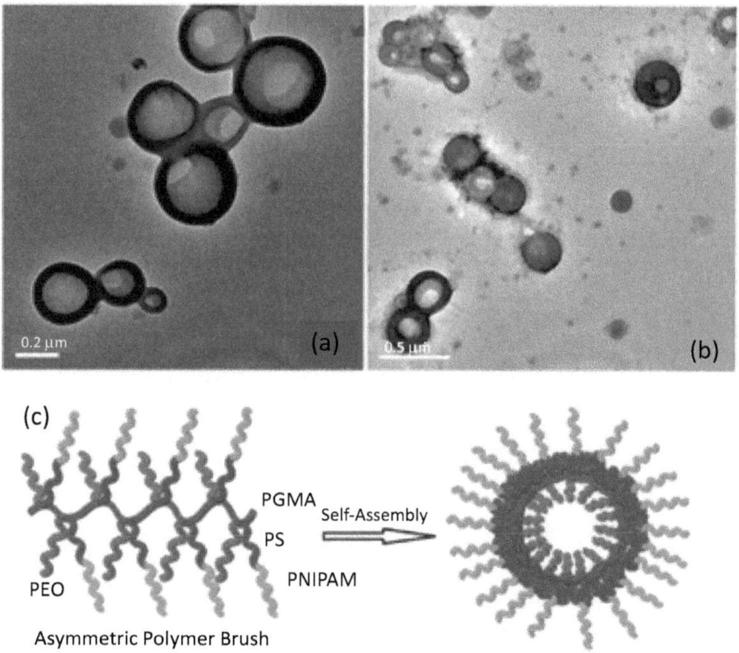

Figure 3.6 TEM images of self-assembly aggregates of PGMA-graft PEO/PS_{31}-*b*-$PNIPAM_{53}$) in water at 20 (a) and 40 °C (b), and a schematic representation for the self-assembly of the asymmetric macromolecular brush into vesicle structure (c). Reproduced with permission from Ref. [67], Copyright 2010, American Chemical Society.

3.2.5 Self-Assembly Approaches of Proteins

Recently, a new class of capsules which are prepared by the assembly of protein–polymer conjugates [76] or amphiphilic recombinant proteins [77] have been reported [76]. For instance, protein–polymer conjugates assembled at water droplet/oil interfaces into polymer capsules which are termed proteinosomes. Through tuning the chemical structure of protein–polymer building blocks, proteinosome can be synthesized with special structures and functions, such as stimulus-response, semipermeability, and enzymatic activity. A large number of guest molecules (~500 nm in diameter) could be encapsulated at high efficiency inside the proteinosomes. Furthermore, the protein–polymer capsules are robust enough to undergo dehydration and rehydration and remain structurally intact even when heated to 70 °C for 90 min. It can be found that the proteinosomes can be used to develop artificial cells which are capable of undertaking temperature cycling procedures, such as PCR-mediated amplification of entrapped genetic polymers.

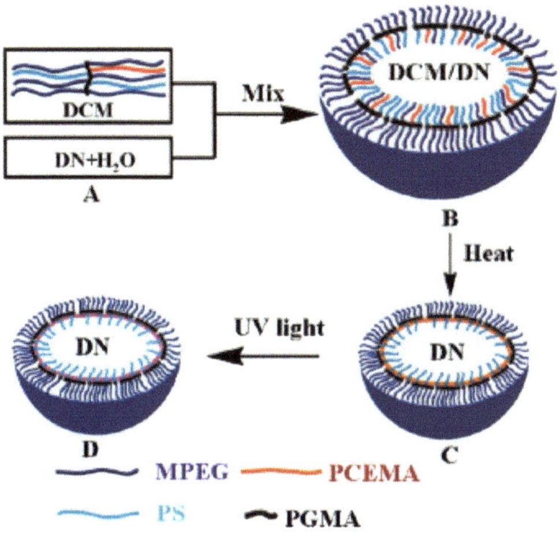

Scheme 3.3 Schematic illustration of the preparation process for the capsules. Reproduced with permission from Ref. [10], Copyright 2013, American Chemical Society.

3.2.6 Dendrimers

Dendrimers are highly branched molecules emanating from a central multivalent core and extending radially outward through a stepwise repetitive reaction sequence [78]. The first dendrimer was synthesized by Vögtle and coworkers thirty years ago [79]. Subsequently dendrimers were identified as a new class of compounds by Tomalia and Newkome [80, 81]. From then on, the field is developing rapidly [82–84]. Dendrimers consist of two topologically different parts: a little initiator core and an increasingly dense surface. Thus, at some stage dendrimers can be considered as a distinct core–shell structure with an internal cavity and an external shell.

Generally, dendrimers can be synthesized by the iterative procedure of repeating reaction units and every completed cycle means a new generation of dendrimers. There are two methods to fabricate dendrimers: the divergent and the convergent method (Fig. 3.7) [85]. Using the divergent system, dendrimers are synthesized from the core and stepwise outward method leading to a treelike structure. This method can be used in the large scale production of dendrimers, such as poly(amido amine) (PAMAM) and poly(propylene imine) (PPI) dendrimers which are even commercially available. In the convergent method, branches are synthesized separately, and the synthesis starts from the peripheral moieties and gradually moves inward towards a focal point. Subsequently the independent branches are connected to a core molecule [86, 87]. For both methods, the synthetic dendrimers have a well-defined and regular structure and a very low polydispersity which is beneficial in biological applications. However, the synthesis process is relatively tedious and costly.

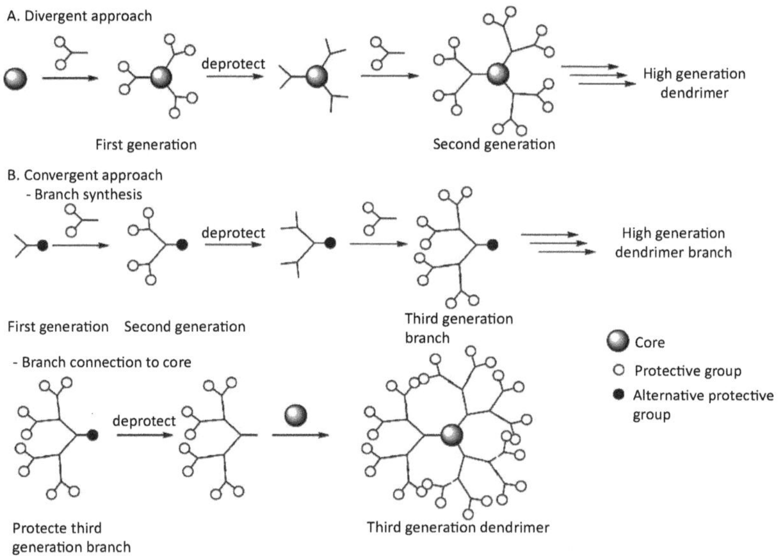

Figure 3.7 (A) Divergent approach to dendrimer synthesis; the dendrimer grows outward from the core molecule. (B) Convergent approach; separately synthesized dendrimer branches are coupled to a core. Reproduced with permission from Ref. [78], Copyright 2009, American Chemical Society.

Due to their versatility, dendrimers have attracted great attention for applications in the field of biomedicine such as drug and gene delivery [88] and imaging. Many researchers modified the dendrimer surface leading to molecules which can be reversibly imprisoned in internal cavities and then site-selectively liberated by suitable external or internal stimuli. For example, You and coworkers [89] prepared a dendritic core–shell nanostructure by functionalizing a hydroxyl-terminated dendrimer with RAFT polymerization chain-transfer agent. RAFT polymerization of *N*-isopropylacrylamide produces a shell around the dendrimer. This dendritic core–shell nanostructure with a temperature-sensitive shell may have potential applications in drug delivery. In another example, Cheng and coworkers [88] synthesized heptafluorobutyric acid-modified generation 5 (G5) poly(amidoamine) (PAMAM) dendrimers which shows excellent serum resistance and exhibits high gene transfection efficacy.

3.2.7 Layer-by-Layer Assembly Approach

Another commonly used method for the preparation of polymeric capsules is to coat a polymeric shell around a prepared sacrificial template which can subsequently be removed [3]. In the last 20 years, the layer-by-layer (LbL) assembly approach has attracted great attention due to its facile and accurate control of the size, morphology, functions and further applications of the prepared capsules. The LbL technique fabricates multilayered films on substrates and polymer capsules are prepared by removal of the core template (Fig. 3.8) [4, 91–96]. The rapid development of the LbL technique has led to the extensive applications of multilayer capsules, especially in biomedical fields such as drug delivery and biosensors. Plentiful surface coating materials, templates, strategies, and a range of interactions have been reported for the LbL technique due to its versatility. While there are still some issues about mechanical stability and aggregation in the LbL technique, in recent years researchers have devoted their efforts in finding methods to overcome these shortcomings. The increasing number of templates, layering materials, the widely exploited driving force and the extensive emerged assembly methods make the layer-by-layer technique a promising method to fabricate multilayer stimulus-responsive polymer capsules.

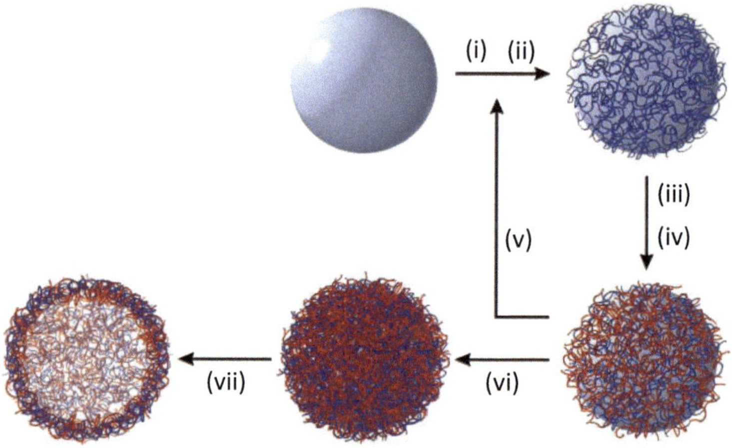

Figure 3.8 LbL assembly on colloidal substrates; (i) Immersion in a positively charged polyelectrolyte solution; (ii) rinse; (iii) immersion in a negatively charged polyelectrolyte solution; (iv) rinse; (v) repeat steps i–iv until desired number of layers has been deposited; (vi) core–shell particles can be the end of the process, or (vii) the template is dissolved away to leave a hollow polyelectrolyte capsule. Reproduced with permission from Ref. [90], Copyright 2010, John Wiley & Sons Ltd.

3.2.8 Surface/Interfacial Polymerization Approaches

In recent years, the development of polymer chemistry has led to the emergence of a class of facile methods to synthesize polymer capsules with various properties. Surface-initiated polymerization (SIP) has attracted great attention due to its diversity of polymerization methods, for example, nitroxide-mediated radical polymerization (NMRP) [97], ATRP [98–100], RAFT polymerization [101, 102], precipitation polymerization, and photopolymerization [103] are among the most popular polymerization techniques to synthesize polymer capsules via SIP. Meanwhile, the template-based polymerization strategy is also popular in the preparation of polymer capsules [3, 104]. In this method, a prepared sacrificial inorganic/organic template is used to be coated with polymer which can form a polymer shell by crosslinking reaction; subsequently the template can be removed thus leaving a polymeric capsule [97, 105, 106]. Thus, by combining this method with surface-initiated polymerization (SIP) technique, a versatile approach for the preparation of multifunctional polymer capsules has been widely studied.

Tailoring morphology and physicochemical properties is of great interest to improve and broaden the potential of polymeric capsules for advanced applications in drug delivery, nanoreactors, and biotechnology. There are obvious advantages of this versatile approach involving template-based polymerization strategy and surface-initiated polymerization for the preparation of polymer capsules: (1) The size and morphology of polymer capsules are easily controlled quantificationally by choosing different templates and controlling the degree of polymerization. (2) The composition of capsules is easily tuned by choosing different monomers. (3) Compared with the "grafting to" processes in which polymer chains are adsorbed to the template, resulting in thin polymer film, surface-initiated polymerization, namely, the "grafting from" processes, anchor the initiating group to the template thus leading to higher brush densities and thicker films [107]. Moreover, the anchored initiating group can continue to react with other groups when detaching from the template.

In the following sections, various polymerization methods accompanied by typical examples in the preparation of polymer capsules are presented. Different kinds of templates are also briefly summarized.

3.2.8.1 NMRP techniques

The pioneering example of controlled radical polymerization methods for the preparation of polymer capsules via SIP was shown by Hawker *et al.* [97]. In their paper, surface-initiated controlled NMRP was employed to synthesize polystyrene using silica nanoparticles as the template; trichlorosilyl-substituted alkoxyamine was employed as initiating groups and anchored to the surface of silica templates. Then the chain growth polymerization of monomer units proceeded from the initiating groups on the surface, and active ester and benzocyclobutene groups were utilized as the monomer units, under chemical or thermal conditions. The reaction of these monomer resulted in a crosslinked polymeric shell. After removing the silica template with HF, the hollow polymer capsules obtained were stable under solvent or thermal stimuli due to their crosslinked feature.

3.2.8.2 ATRP techniques

After the surface-initiated controlled NMRP to prepare polymer capsules via SIP, surface-initiated ATRP was also successfully developed to synthesize polymeric capsules by Kang *et al.* [106] and Fukuda [105]. For example, by the surface-initiated ATRP method, Kang and coworkers synthesized well-defined SiO_2-*g*-PS-*b*-PMMA core–shell hybrids, crosslinked PS hollow polymer capsules obtained by UV exposure of the hybrid particles and HF corrosion of the silica template [106]. The shell thickness can be controlled by tuning the PMMA and PS chain lengths in the ATRP process and the obtained capsules are stable in organic solvents. In the following study, the authors also demonstrated that the membrane of the PS hollow capsules can be functionalized by hydrophilic poly(ethylene glycol) with the terminal thiol group (PEG-SH) and hydrophobic polystyrene with the terminal azide groups (PS-N_3) via alkyne-azide and thiol-ene surface click reactions, resulting in hairy hollow capsules with different polymer brushes on the surface [108].

Based on surface-initiated ATRP strategy, Fukuda and coworkers prepared an oxetane group-carrying methacrylate, 3-ethyl-3-(methacryloyloxy)methyloxetane (EMO) initiated from the surface of monodisperse silica templates [105]. Taking advantage of the cationic ring-opening reaction of the oxetane groups of the EMO moieties, the PEMO layer of (PEMO-*b*-PMMA)-SiO_2 was crosslinked, then a uniform and monodisperse hollow capsule was obtained by dissolving the SiO_2 core with HF (Fig. 3.9).

3.2.8.3 RAFT techniques

RAFT polymerization, is one of several controlled radical polymerization techniques. It makes use of a chain transfer agent (RAFT agent) to afford control over the generated molecular weight and polydispersity during free-radical polymerization. RAFT polymerization can be performed in a large range of solvents (including water), within a wide temperature range, high functional group tolerance and absence of metals for polymerization. Furthermore, the range of commercially available RAFT agents covers close to all the monomer classes that can undergo radical polymerization. Due to its flexibility with respect to the choice of monomers and reaction conditions, the RAFT process competes favorably with other forms of living polymerization for the generation of biomaterials [100].

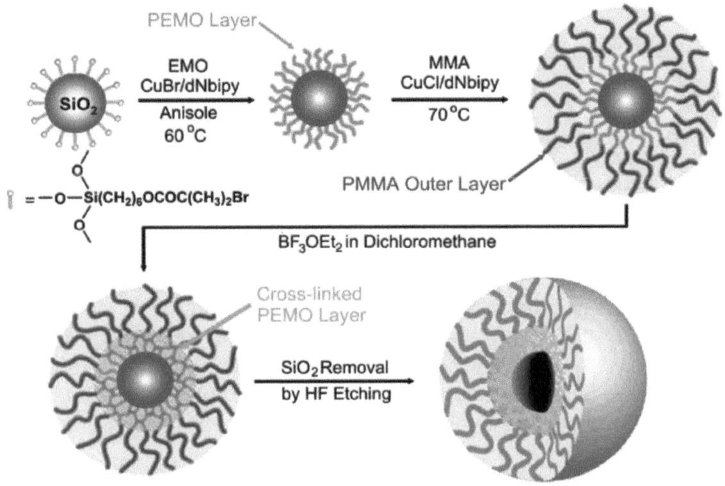

Figure 3.9 Schematic representation for the synthesis of polymeric capsules by SIP. Employing 3-ethyl-3-(methacryloyloxy)-methyloxetane (EMO), and methyl methacrylate (MMA) as monomers, PEMO-*b*-PMMA grafted silica nanoparticles (SNPs) were obtained via surface-initiated ATRP. After crosslinking the PEMO layer of (PEMO-*b*-PMMA)-SNP by cationic ring-opening reaction of the oxetane groups of the EMO moieties, the core is removed by HF etching. This process gave crosslinked (PEMO-*b*-PMMA) polymeric capsules. Reprinted with permission from Ref. [109], Copyright 2007, American Chemical Society.

As shown in a recent study by Voit and coworkers, by surface-initiated RAFT polymerization, polymeric capsules can be successfully synthesized, and the general procedure is shown in Fig. 3.10 [110]. After anchoring a dithiocarbonate RAFT agent onto the silica nanoparticle, block copolymers PtBMA-*co*-PDMIPM-*b*-PHPMA were developed from the surface. A photo-crosslinker 2,3-dimethylmaleic imidopropyl methacrylate was incorporated in the polymer to form a crosslinking shell, and after removal of silica templates by the HF buffer solution (pH 5.0) well-dispersed polymeric capsules could be obtained.

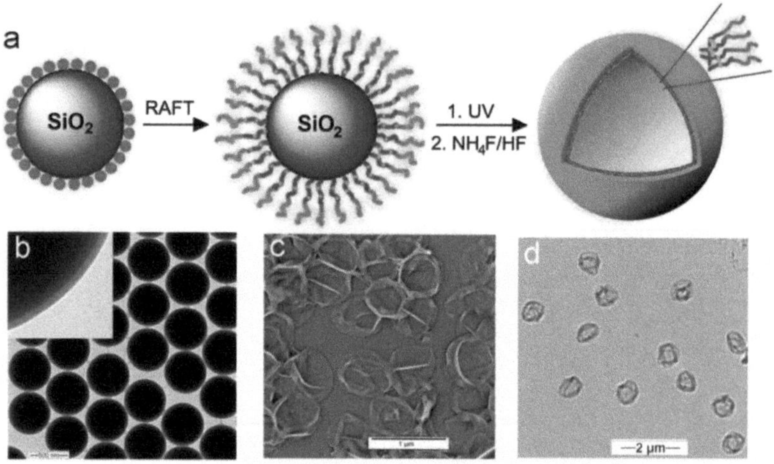

Figure 3.10 (a) Schematic illustration of the procedure to form polymeric capsules based on the surface-initiated RAFT polymerization from silica nanoparticles as templates; (b) TEM image of polymer grafted silica nanoparticles (inset is a magnification of (b)); (c) SEM and (d) TEM images of the synthesized polymeric capsules. Reprinted with permission from Ref. [110], Copyright 2011, American Chemical Society.

In the following study, based on surface-initiated RAFT methods, an intelligent polymeric capsule with an ultrathin membrane was developed by using a pH sensitive monomer diethylaminoethyl methacrylate (DEAEMA) as main units and pyridyldisulfide ethyl methacrylate (PDSM) containing a degradable disulfide bond to crosslink the membrane. Stable polymer capsules were obtained after dissolving the silicon nanoparticles. Also, the capsules can be degraded by a reducing agent such as glutathione which was obviously facilitated in a more acidic environment [111].

3.2.8.4 Precipitation polymerization

Apart from the controlled radical polymerization techniques, precipitation polymerization is also widely used to prepare polymer capsules. Precipitation polymerization is a heterogeneous polymerization process that begins initially as a homogeneous system in the continuous phase, where the monomer and initiator are completely soluble, but upon initiation the formed polymer is insoluble and thus precipitates. After precipitation, the polymerization proceeds by absorption of the monomer and initiator into the polymer particles.

Based on precipitation polymerization, You and coworkers [112] synthesized redox- and temperature-responsive polymer capsules prepared by growing a temperature-responsive shell from 3-(trimethoxysilyl)propyl methacrylate–modified silica nanoparticles (SiO_2-MPS), the shell crosslinked with bioreducible chemical linkages, and by subsequently removing the silica core by hydrofluoric acid etching. The lower critical solution temperature (LCST) of the polymer capsules can be tuned precisely. Experiment results show that the novel nanocapsules are biocompatible and have great potential in controlled drug release application.

3.2.8.5 Photopolymerization

Among these studies, the surface-initiated photopolymerization (SIPP) has also been broadly employed as another facile and versatile strategy to generate polymeric capsules [103, 113]. As shown in Fig. 3.11, based on SIPP, Yin and coworkers synthesized polymer capsules with uniform PMMA brushes by using silica nanoparticles as the template. The crosslinked PDMAEMA layer was immobilized on the surface of silica particles; subsequently under UV-irradiation, methacrylate (MMA) was employed to generate PMMA brushes using thioxanthone (TX) as the photoinitiator [103]. Polymer capsules with uniform PMMA brushes were obtained by dissolving silica templates. The surface initiating polymerization method is facile and robust in the preparation of various polymer capsules by replacing the used monomer.

Another example shown by Yang *et al.* was that a host of optically active polymer can be inserted into the polymer shell by copolymerization of optically active *N*-propargylamide copolymers and maleic anhydride and vinyl acetate [113]. Therefore, each obtained polymer capsule can be supposed as a single chiral resolution vessel because of its inner cavity and the helical polymer chains grafted in the shell.

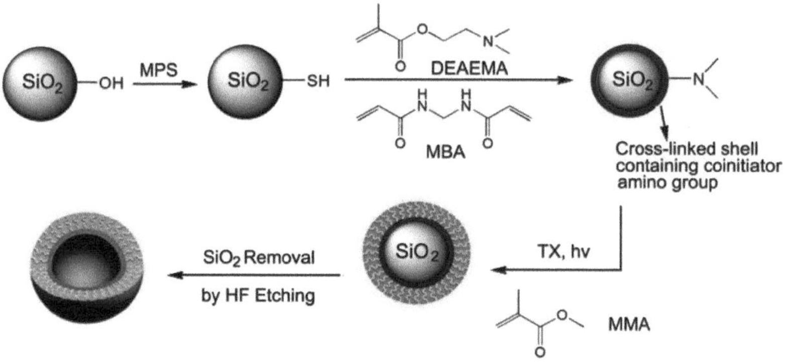

Figure 3.11 Process for synthesis of polymeric vesicles with well-defined PMMA brushes from a SiO_2 nanoparticle. Reproduced from Ref. [103], Copyright 2010, with permission from The Royal Society of Chemistry.

3.2.9 Single-Step Adsorption Approaches

Although the layer-by-layer technique is widely used to fabricate polymer capsules with a tailored size, morphology and functions, multiple and repetitive polymer adsorption procedures make it time-consuming and material-consuming. Moreover, the surface/interfacial polymerization approaches need premodifying of the surface or they suffer from very complicated procedures. On the other hand, it is difficult to remove the shell on demand in some cases. It is thus highly desirable to develop a simple and versatile approach to grow a polymer shell on nanoparticles without premodifying the nanoparticles, and removing the shell on demand. For example, You and coworkers [115] reported a new and easy method of stimuli-triggered growth and removal of a bioreducible nanoshell on silica nanoparticles (Fig. 3.12). pH or temperature could induce the aggregation of disulfide-contained branched polymers on the surface of silica nanoparticles; subsequently, the aggregated polymers could undergo an intermolecular disulfide exchange to crosslink the aggregated polymers, forming a bioreducible polymer shell around silica nanoparticles. When these nanoparticles with a polymer shell are treated with glutathione (GSH) or D,L-dithiothreitol (DTT), the polymer shell could be easily removed from the nanoparticles. The potential application of this method is demonstrated by easily growing and removing a bioreducible shell from liposomes, and improvement of *in vivo* gene transfection activity of liposomes with a bioreducible PEG shell.

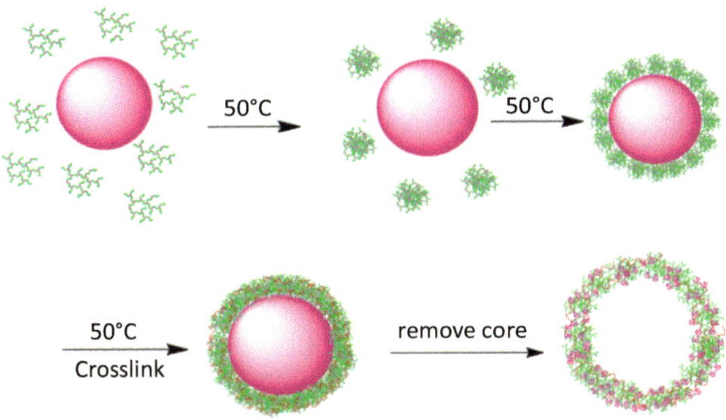

Figure 3.12 The schematic procedure for stimulus-induced growth and removal of a bioreducible shell on silica nanoparticles and the formation of bioreducible nanocapsules. Reprinted with permission from Ref. [114], Copyright 2014, American Chemical Society.

In another example, Caruso *et al.* have reported a facile method to synthesize polymer capsules via the single-step adsorption method. The polymers are adsorbed in a solid core/mesoporous shell (SC/MS) silica particle template in single-step; the polymer capsules are fabricated by crosslinking the polymer shell and removing the silica templates [116].

In addition, a great number of polymer capsules such as polyelectrolytes (i.e., PAH), polypeptide-drug conjugates (PGA-Dox), and polypeptides (i.e., PLL and PGA) have been synthesized by the single-step adsorption approach. Furthermore, Mertz *et al.* have reported the fabrication of biopolymer capsules via the adsorption of biopolymers on the surface of bromoisobutyramide (BrIBAM)-functionalized templates. The noncovalent interaction between biopolymers and BrIBAM is used as the assembly driving force [117].

3.2.10 Polymerization and Self-Assembly Approaches in Nanodroplet

Among the template-based strategies for polymer capsule preparation, a class of soft templates including emulsion droplets or vesicles have gained wide attention due to its removal under mild conditions and the easy encapsulation of compounds [118]. Nanodroplets, which have a uniform nano-environment, can control many internal chemical reactions. For example, they can act as a nanoreactor to induce the region-selectivity of organic reactions and accelerate the reaction rates [119]. Thus, emulsion droplets can be used to generate polymer capsules.

Emulsion droplets are obtained by shearing a mixture comprising two immiscible liquid phases (e.g., oil and water), one or more surfactants or co-surfactant. The shearing proceeds usually via exposure to high power ultrasound of the mixture or with a high-pressure homogenizer which are high-shearing processes. The size of the obtained droplets ranges between 50 and 500 nm, depending on the type and amount of the surfactant or co-surfactant.

Miniemulsion polymerization is a promising approach to form polymer capsules which is compatible with almost all kinds of monomers/polymers and chemical reactions [120, 121]. Moreover, it is easy to modify the surface of polymer capsules with specific molecules such as biomolecules for sensing or targeting delivery [20]. Up to now, a host of excellent reports have shown the versatility of miniemulsion polymerizations which can be used to prepare polymer capsules involving various polymerization methods ranging from radical [122] and anionic [123] polymerizations to polyaddition [124] or polycondensation methods [109, 125]. For example, Landfester and coworkers reported that a polymeric shell, for example, PMMA, PMA, or PCL were synthesized from nanodroplets by changing the ratio of the solvent/non-solvent phase (dichloromethane/cyclohexane) under mild evaporation.)

In addition to miniemulsion polymerization, the self-assembly of polymers in nanodroplets can also be used to form polymer capsules [126]. For example, You and coworkers [126] reported a temperature-responsive PEG-based branched polymer with disulfide bonds in its backbone has been synthesized via reversible RAFT polymerization of 2-(2-methoxyethoxy) ethyl methacrylate, oligo(ethylene glycol) methacrylate, and *N,N'*-cystamine bisacrylamide. Subsequently, the polymer can self-assemble into nanocapsules in nanodroplets upon heating. The produced nanocapsules are bioreducible due to which the shells are crosslinked by disulfide bonds (Fig. 3.13). These nanocapsules can load drug molecules and deliver them into a cell.

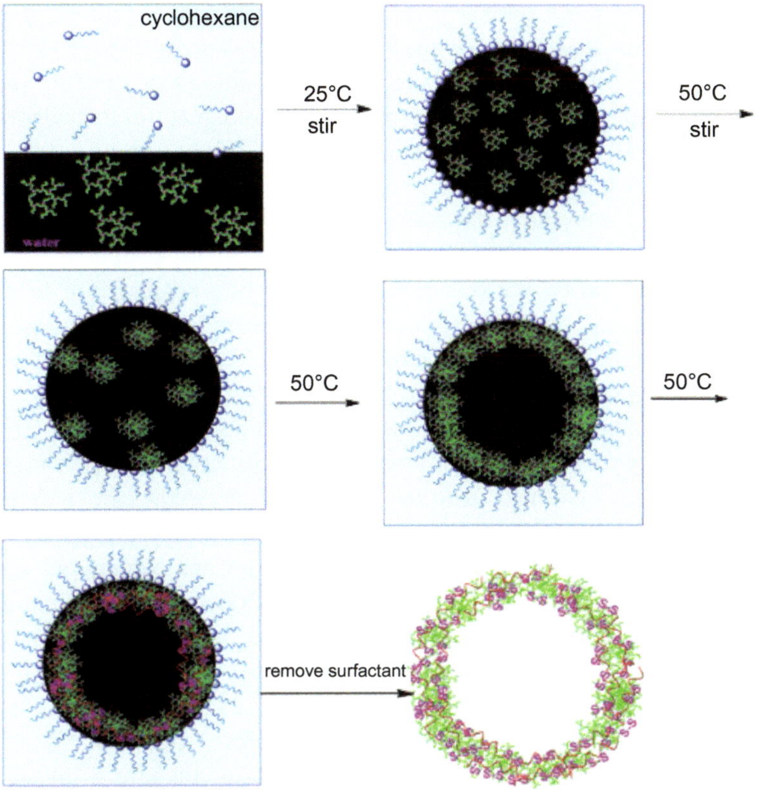

Figure 3.13 The possible mechanism for the assembly of PEG-based branched polymer in nanodroplets. Reprinted with permission from Ref. [126], Copyright 2014, John Wiley and Sons.

3.3 Application

Capsules are artificial compartmentalization inspired from natural cells. Due to its controlled size and morphology, various fabrication methods and versatile properties, polymer capsules have extensive applications in drug delivery, gene delivery, smart biomimetic microreactors, sensing, and so on. In the subsequent chapter these applications will be discussed in detail.

3.3.1 Drug Delivery

A major promising application of synthetic capsules is the controlled encapsulation and release of pharmaceuticals. Several works about capsules for drug delivery application were reported [127, 128]. A smart drug delivery carrier should possess some key abilities: high loading capacity, efficient delivery in physiological fluids, triggered release in the target sites, degradability and low immunogenicity, and toxicity [129–131].

Hollow polymer capsules are promising candidates for drug delivery due to their easily controlled properties and multifunctionality. Various molecules can be encapsulated in the cavity of capsules and drug release can be realized by external stimuli. In fact, stimuli can be divided into several kinds, for example, chemical stimuli (electrochemical stimuli, pH, ion strength, etc.), physical stimuli (ultrasound, magnetic field, mechanical deformation, etc.), and biological stimuli (receptor, enzyme-induction, etc.). Based on these stimuli, loading of drug and therapeutic molecules into polymer capsules can be realized by three strategies, including co-assembly which means drugs or therapeutic molecules are conjugated with polymers and then assembled together with polymers, pre-loading [132], which means materials are adsorbed onto or encapsulated into functional templates before the LbL assembly, and post-loading [133, 134] which means molecules are encapsulated into prepared capsules by altering the permeability of polymer capsules.

In the following sections, various stimuli including physical, chemical and biological stimuli which can trigger drug encapsulation and release are briefly summarized. Selective examples are presented.

3.3.1.1 Physical stimuli for drug delivery

Physical stimuli can affect the capsule permeability leading to controlled encapsulation and release. Generally physical stimuli include: light, temperature, magnetic fields, ultrasound, and mechanical deformation.

(i) Laser light

Polymer capsules formed by light-responsive polymers can be utilized to mask drugs and is disassembled by laser light when the particles reach the targeting sites in *in vivo* drug delivery. It is beneficial to avoid the undesired release of drugs. Near-infrared (near-IR) light is the most common laser light used in drug delivery due to the less harm it causes and much deeper penetrability in tissue [135]. Several groups have reported on the light-responsive polymer capsules which are used in cancer therapy [136–139]. Laser light is utilized not only in controlled release but also in encapsulation. For example, Bedard and coworkers have reported the light induced encapsulation of a fluorescently labeled polymer in a (PAH/poly[1-[4-(carboxy-4-hydroxyphenylazo) benzenesulfoamido]-1,2-ethanediyl, sodium salt] (PAzo))$_3$/PAH/poly (vinylsulfonate) (PVS) microcapsule [141]. In another example, Liu and coworkers developed a stimuli-triggered crosslinking strategy on block copolymer capsules containing light-labile carbamate-caged primary amines, to produce bilayer hydrophobicity-to-hydrophilicity transition and membrane permeabilization for the co-release of both hydrophobic and hydrophilic substances and light-switchable biocatalysis of enzyme-entrapped vesicle nanoreactors (see Fig. 3.14) [140].

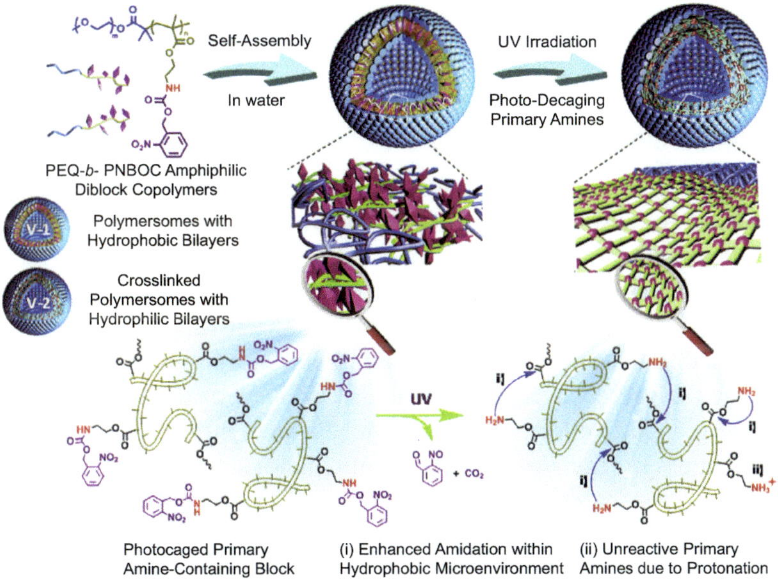

Figure 3.14 Design of vesicles exhibiting concurrent photo-triggered "traceless" crosslinking and vesicle membrane permeabilization. Reprinted with permission from Ref. [140], Copyright 2014, John Wiley and Sons.

(ii) Temperature

Temperature is another important physical stimulus in drug delivery. Temperature can not only be used to controllably encapsulate drugs, but also be useful in cancer therapy through the combination of thermal therapy with chemotherapy. Temperature has influence on the capsule fabrication process and material properties [142]. Köhler *et al.* have reported thermal responsive LbL assembled capsules with enhanced mechanical properties which can be used to encapsulate materials [134]. In another example, Hong and coworkers [143] reported a novel nanocontainer which could regulate the release of payloads, based on zwitterionic sulfobetaine copolymer-functionalized MSNs [MSNPoly(DMAEMA-*co*-DMAPS)] via surface RAFT polymerization. The dual-thermo-responsive copolymer shell has an interesting temperature-dependent behavior in aqueous solutions. The release of RhB from MSN-poly(DMAEMA-*co*-DMAPS) could be controlled by adjusting the temperature in a normal saline solution. Heating-induced volume contraction of polymer shells and accelerated diffusion synergistically make the guest molecules release faster (Fig. 3.15). This drug delivery carrier should have potential applications in cancer therapy through the combination of thermal therapy with chemotherapy.

(iii) Magnetic fields

Magnetic nanoparticles (MNPs) are a particular class of nanoparticles which are extensively utilized in biomedical fields. The so-called "magnetic drug targeting" has attracted great attention as a prospective method for drug delivery. Polymer capsules incorporated with MNPs can be employed to deliver the loadings to the target sites through an external magnetic field. Moreover, the release of drugs triggered by magnetic fields can lead to the enhancement of therapeutic effects [144, 145]. MNPs are synthesized by coating polymers on the surface of magnetic nanoparticles such as Fe_3O_4. The coating materials are required to be biocompatible. For example, Yukio Nagasaki group [146] reported the poly(ethyleneglycol)-block-poly(4-vinylbenzylphosphonate) magnetic nanoparticles containing doxorubicin as a potential targeted drug delivery system. In another example, Lu and coworkers demonstrated that the permeability of capsules can be changed by magnetic field triggered aggregates of NP [147].

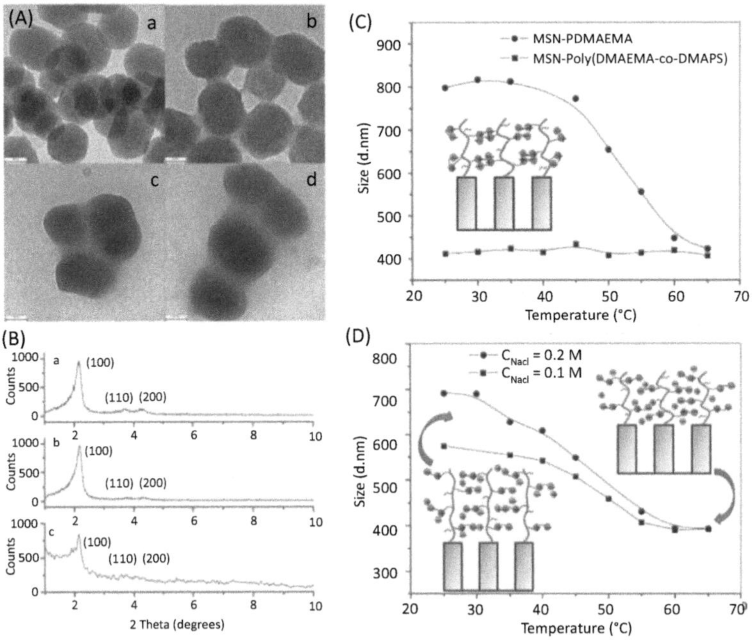

Figure 3.15 (A) HRTEM images of pristine MSNs (a) and poly(DMAEMA-*co*-DMAPS)-coated MSNs with different thickness of polymer layers (b, c, and d). All scale bars are 50 nm. (B) Powder X-ray diffraction patterns of MSN-OH (a), MSN-RAFT (b), and MSN poly(DMAEMA-*co*-DMAPS) (c). (C) Temperature dependence of hydrodynamic diameter (Dh) of MSN-PDMAEMA and MSN-poly(DMAEMA-*co*-DMAPS) in pure water. The insert represents the collapsed copolymer shell due to the inter- or intra-chain associations. (D) The relationship of Dh of MSN-poly(DMAEMA-*co*-DMAPS) with temperature in aqueous solution at two ionic strengths (0.1 M and 0.2 M NaCl). The inserts represent the possible conformations of copolymer chains below and above the LCST. Reproduced with permission from Ref. [143], Copyright 2012, John Wiley & Sons Ltd.

(iv) Ultrasound

As a strict definition, ultrasound is sound with frequencies from 20 kHz up to several gigahertz beyond the upper audible limit of human hearing. Ultrasound is widely used in chemical synthesis and biomedicine fields such as coating carbon nanotubes, polymer destruction, mixing, acceleration chemical processes, ultrasound imaging as well as drug delivery [148]. Ultrasound possesses strong power (20 kHz, 100–500 W) which are able to destroy polyelectrolyte multilayer capsules [149, 150]. Ultrasound is also used to control the permeability [151] of polymer capsules thus resulting in controlled release. Furthermore, ultrasound also has potential applications in encapsulation [152]. For example, by means of sonicating oil in protein solution, two drugs (rifampicin and indomethacin) were successfully encapsulated in prepared protein nanocarriers.

(v) Mechanical deformation

With the development of polymer chemistry, a variety of polymer capsules have emerged. A simple method to sort these capsules is through mechanical properties, namely the relative stiffness. In turn, the mechanical properties are employed to evaluate polymer capsules [153]. Moreover, the controlled encapsulation and release can be realized by the means of membrane mechanical deformation. For example, based on this method, release of fluorescently labeled dextran molecules was studied, combining fluorescence microscopy with AFM force spectroscopy [154].

3.3.1.2 Chemical stimuli for drug delivery

Chemical stimuli can change the capsule permeability leading to controlled encapsulation and release. Generally physical stimuli include: pH, ionic strength, chemical and electrochemical response, and solvent.

(i) pH

Due to the pH levels of different kinds of cells, tissues, organs, and pathophysiological states being quite different, including as in the acidic stomach interior (pH 1–3) [155], alkaline duodenum (pH 7.0–8.5) [156], and blood (pH 7.35–7.45), pH stimulus has been widely used in biomedical fields. A large number of pH-responsive nanoparticles have been synthesized. The structures and surface features of these materials may change in response to pH stimulus. These changes include deformation, disassembly, and charge switching which are beneficial in targeting drug release. The permeability of polymer capsules formed from pH-responsive polymers may increase or decrease along with the pH variation. For example, the permeability properties of polyelectrolyte multilayer capsules have been investigated as a function of pH and salt concentration [151]. The polyelectrolyte multilayer capsules can be switched between an open and a closed state by tuning the pH-value. The capsules are closed at a pH value of 8 and higher, but at a pH lower than 6 the macromolecules permeate into the capsule interior. Furthermore, this transition is reversible. This property provides an opportunity to encapsulate different materials under mild conditions. In another example, Pan and coworkers [157] reported the hybrid poly(2-(diethylamino)ethyl methacrylate)-coated mesoporous silica nanoparticles (MSN-PDEAEMA) which are pH-responsive (Scheme 3.4). The pH-responsive PDEAEMA brushes anchored on MSNs could serve as a switch to control the opening and closing of the pores. Release of guest molecules was conducted at different pHs, and the results showed rapid release in an acidic aqueous solution but very little leakage in an alkaline solution. By adjusting the pH of the solution repeatedly, the release of encapsulated molecules could be switched on and off at will (Fig. 3.14). It can be noted that these pH-sensitive nano-systems are promising in drug and gene delivery.

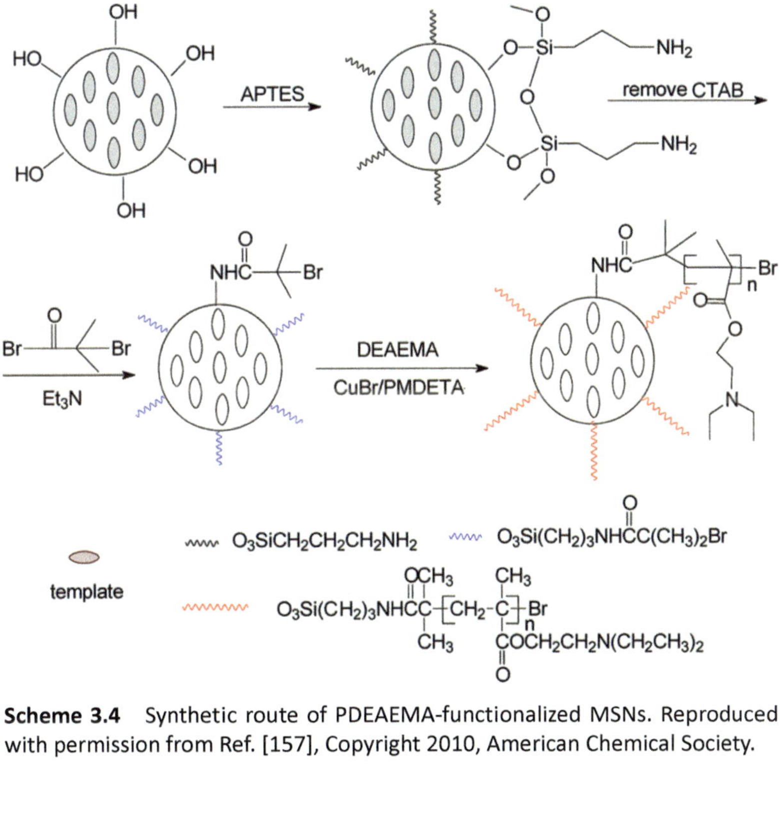

Scheme 3.4 Synthetic route of PDEAEMA-functionalized MSNs. Reproduced with permission from Ref. [157], Copyright 2010, American Chemical Society.

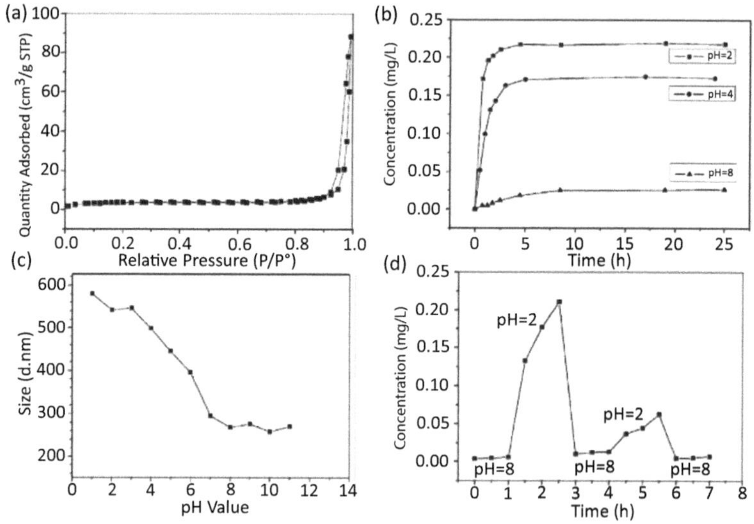

Figure 3.16 (A) Adsorption/desorption isotherm for MSN-PDEAEMA. (B) The relationship of concentration of the released RhB with time at different pH values. (C) pH dependence of hydrodynamic diameter of MSNPDEAEMA. (D) Release profile of RhB loaded in MSN-PDEAEMA nanocontainers with pH of the release medium alternately changed between 8 and 2 at ambient temperature. Reproduced with permission from Ref. [157], Copyright 2010, American Chemical Society.

(ii) Ionic strength

There are a large number of inorganic salts in physiological conditions, such as sodium chloride (NaCl) with the concentration of 0.15 M in blood, which have significant influence in the permeability of polymer capsules. Therefore, salt-induced permeability and changes of polyelectrolyte multilayer capsules have been deeply investigated [142, 158, 159] With the increase of ion strength, the permeability of polyelectrolyte multilayer capsules increase. Subsequently, the encapsulated molecules can be released with the same salt treatment. Nevertheless, an opposite effect was reported from the behavior of multilayer films at increasing salt concentration. For poly(styrenesulfonate, sodium salt) (PSS), with the increase of salts strength, the electrostatic interaction between the charged groups decreases, leading to the capsules shrinkage through hydrophobic groups of PSS. For example, Gao *et al.* synthesized capsules formed by PSS and poly(diallyldimethylammonium chloride) (PDADMAC) template on melamino-formaldehyde (MF) particles. The controlled encapsulation of dextran biomacromolecules can be achieved by opening and reclosing capsules [160].

(iii) Electrochemical stimuli

Electrical and electrochemical redox stimuli have attracted considerable attention in both research and applications due to its facile trigger mode [161]. Electrochemical impedance spectroscopy is used to measure the redox of polyelectrolyte multilayer capsules which can protect ferrous metals [162].

In the preparation of electrochemical-responsive nanoparticles, the linkers are quite important, such as the *b*-cyclodextrin-ferrocene (β-CD-Fc) linker. β-cyclodextrin (β-CD) can strongly interact with ferrocene (Fc) derivatives in a reduced state, but hardly binds the oxidized state [163]. Hence, the host–guest interaction and response to electrochemical stimuli between β-CD and Fc can be used in the synthesis of electrochemical active polymers and controlled drug release [164].

(iv) Solvent

Solvent is also an important factor in the fabrication and research of polymer capsules. Water is the most commonly used solvent and organic solvents are also used in this field. Solvent also plays an important role in the fabrication of hydrogen-bonded films [165]. For example, Zhang and coworkers [165] reported the build-up of hydrogen-bonding-directed poly(4-vinylpyridine)/poly(4-vinylphenol) (PVPy/PVPh) multilayer film that was fabricated by the LbL assembly of PVPy and PVPh from an ethanol solution. It can be found that with the increase of the ratio of *N,N*-dimethylformamide (DMF) in the mixed ethanol/DMF solvents, a significant decrease of the amount of polymers adsorbed takes place. The phenomenon may derive from the increased polarity of the adsorption solutions. As a result, solvent can be used to tune the structure and permeability of hydrogen-bonding directed multilayer polymer capsules.

3.3.1.3 Biological stimuli for drug delivery

Biological stimuli can be used in targeting delivery and release. In this section, receptor-implementing targeting and enzyme-triggering release will be introduced.

(i) Receptor-implementing targeting

To achieve efficient drug delivery, the first step is to deliver the encapsulated molecules to a precise targeting site. Otherwise, many side effects will occur if the drugs are delivered to undesired cells or organs. Receptors are the only category of stimuli to interact with the targeted capsules. Thus, capsules can be designed with proper properties or structures which can be recognized by special receptors. Moreover, only nanometer-sized delivery carriers can be employed in circulation. For example, targeting and uptake of polyelectrolyte multilayered coated 500 nm particles to colorectal cancer cells has been investigated [166].

(ii) Enzyme-triggering release

Another route to releasing encapsulated materials inside cells is by using biological stimuli elicited by enzymes. Biodegradable polyelectrolyte multilayer capsules could be attractive for delivery of drugs with an intracellular target such as nucleic acids and proteins [167]. Destroying the capsules with enzymes is of high practical importance. Release of oligonucleotide sequences has been demonstrated for microcapsules assembled by the polycation-free method [168]. An example of biodegradable microcapsules was demonstrated by Itoh *et al.* [169] who constructed DEXS/Chitosan microcapsules responsive to chitosanase; the main disadvantage of this approach is that chitosanase does not exist in mammalian cells. Release of fluorescently labeled transferrin from disulfide crosslinked polymer capsules made of poly(vinylpyrrolidone) and poly(methacrylic acid) functionalized with cysteamine was demonstrated by the addition of dithiothreitol (DTT) [169]. This approach has potential for *in vivo* applications as thiol–disulfide exchange leading to deconstruction of the capsules can occur with the assistance of intracellular proteins.

3.3.2 Gene Delivery

Gene therapy has attracted extensive attention over the past three decades as a promising approach in the treatment of a broad variety of human diseases, ranging from viral infections [170] to genetic disorders [171] and cancer [172]. The efficient delivery of DNA into cells is critical to the success of gene therapy. Hollow polymer capsules have been widely used in drug delivery, and similarly, it can also be used to deliver genes. Although some studies demonstrate polymer capsules can encapsulate DNA molecules, encapsulation of oligonucleotides are highly susceptible to degradation by endonucleases in physiological fluids and in the extracellular environment [173]. Thus, there is a higher requirement for polymer capsules to encapsulate and protect genes in a harsh biological environment. For example, Caruso and coworkers [168] reported a polycation-free encapsulation method to obtain high concentrations of short oligonucleotide chains confined within monodisperse, degradable microcapsules. The encapsulation method exploits amine-functionalized silica (SiO_2+) particles to adsorb oligonucleotides, followed by the assembly of thiol-functionalized poly(methacrylic acid) (PMASH) and poly(vinylpyrrolidone) (PVPON) multilayers [174]. Removal of the template particles produces degradable capsules filled with oligonucleotides (Scheme 3.5).

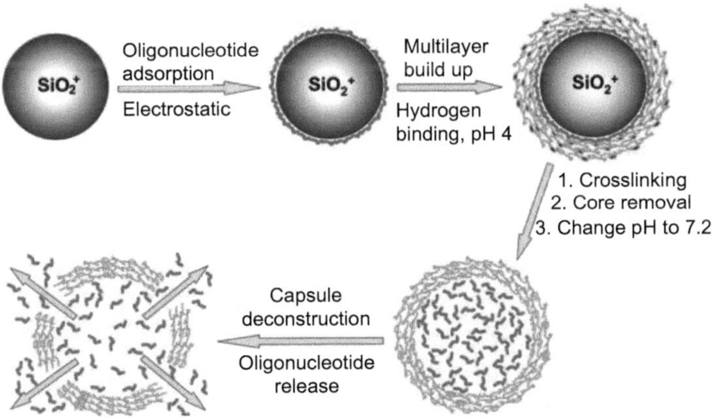

Scheme 3.5 Schematic representation of the encapsulation of short oligonucleotide sequences within polyelectrolyte capsules [168]. Reproduced with permission from Ref. [146], Copyright 2006, John Wiley & Sons Ltd.

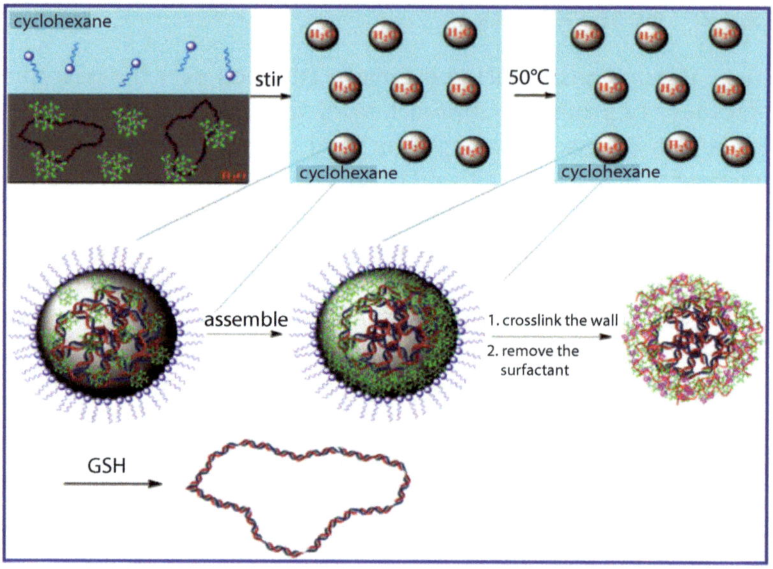

Figure 3.17 The outline for encapsulation of large-sized DNA into small-sized nanocapsules and the release of DNA from the nanocapsules in the presence of glutathione (GSH). Reprinted with permission from Ref. [114], Copyright 2014, American Chemical Society.

In another example, You and coworkers [114] reported a versatile method for encapsulating large-sized plasmid DNAs into small-sized bioreducible nanocapsules in which shearing force and surfactant can fold large-sized plasmid DNAs into small-sized emulsion droplets containing bioreducible branched polymers. Subsequently, temperature triggers the bioreducible branched polymers to aggregate and crosslink at the water/oil interface of the emulsion nanodroplet, forming a bioreducible shell around the nanodroplet. Thus, a small-sized nanocapsule (~110 nm) containing large-sized plasmid DNA (~1900 nm long) forms by the removal of the surfactant, as shown in Fig. 3.17. The nanocapsules are bioreducible and biocompatible and can act as a vector to deliver DNA into the cell.

3.3.3 Biomimetic Microreactors

In the bio-system, compartmentalization is important for metabolism, such as enzymatic reaction and protein folding.

Compartmentalization provides geometrical restriction which is beneficial in regulating cell growth and in the exchange of materials between cells and their environment [20]. Inspired by nature, a wide range of artificial polymer capsules emerge to mimic the natural counterparts, which can be used as smart nanoreactors and even artificial organelles [175, 176]. In the following subsequent chapter, the applications of polymer capsules as microreactors in enzyme catalysis, polymerization, nanoparticles synthesis, and artificial organelles are introduced.

3.3.3.1 Enzyme catalysis

A quite important application of polymer capsule micoreactors is enzyme catalysis. Enzyme can be encapsulated within the cavities of capsules, and the enzymatic reactions are performed *in situ* through the membrane diffusion of reactants and products. The permeability of polymer capsules can be tuned by the above-mentioned stimuli response strategies. For example, Voit and coworkers reported a pH sensitive poly(diethylaminoethyl methacrylate) (PDEAEM) capsules [177, 178] loaded with myoglobin. The catalytic activity of these capsules can be tuned by pH. When the pH value is 8.0, the PDEAEM membrane is deprotonated and fully hydrophobic, thus leading to difficult diffusion of hydrophilic hydrogen peroxide to the core, with the myoglobin showing a relatively low catalytic activity. When the pH value is adjusted to 6.0, the PDEAEM membrane is swollen and totally hydrophilic, thus resulting in the easy diffusion of hydrophilic substrates to the core and an acceleration of the reaction process.

3.3.3.2 Polymerization

Apart from enzyme catalysis, another important application of polymer capsule microreactors is polymerization [179–181]. The interior cavities of polymer capsules are a good confined reaction space for polymerization. For example, encapsulation of reactants into the hollow polymer capsules is beneficial to controlled/living radical polymerization on account of the persistent radical effect.

For example, Zhang and coworkers [181] reported a study on RAFT polymerization of a typical hydrophobic monomer of styrene within a hollow capsule suspended in water (Fig. 3.18). The hydrophobic monomer of styrene, the RAFT agent *S*-benzyldithiobenzoate, and the initiator 2,2′-azobis(isobutyronitrile) were encapsulated into the hollow polymer capsules, and RAFT polymerization of styrene took place in a well-controlled manner in water. Compared to the homogeneous bulk polymerization, the obtained polymer of polystyrene has a competitively low dispersity index.

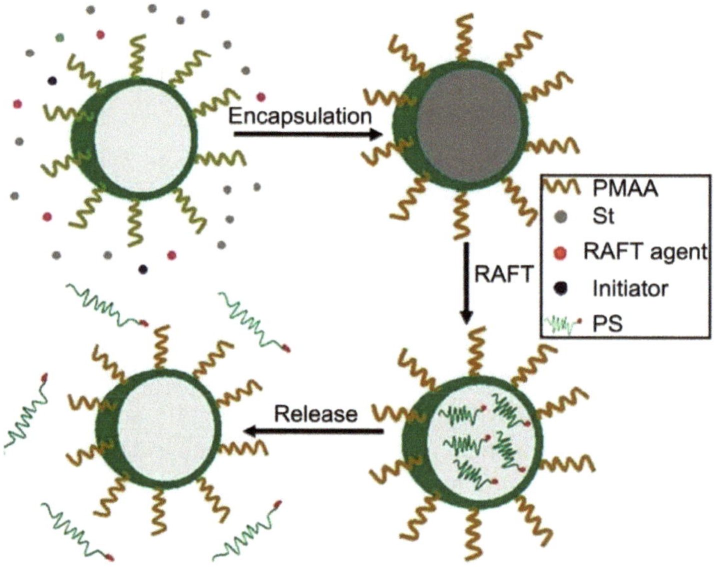

Figure 3.18 Schematic nanoreactors formation when-RAFT polymerization of styrene in water is used. Reproduced with permission from Ref. [181], Copyright 2010, John Wiley & Sons Ltd.

3.3.3.3 Nanoparticles synthesis

The hollow polymer capsule microreactors have been reported to promote nanoparticles synthesis [182–184]. The core cavities of polymer capsules are able to provide a good environment to optimize the reaction. Furthermore, the polymer capsules are semipermeable which allows small ions to penetrate and stops large particles to penetrate. Thus, the property can be utilized to synthesize inorganic crystals inside the core domain of polymer capsules. For example, the growth of calcium carbonate and barium carbonate particles only inside the polymer capsules was obtained.

3.3.3.4 Artificial organelles

In cell biology, organelles is a specialized membrane-delineated component within a cell that has a specific function [185–187]. There are a specific class of enzymes inside the cell organelles which allow the subcellular spatial control of different functions. Various researchers are dedicated to designing artificial organelles. Recently Hest and coworkers made a great effort to exploit enzyme-loaded polymersomes into artificial organelles [187].

3.3.4 Sensing

Except for the above-mentioned drug delivery and enzymatic catalysis, polymeric capsules also have a potential in sensing applications. Miniaturization and integration of sensing devices may be used for *in situ* detection of various molecules with low concentrations. LbL-assembled capsules have been used to sense and detect small ions and oxygen through fixing some special fluorescent indicators into the capsules [188]. Biomolecules [such as DNA (deoxyribonucleic acid)] may also be detected by means of immobilizing a molecular beacon into capsules [189]. Obviously, the permeability of the capsule wall is of vital importance for the sensing process. This can be tuned through changing the film thickness, assembly conditions, composition, etc. [190–193] so that the capsule wall (acting as a selective barrier) effectively impedes the sensing molecules from diffusing. The molecular beacon approach revealed that the capsule shell (PSS/PAH bilayers) can exclude DNA molecules larger than 4.7 nm and 6.0 nm in diameter after a 3 h and 75 h incubation, respectively [189]. Some capsules loaded with fluorescent compounds have also been utilized in immunoassays. After bio-specific binding of the analyte to the capsules, the fluorophore encapsulated into the capsule was gradually solubilized and diffused out. It emitted an obvious fluorescent signal after exposure to the ethanol environment, and thus exhibited significant enhancement of sensitivity for immunoassay applications [194].

3.4 Conclusion

Many different kinds of polymer capsules have so far been developed, each with both advantages and drawbacks. A variety of applications of hollow polymer capsules have been discovered and each application is affected by many factors. Furthermore, many polymer capsules are responsible for a multitude of stimuli, which can be used in controlled encapsulation and release. These polymer capsules can be called smart polymer capsules. In further developments, more robust and facile techniques to fabricate polymer capsules will be synthesized and new opportunities for polymer capsules will be applied in biomimetic catalysis, drug delivery, and other diverse fields.

References

1. Schwille P., *Science*, 333 (2011) 1252–1254.
2. Szostak J. W., Bartel D. P., Luisi P. L., *Nature*, 409 (2001) 387–390.
3. Meier W., *Chemical Society Reviews*, 29 (2000) 295–303.
4. Caruso F., Caruso R. A., Möhwald H., *Science*, 282 (1998) 1111–1114.
5. Utada A., Lorenceau E., Link D., Kaplan P., Stone H., Weitz D., *Science*, 308 (2005) 537–541.
6. Zhu Y., Fan L., Yang B., Du J., *ACS Nano*, 8 (2014) 5022–5031.
7. Kukula H., Schlaad H., Antonietti M., Förster S., *Journal of the American Chemical Society*, 124 (2002) 1658–1663.
8. Holowka E. P., Pochan D. J., Deming T. J., *Journal of the American Chemical Society*, 127 (2005) 12423–12428.
9. Kurniasih I. N., Keilitz J., Haag R., *Chemical Society Reviews*, 44 (2015) 4145–4164.
10. Liu F., Hu J., Liu G., Hou C., Lin S., Zou H., Zhang G., Sun J., Luo H., Tu Y., *Macromolecules*, 46 (2013) 2646–2657.
11. Zhou Y., Huang W., Liu J., Zhu X., Yan D., *Advanced Materials*, 22 (2010) 4567–4590.
12. Zhang L., Eisenberg A., *Science*, 268 (1995) 1728.
13. Jenekhe S. A., Chen X. L., *Science*, 279 (1998) 1903–1907.
14. Discher B. M., Won Y. Y., Ege D. S., Lee J. C. M., Bates F. S., Discher D. E., Hammer D. A., *Science*, 284 (1999) 1143–1146.

15. Pearson R. T., Warren N. J., Lewis A. L., Armes S. P., Battaglia G., *Macromolecules*, 46 (2013) 1400–1407.
16. Discher B. M., Won Y.-Y., Ege D. S., Lee J. C., Bates F. S., Discher D. E., Hammer D. A., *Science*, 284 (1999) 1143–1146.
17. Messager L., Gaitzsch J., Chierico L., Battaglia G., *Current Opinion in Pharmacology*, 18 (2014) 104–111.
18. Letchford K., Burt H., *European Journal of Pharmaceutics and Biopharmaceutics*, 65 (2007) 259–269.
19. LoPresti C., Lomas H., Massignani M., Smart T., Battaglia G., *Journal of Materials Chemistry*, 19 (2009) 3576–3590.
20. Gaitzsch J., Huang X., Voit B., *Chemical Reviews*, 116 (2016) 1053–1093.
21. Battaglia G., Ryan A. J., *The Journal of Physical Chemistry B*, 110 (2006) 10272–10279.
22. Battaglia G., Ryan A. J., *Journal of the American Chemical Society*, 127 (2005) 8757–8764.
23. Battaglia G., Ryan A. J., *Angewandte Chemie International Edition*, 45 (2006) 2052–2056.
24. Shen H., Eisenberg A., *Macromolecules*, 33 (2000) 2561–2572.
25. Lee H. J., Yang S. R., An E. J., Kim J.-D., *Macromolecules*, 39 (2006) 4938–4940.
26. Sun J.-T., Hong C.-Y., Pan C.-Y., *Soft Matter*, 8 (2012) 7753–7767.
27. Blanazs A., Ryan A., Armes S., *Macromolecules*, 45 (2012) 5099–5107.
28. Battaglia G., Ryan A. J., *Nature Materials*, 4 (2005) 869–876.
29. Discher D. E., Eisenberg A., *Science*, 297 (2002) 967–973.
30. Discher B. M., Hammer D. A., Bates F. S., Discher D. E., *Current Opinion in Colloid & Interface Science*, 5 (2000) 125–131.
31. McCormick C. L., Sumerlin B. S., Lokitz B. S., Stempka J. E., *Soft Matter*, 4 (2008) 1760–1773.
32. Cui J. W., van Koeverden M. P., Mullner M., Kempe K., Caruso F., *Advances in Colloid and Interface Science*, 207 (2014) 14–31.
33. Qiao Z.-Y., Ji R., Huang X.-N., Du F.-S., Zhang R., Liang D.-H., Li Z.-C., *Biomacromolecules*, 14 (2013) 1555–1563.
34. Ruysschaert T., Sonnen A. F., Haefele T., Meier W., Winterhalter M., Fournier D., *Journal of the American Chemical Society*, 127 (2005) 6242–6247.

35. Marsden H. R., Quer C. B., Sanchez E. Y., Gabrielli L., Jiskoot W., Kros A., *Biomacromolecules,* 11 (2010) 833–838.

36. Du J., Tang Y., Lewis A. L., Armes S. P., *Journal of the American Chemical Society,* 127 (2005) 17982–17983.

37. Li Y., Lokitz B. S., McCormick C. L., *Angewandte Chemie International Edition,* 45 (2006) 5792–5795.

38. Wang G., Tong X., Zhao Y., *Macromolecules,* 37 (2004) 8911–8917.

39. Borchert U., Lipprandt U., Bilang M., Kimpfler A., Rank A., Peschka-Süss R., Schubert R., Lindner P., Förster S., *Langmuir,* 22 (2006) 5843–5847.

40. Hong C.-Y., You Y.-Z., Pan C.-Y., *Journal of Polymer Science Part A: Polymer Chemistry,* 42 (2004) 4873–4881.

41. Matyjaszewski K., Müller A. H., *Controlled and Living Polymerizations: from Mechanisms to Applications,* (2009), John Wiley & Sons.

42. Zou P., Pan C.-Y., *Macromolecular Rapid Communications,* 29 (2008) 763–771.

43. Wan W.-M., Hong C.-Y., Pan C.-Y., *Chemical Communications,* (2009) 5883–5885.

44. Zheng Q., Zheng G.-H., Pan C.-Y., *Polymer International,* 55 (2006) 1114–1123.

45. Blanazs A., Armes S. P., Ryan A. J., *Macromolecular Rapid Communications,* 30 (2009) 267–277.

46. Esser-Kahn A. P., Odom S. A., Sottos N. R., White S. R., Moore J. S., *Macromolecules,* 44 (2011) 5539–5553.

47. Shi Z.-Q., Zhou Y.-F., Yan D.-Y., *Macromolecular Rapid Communications,* 29 (2008) 412–418.

48. Cha J. N., Birkedal H., Euliss L. E., Bartl M. H., Wong M. S., Deming T. J., Stucky G. D., *Journal of the American Chemical Society,* 125 (2003) 8285–8289.

49. Arumugam S., Vutukuri D. R., Thayumanavan S., Ramamurthy V., *Journal of the American Chemical Society,* 127 (2005) 13200–13206.

50. Savariar E. N., Aathimanikandan S. V., Thayumanavan S., *Journal of the American Chemical Society,* 128 (2006) 16224–16230.

51. Kale T. S., Klaikherd A., Popere B., Thayumanavan S., *Langmuir,* 25 (2009) 9660–9670.

52. Zhu Y.-Q., Liu L., Du J.-Z., *Macromolecules,* 46 (2012) 194–203.

53. Wang L.-H., Zhang Z.-D., Hong C.-Y., He X.-H., You W., You Y.-Z., *Advanced Materials,* 27 (2015) 3202–3207.

54. Wasio N. A., Quardokus R. C., Forrest R. P., Lent C. S., Corcelli S. A., Christie J. A., Henderson K. W., Kandel S. A., *Nature*, 507 (2014) 86–89.
55. Bruns C. J., Fujita D., Hoshino M., Sato S., Stoddart J. F., Fujita M., *Journal of the American Chemical Society*, 136 (2014) 12027–12034.
56. Zhang M., Yan X., Huang F., Niu Z., Gibson H. W., *Accounts of Chemical Research*, 47 (2014) 1995–2005.
57. Yan D.-Y., Gao C., Frey H., *Hyperbranched Polymers: Synthesis, Properties, and Applications*, (2011) John Wiley & Sons.
58. Zhou Y.-F., Yan D.-Y., *Angewandte Chemie International Edition*, 43 (2004) 4896–4899.
59. Wang Y.-L., Li B., Jin H.-B., Zhou Y.-F, Lu Z.-Y, Yan D.-Y., *Chemistry – An Asian Journal*, 9 (2014) 2281–2288.
60. Zhou Y.-F., Yan D.-Y., Dong W.-Y., Tian Y., *The Journal of Physical Chemistry B*, 111 (2007) 1262–1270.
61. Zou J., Ye X., Shi W., *Macromolecular Rapid Communications*, 26 (2005) 1741–1745.
62. Guo B., Shi Z.-Q., Yao Y., Zhou Y.-F., Yan D.-Y., *Langmuir*, 25 (2009) 6622–6626.
63. Guo Z.-P., Li Y.-H., Tian H.-Y., Zhuang X.-L., Chen X.-S., Jing X.-B., *Langmuir*, 25 (2009) 9690–9696.
64. Mai Y.-Y., Zhou Y.-F., Yan D.-Y., *Small*, 3 (2007) 1170–1173.
65. Tao W., Liu Y., Jiang B.-B., Yu S.-R., Huang W., Zhou Y.-F., Yan D.-Y., *Journal of the American Chemical Society*, 134 (2011) 762–764.
66. Liu Y., Yu C.-Y., Jin H.-B., Jiang B.-B., Zhu X.-Y., Zhou Y.-F., Lu Z.-Y., Yan D.-Y., *Journal of the American Chemical Society*, 135 (2013) 4765–4770.
67. Lian X.-M., Wu D.-X., Song X.-H., Zhao H.-Y., *Macromolecules*, 43 (2010) 7434–7445.
68. Dziezok P., Fischer K., Schmidt M., Sheiko S. S., Möller M., *Angewandte Chemie International Edition in English*, 36 (1997) 2812–2815.
69. Schappacher M., Deffieux A., *Macromolecules*, 38 (2005) 7209–7213.
70. Wu D.-X., Yang Y.-Y., Cheng X.-H., Liu L., Tian J., Zhao H.-Y., *Macromolecules*, 39 (2006) 7513–7519.
71. Lanson D., Ariura F., Schappacher M., Borsali R., Deffieux A., *Macromolecules*, 42 (2009) 3942–3950.
72. Stephan T., Muth S., Schmidt M., *Macromolecules*, 35 (2002) 9857–9860.

73. Vasilevskaya V., Gusev L., Khokhlov A., Ikkala O., G. Ten Brinke, *Macromolecules*, 34 (2001) 5019–5022.
74. Gu L.-N., Shen Z., Zhang S., Lu G.-L., Zhang X.-H., Huang X.-Y., *Macromolecules*, 40 (2007) 4486–4493.
75. Zheng R.-H., Liu G.-J., *Macromolecules*, 40 (2007) 5116–5121.
76. Huang X., Li M., Green D. C., Williams D. S., Patil A. J., Mann S., *Nature Communications*, 4 (2013) 2239.
77. Vargo K. B., Parthasarathy R., Hammer D. A., *Proceedings of the National Academy of Sciences*, 109 (2012) 11657–11662.
78. van Dongen S. F., de Hoog H.-P. M., Peters R. J., Nallani M., Nolte R. J., van Hest J. C., *Chemical Reviews*, 109 (2009) 6212–6274.
79. Vögtle F., Buhleier E., Wehner W., *Synthesis*, 2 (1978) 155–158.
80. Newkome G. R., Yao Z., Baker G. R., Gupta V. K., *The Journal of Organic Chemistry*, 50 (1985) 2003–2004.
81. Tomalia D., Baker H., Dewald J., Hall M., Kallos G., Martin S., Roeck J., Ryder J., Smith P., *Polymer Journal*, 17 (1985) 117–132.
82. Tomalia D. A., Naylor A. M., Goddard W. A., *Angewandte Chemie International Edition in English*, 29 (1990) 138–175.
83. Bosman A., Janssen H., Meijer E., *Chemical Reviews*, 99 (1999) 1665–1688.
84. Carlmark A., Hawker C., Hult A., Malkoch M., *Chemical Society Reviews*, 38 (2009) 352–362.
85. Jansen J. F., de Brabander-van den Berg E., Meijer E., *Science*, 1994, 266, 1226.
86. Hawker C. J., Frechet J. M., *Journal of the American Chemical Society*, 112 (1990) 7638–7647.
87. Hawker C., Fréchet J. M., *Journal of the Chemical Society, Chemical Communications*, (1990) 1010–1013.
88. Wang M., Liu H., Li L., Cheng Y., *Nature Communications*, 5 (2014) 5293.
89. You Y.-Z., Hong C.-Y., Pan C.-Y., Wang P.-H., *Advanced Materials*, 16 (2004) 1953–1957.
90. Becker A. L., Johnston A. P., Caruso F., *Small*, 6 (2010) 1836–1852.
91. Decher G., Hong J. D., *Makromol. Chem., Macromol. Symp.*, 46 (1991) 321–327.
92. Decher G., Hong J., Schmitt J., *Thin Solid Films*, 210 (1992) 831–835.
93. Decher G., *Science*, 277 (1997) 1232–1237.

94. Donath E., Sukhorukov G. B., Caruso F., Davis S. A., Möhwald H., *Angewandte Chemie International Edition*, 37 (1998) 2201–2205.

95. Decher G., Schlenoff J. B., *Multilayer Thin Films: Sequential Assembly of Nanocomposite Materials*, (2006) John Wiley & Sons.

96. Decher G., Hong J., *Berichte Der Bunsengesellschaft Für Physikalische Chemie*, 95 (1991) 1430–1434.

97. Blomberg S., Ostberg S., Harth E., Bosman A. W., Van Horn B., Hawker C. J., *Journal of Polymer Science Part A: Polymer Chemistry*, 40 (2002) 1309–1320.

98. Wang J.-S., Matyjaszewski K., *Journal of the American Chemical Society*, 117 (1995) 5614–5615.

99. Matyjaszewski K., Xia J., *Chemical Reviews*, 101 (2001) 2921–2990.

100. Ouchi M., Terashima T., Sawamoto M., *Chemical Reviews*, 109 (2009) 4963–5050.

101. Chiefari J., Chong Y., Ercole F., Krstina J., Jeffery J., Le T. P., Mayadunne R. T., Meijs G. F., Moad C. L., Moad G., *Macromolecules*, 31 (1998) 5559–5562.

102. Moad G., Rizzardo E., Thang S. H., *Australian Journal of Chemistry*, 62 (2009) 1402–1472.

103. Chen F.-T., Jiang X.-S., Liu R., Yin J., *Polymer Chemistry*, 2 (2011) 614–618.

104. Huang X., Voit B., *Polymer Chemistry*, 4 (2013) 435–443.

105. Morinaga T., Ohkura M., Ohno K., Tsujii Y., Fukuda T., *Macromolecules*, 40 (2007) 1159–1164.

106. Fu G., Shang Z., Hong L., Kang E., Neoh K., *Macromolecules*, 38 (2005) 7867–7871.

107. Barbey R., Lavanant L., Paripovic D., Schüwer N., Sugnaux C., Tugulu S., Klok H.-A., *Chemical Reviews*, 109 (2009) 5437–5527.

108. Li G. L., Xu L. Q., Tang X., Neoh K., Kang E., *Macromolecules*, 43 (2010) 5797–5803.

109. Crespy D., Stark M., Hoffmann-Richter C., Ziener U., Landfester K., *Macromolecules*, 40 (2007) 3122–3135.

110. Huang X., Appelhans D., Formanek P., Simon F., Voit B., *Macromolecules*, 44 (2011) 8351–8360.

111. Huang X., Appelhans D., Formanek P., Simon F., Voit B., *ACS Nano*, 6 (2012) 9718–9726.

112. Wang Z.-K., Wang D., Wang H., Yan J.-J., You Y.-Z., Wang Z.-G., *Journal of Materials Chemistry*, 21 (2011) 15950–15956.

113. Zhou K., Tong L., Deng J., Yang W., *Journal of Materials Chemistry*, 20 (2010) 781–789.

114. Wang L.-H., Ding S.-G., Yan J.-J., You Y.-Z., *The Journal of Physical Chemistry B*, 118 (2014) 3893–3898.

115. Han L.-F., Chen Q.-B., Hu Z.-T., Piao J.-G., Hong C.-Y., Yan J.-J., You Y.-Z., *Macromolecular Rapid Communications*, 35 (2014) 649–654.

116. Wang Y., Bansal V., Zelikin A. N., Caruso F., *Nano Letters*, 8 (2008) 1741–1745.

117. Mertz D., Tan P., Wang Y., Goh T. K., Blencowe A., Caruso F., *Advanced Materials*, 23 (2011) 5668–5673.

118. Zhang Y., Yang S., Guan Y., Cao W., Xu J., *Macromolecules*, 36 (2003) 4238–4240.

119. Song H., Chen D. L., Ismagilov R. F., *Angewandte Chemie International Edition*, 45 (2006) 7336–7356.

120. Musyanovych A., Landfester K., *Macromolecular Bioscience*, 14 (2014) 458–477.

121. Landfester K., Musyanovych A., Mailänder V., *Journal of Polymer Science Part A: Polymer Chemistry*, 48 (2010) 493–515.

122. Tiarks F., Landfester K., Antonietti M., *Langmuir*, 17 (2001) 908–918.

123. Musyanovych A., Landfester K., *Surface and Interfacial Forces–From Fundamentals to Applications*, (2008) Springer, pp. 120–127.

124. Jagielski N., Sharma S., Hombach V., Mailänder V., Rasche V., Landfester K., *Macromolecular Chemistry and Physics*, 208 (2007) 2229–2241.

125. Torini L., Argillier J., Zydowicz N., *Macromolecules*, 38 (2005) 3225–3236.

126. Wang L.-H., Ding S.-G., Yan J.-J., You Y.-Z., *Macromolecular Rapid Communications*, 35 (2014) 298–302.

127. Antipov A. A., Sukhorukov G. B., Donath E., Möhwald H., *The Journal of Physical Chemistry B*, 105 (2001) 2281–2284.

128. Gaponik N., Radtchenko I. L., Gerstenberger M. R., Fedutik Y. A., Sukhorukov G. B., Rogach A. L., *Nano Letters*, 3 (2003) 369–372.

129. Ferrari M., *Nature Reviews Cancer*, 5 (2005) 161–171.

130. Farokhzad O. C., Langer R., *ACS Nano*, 3 (2009) 16–20.

131. Devadasu V. R., Bhardwaj V., Kumar M. R., *Chemical Reviews*, 113 (2012) 1686–1735.

132. Sukhorukov G. B., Volodkin D. V., Günther A. M., Petrov A. I., Shenoy D. B., Möhwald H., *Journal of Materials Chemistry*, 14 (2004) 2073–2081.

133. Déjugnat C., Sukhorukov G. B., *Langmuir*, 20 (2004) 7265–7269.

134. Köhler K., Shchukin D. G., Möhwald H., Sukhorukov G. B., *The Journal of Physical Chemistry B*, 109 (2005) 18250–18259.

135. Friebel M., Do K., Hahn A., Mu G., *Journal of Biomedical Optics*, 4 (1999) 36–46.

136. Tao X., Li J., Möhwald H., *Chemistry–A European Journal*, 10 (2004) 3397–3403.

137. Skirtach A. G., Antipov A. A., Shchukin D. G., Sukhorukov G. B., *Langmuir*, 20 (2004) 6988–6992.

138. Radt B., Smith T. A., Caruso F., *Advanced Materials*, 16 (2004) 2184–2189.

139. Koo H. Y., Lee H.-J., Kim J. K., San Choi W., *Journal of Materials Chemistry*, 20 (2010) 3932–3937.

140. Wang X.-R., Liu G.-H, Hu J.-M., Zhang G.Y., Liu S.-Y., *Angewandte Chemie International Edition*, 53 (2014) 3138–3142.

141. Bédard M., Skirtach A. G., Sukhorukov G., *Macromolecular Rapid Communications*, 28 (2007) 1517–1521.

142. Büscher K., Graf K., Ahrens H., Helm C. A., *Langmuir*, 18 (2002) 3585–3591.

143. Sun J.-T., Yu Z.-Q., Hong C.-Y., Pan C.-Y., *Macromolecular Rapid Communications*, 33 (2012) 811–818.

144. Wang W., Liu L., Ju X. J., Zerrouki D., Xie R., Yang L., Chu L. Y., *ChemPhysChem*, 10 (2009) 2405–2409.

145. Hu S.-H., Tsai C.-H., Liao C.-F., Liu D.-M., Chen S.-Y., *Langmuir*, 24 (2008) 11811–11818.

146. Hałupka-Bryl M., Asai K., Thangavel S., Bednarowicz M., Krzyminiewski R., Nagasaki Y., *Colloids and Surfaces B: Biointerfaces*, 118 (2014) 140–147.

147. Lu Z., Prouty M. D., Guo Z., Golub V. O., Kumar C. S., Lvov Y. M., *Langmuir*, 21 (2005) 2042–2050.

148. Unger E., *The Leading Edge in Diagnostic Ultrasound, Atlantic City, NJ: May*, (1997) 13–16.

149. Shchukin D. G., Gorin D. A., Möhwald H., *Langmuir*, 22 (2006) 7400–7404.

150. Skirtach A. G., De Geest B. G., Mamedov A., Antipov A. A., Kotov N. A., Sukhorukov G. B., *Journal of Materials Chemistry*, 17 (2007) 1050–1054.

151. Antipov A. A., Sukhorukov G. B., Leporatti S., Radtchenko I. L., Donath E., Möhwald H., *Colloids and Surfaces A: Physicochemical and Engineering Aspects*, 198 (2002) 535–541.

152. Han Y., Shchukin D., Yang J., Simon C. R., Fuchs H., Möhwald H., *Acs Nano*, 4 (2010) 2838–2844.

153. Smith K. A., Alexeev A., Verberg R., Balazs A. C., *Langmuir*, 22 (2006) 6739–6742.

154. Fernandes P. A., Delcea M., Skirtach A. G., Möhwald H., Fery A., *Soft Matter*, 6 (2010) 1879–1883.

155. Dressman J. B., Berardi R. R., Dermentzoglou L. C., Russell T. L., Schmaltz S. P., Barnett J. L., Jarvenpaa K. M., *Pharmaceutical Research*, 7 (1990) 756–761.

156. Kararli T. T., *Biopharmaceutics & Drug Disposition*, 16 (1995) 351–380.

157. Sun J.-T., Hong C.-Y., Pan C.-Y., *The Journal of Physical Chemistry C*, 114 (2010) 12481–12486.

158. Ibarz G., Dähne L., Donath E., Möhwald H., *Advanced Materials*, 13 (2001) 1324–1327.

159. Fery A., Schöler B., Cassagneau T., Caruso F., *Langmuir*, 17 (2001) 3779–3783.

160. Gao C., Möhwald H., Shen J. C., *ChemPhysChem*, 5 (2004) 116–120.

161. Mendes P. M., *Chemical Society Reviews*, 37 (2008) 2512–2529.

162. Völker E., Calvo E. J., Williams F. J., *Israel Journal of Chemistry*, 48 (2008) 305–312.

163. Matsue T., Evans D. H., Osa T., Kobayashi N., *Journal of the American Chemical Society*, 107 (1985) 3411–3417.

164. Peng L., Wang Z., Feng A., Huo M., Fang T., Wang K., Wei Y., Yuan J., *Polymer*, 88 (2016) 112–122.

165. Zhang H., Wang Z., Zhang Y., Zhang X., *Langmuir*, 20 (2004) 9366–9370.

166. Cortez C., Tomaskovic-Crook E., Johnston A. P., Radt B., Cody S. H., Scott A. M., Nice E. C., Heath J. K., F. Caruso, *Advanced Materials*, 18 (2006) 1998–2003.

167. Shu S., Zhang X., Wu Z., Wang Z., Li C., *Biomaterials*, 31 (2010) 6039–6049.

168. Zelikin A. N., Li Q., Caruso F., *Angewandte Chemie International Edition*, 45 (2006) 7743–7745.

169. Itoh Y., Matsusaki M., Kida T., Akashi M., *Biomacromolecules*, 7 (2006) 2715–2718.

170. Kumar P., Ban H.-S., Kim S.-S., Wu H., Pearson T., Greiner D. L., Laouar A., Yao J., Haridas V., Habiro K., *Cell*, 134 (2008) 577–586.

171. Cavazzana-Calvo M., Hacein-Bey S., de Saint Basile G., Gross F., Yvon E., Nusbaum P., Selz F., Hue C., Certain S., Casanova J.-L., *Science*, 288 (2000) 669–672.

172. Shen H., Sun T., Ferrari M., *Cancer Gene Therapy*, 19 (2012) 367–373.

173. Yin H., Kanasty R. L., Eltoukhy A. A., Vegas A. J., Dorkin J. R., Anderson D. G., *Nature Reviews Genetics*, 15 (2014) 541–555.

174. Zelikin A. N., Quinn J. F., Caruso F., *Biomacromolecules*, 7 (2006) 27–30.

175. Kim K. T., Meeuwissen S. A., Nolte R. J., van Hest J. C., *Nanoscale*, 2 (2010) 844–858.

176. Palivan C. G., Fischer-Onaca O., Delcea M., Itel F., Meier W., *Chemical Society Reviews*, 41 (2012) 2800–2823.

177. Boyer C., Bulmus V., Davis T. P., Ladmiral V., Liu J., Perrier S., *Chemical Reviews*, 109 (2009) 5402–5436.

178. Gräfe D., Gaitzsch J., Appelhans D., Voit B., *Nanoscale*, 6 (2014) 10752–10761.

179. Valade D., Jeon Y., Kessel S., Monteiro M. J., *Journal of Polymer Science Part A: Polymer Chemistry*, 50 (2012) 4762–4771.

180. Zetterlund P. B., *Polymer Chemistry*, 2 (2011) 534–549.

181. Zhang M., Zhang W., *Journal of Polymer Science Part A: Polymer Chemistry*, 48 (2010) 5446–5455.

182. Landfester K., Willert M., Antonietti M., *Macromolecules*, 33 (2000) 2370–2376.

183. Sukhorukov G. B., Susha A. S., Davis S., Leporatti S., Donath E., Hartmann J., Möhwald H., *Journal of Colloid and Interface Science*, 247 (2002) 251–254.

184. Li W., Sha X., Dong W., Wang Z., *Chemical Communications*, (2002) 2434–2435.

185. Tanner P., Baumann P., Enea R., Onaca O., Palivan C., Meier W., *Accounts of Chemical Research*, 44 (2011) 1039–1049.

186. Tanner P., Egli S., Balasubramanian V., Onaca O., Palivan C. G., Meier W., *FEBS Letters*, 585 (2011) 1699–1706.

187. Peters R. J., Louzao I., van Hest J. C., *Chemical Science*, 3 (2012) 335–342.

188. McShane M. J., Brown J., Guice K. B., Lvov Y. M., *Journal of Nanoscience and Nanotechnology*, 2 (2002) 411–416.

189. Johnston A. P., Caruso F., *Journal of the American Chemical Society*, 127 (2005) 10014–10015.

190. Dubas S. T., Schlenoff J. B., *Macromolecules*, 32 (1999) 8153–8160.

191. Cho J., Quinn J. F., Caruso F., *Journal of the American Chemical Society*, 126 (2004) 2270–2271.

192. Quinn J. F., Yeo J. C., Caruso F., *Macromolecules*, 37 (2004) 6537–6543.

193. Yoo D., Shiratori S. S., Rubner M. F., *Macromolecules*, 31 (1998) 4309–4318.

194. Trau D., Yang W., Seydack M., Caruso F., Yu N.-T., Renneberg R., *Analytical Chemistry*, 74 (2002) 5480–5486.

Chapter 4

On the Use of Complex Coacervates for Encapsulation

Agata Maria Brzozowska
*Institute of Materials Research and Engineering, Agency for Science, Technology and Research (A*STAR), 2 Fusionopolis Way, Innovis, #08-03, Singapore 138634*
brzozowskaam@imre.a-star.edu.sg

4.1 Introduction

In very general terms, the process of encapsulation refers to a method of protecting molecules from environmental factors, and/or improving their compatibility with the surrounding medium. The molecules of interest may be sensitive to oxidation (e.g., vitamins, fatty acids), light (e.g., pigments, dyes), pH (e.g., proteins, enzymes), temperature (e.g., proteins, drugs), and the presence of reactive compounds or solvents (e.g., enzymes, dyes). Encapsulation offers a means of control over the release of the encapsulated compounds with respect to location, trigger, and duration. For example, in the food industry encapsulation is used to increase fluidity and flavor sensation by prolonged, controlled release of lipids [1, 2] achieved

Polymer Capsules
Edited by Ye Liu and Xian Jun Loh

ISBN 978-981-4774-60-4 (Hardcover), 978-0-429-42873-9 (eBook)
www.jennystanford.com

by control over their physical state after emulsification. In the pharmaceutical industry encapsulation allows for targeted drug delivery and controlled drug release [3] achieved by selection of encapsulation material that increases its permeability at specific pH; in the coating industry lyophobic biocides and dyes can be effectively incorporated into formulations once encapsulated [4]; in agriculture, encapsulation allows the development of enhanced delivery fertilizers that release nutrients gradually during the entire crop period, preventing over-fertilization and contamination of ground water [5]. The exact method of encapsulation depends on the physicochemical properties of the compound of interest as well as on the chosen application.

The process of encapsulation is commonly characterized in the literature from the perspective of (i) the particular physicochemical process used (complex coacervation, solvent evaporation, heat treatment, etc.), (ii) the morphology of the carriers (vesicles, hard shells, soft capsules, hydrogels, etc.), (iii) the general characteristics of the encapsulated compounds (water soluble compounds, hydrophobic compounds, biopolymers, polysaccharides, enzymes, proteins, cells, etc.), or (iv) the application of the designed encapsulation system (drug delivery, food additives, personal care products, plasticizers, paint additives, protective functional coatings, etc.). There are many excellent publications available providing detailed overviews of various approaches, for instance, by O'Donnell *et al.* [6] on encapsulation using the solvent evaporation technique; by Walde *et al.* [7] on encapsulation of enzymes inside vesicles; or recently published books dedicated to the subject of encapsulation in general, for instance [8–10].

The main focus of this chapter is on encapsulation using the complex coacervate phase, with emphasis on the complex coacervation between polyelectrolytes. This promising method has gained increased attention in recent years.

4.2 Coacervation

The phenomenon of coacervation was first described by Tiebackx [11] who in 1911 discovered complexation between gelatin and gum arabic in acetic acid solution. The term "coacervation" was introduced later by Jong and Kryut [12]. It originates from the Latin word "acervus" (aggregation) and the prefix "co-" (together). Coacervation refers to a phase separation that takes place in a liquid phase. This process results in the formation of two phases existing in equilibrium, as defined by de Jong [12, 13]. One of these phases consists of diluted solute and the other of almost pure solvent. The first theoretical model describing coacervation was proposed by Overbeek and Voorn in 1957 [14], followed by extensive studies and theoretical descriptions by other researchers, for example, [15–21]. In his work, Voorn distinguished two types of coacervation: (i) a simple coacervation that does not involve electrostatic interactions and (ii) a complex coacervation that involves electrostatic interactions between oppositely charged molecules [19].

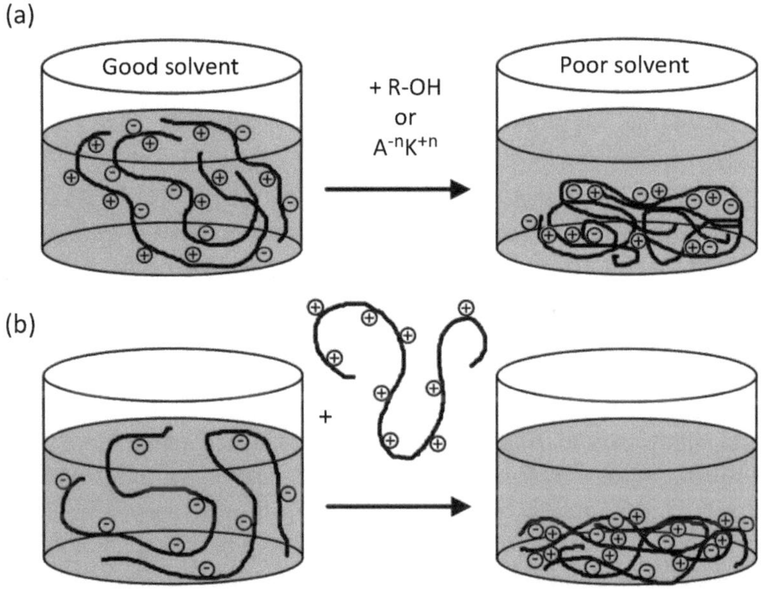

Figure 4.1 A schematic representation of (a) a simple coacervation resulting from a change in the solvent quality, and (b) a complex coacervation resulting from electrostatic interactions between oppositely charged polyelectrolytes.

Simple coacervation takes place when the solubility of molecules decreases due to change in molecule hydration upon addition of alcohol or multivalent ions [22] to the solution. A good example may be coacervation of gelatin as described by, for instance, Bohidar *et al.* [21]. Gelatin is a molecule with an almost equal number of positively and negatively chargeable groups and isoelectric point at approximately pH 5. Changing the quality of the solvent from good to poor, the gelatin chain collapses, shortening the distance between the oppositely chargeable groups, which results in the formation of the coacervate phase. In the experiment described by Bohidar *et al.*, the change of the solvent quality was achieved by the addition of alcohol. The concentration of alcohol required to induce coacervation depended on its molecular weight and thus, its hydrophobicity [23]. The higher the hydrophobicity the lower the concentration of alcohol required. Addition of alcohol beyond the amount necessary for coacervation to take place leads to precipitation of the solute. Another example of simple coacervation may be the coacervation of soy glicinin [24]. Soy glicinin is a protein of plant origin and is considered as a replacement for encapsulation materials of animal origin. The introduction of plant-derived alternatives is of particular importance in view of the health risks related to the use of animal products as well as cultural and lifestyle needs, for example, vegetarian, kosher, and halal products. Lazko *et al.* [24] reported on the simple coacervation of soy glicinin around oil droplets induced by the change of pH of a reaction medium consisting of aqueous solution of the protein, and small amount of hexadecane and HCl. Coacervation was induced by the increase of pH from pH 2 to pH 5 upon the addition of NaOH under agitation. The thus formed coacervate layer was crosslinked by the addition of glutaraldehyde solution. The authors studied various conditions of temperature, pH, and agitation, and found significant effects on the morphology of the resulting capsules, for example, the duration of agitation affected the size distribution and the encapsulation efficiency; the rate of pH change affected the aggregation of formed microcapsules as well as the homogeneity of the protein layer on oil droplets. Seweryn *et al.* [25] discussed an effect of a type of inorganic salt on the simple coacervation of surfactants, using the formulation of dishwashing liquid as an example. The authors observed significant differences between the effects of monovalent and multivalent ions: (i) polymer concentration was higher in the coacervate phase formed in the presence of monovalent ions; (ii) with increasing valence of cations present, the emulsification properties of the coacervate phase decreased; (iii) the presence of monovalent ions resulted in better foaming properties of the formulation; (iv) the formulations containing multivalent ions appeared less irritating on the skin; and (v) the presence of divalent ions resulted in formulations with lower maximum dynamic viscosity.

The application of simple coacervation to encapsulation may be limited by the sensitivity of the molecule to be encapsulated to dehydration or, in case of proteins, to denaturation. Concentrations of alcohol or salt required to induce the coacervation of the capsule material may be sufficiently high to induce precipitation or depletion of the target molecules from the solution before the process of encapsulation is completed.

Complex coacervation, also referred to as an "associative phase separation," results from electrostatic interactions between oppositely charged molecules in aqueous solution. The resulting phase is referred to as a "complex coacervate" phase. One of the earliest studied examples of complex coacervation is a complexation between gelatin and gum arabic as described by, for instance, de Ruiter *et al.* [26]. In their experiments the authors formed the complex coacervate phase by mixing stock solutions of respective polyelectrolytes, at pH 3.7 maintained by sodium acetate–acetic acid buffer. At this pH the majority of the chargeable groups become dissociated, facilitating the complexation.

Complex coacervation can take place between two polyelectrolytes or between a polyelectrolyte and a colloidal particle. Both processes are known and have been extensively studied for decades. Complex coacervation between polyelectrolytes and colloidal particles has been extensively discussed in literature by, amongst others, Kizilay *et al.* [27], and will not be discussed in detail here. However, it is worth mentioning that despite many similarities, complex coacervation between two polyelectrolytes differs from complex coacervation between polyelectrolyte and colloidal particles. This is mainly due to differences in molecules morphology, and thus in their behavior in solution. Polyelectrolyte chains are more flexible than colloidal particles, and can adopt various configurations in response to the changing composition of the solution or quality of the solvent. Kizilay *et al.* notes that (i) due to greater heterogeneity of polyelectrolytes the phase transitions observed in polyelectrolyte–polyelectrolyte systems are broadened and liquid–liquid transition is often masked; (ii) a gain of configurational entropy upon complexation should be greater in systems consisting of polyelectrolytes only; (iii) the complexation between polyelectrolytes and colloid particles can be described as condensation of polymer chains on the surface of a particle, or, in case of small particles, as binding the ligands to a host (polyelectrolyte chain); (iv) upon complexation, a structure of a colloidal particle (e.g., an enzyme) should remain intact whereas the conformation of the polyelectrolyte chain changes [27].

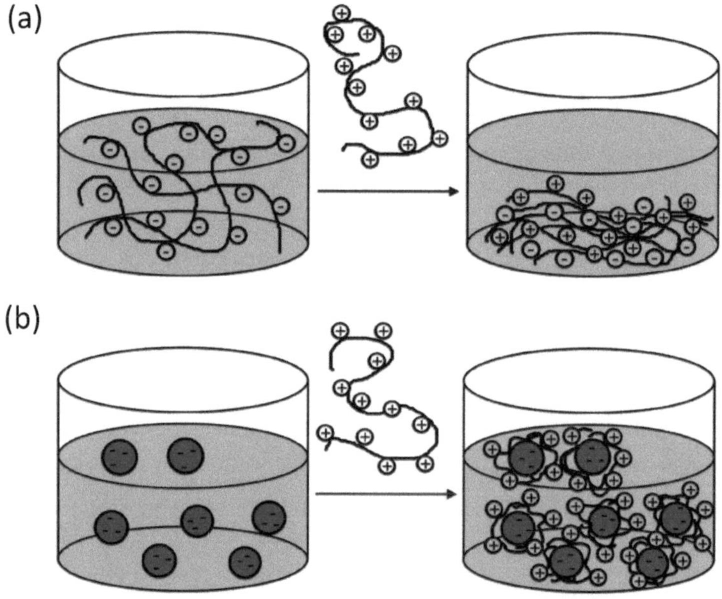

Figure 4.2 A simplified schematic representation of complex coacervation between two polyelectrolyte chains (a), and between polyelectrolyte chains and colloidal particles (b).

Thus, in very general terms, the understanding of complex coacervation between colloidal particles and polyelectrolytes is of particular interest for the immobilization of particles that are unlikely to change shape or conformation during the process, that is, encapsulation of enzymes, formulations for antigen delivery, process of protein purification, stabilization of food products [28], and encapsulation of particles used in diagnostics. The understanding of complex coacervation between polyelectrolytes is crucial for applications that require the formation of a capsule/shell of well-defined morphology around drops or small molecules and/or particles that can change shape in response to changes in the environment.

4.2.1 Conditions for Complex Coacervation

In this section we discuss conditions necessary for complex coacervation to occur. The extent of this discussion is limited to parameters that can and need to be controlled during assembly of capsules, and to a concise explanation of their significance in the process. For clarity of presentation, the focus is on linear polyelectrolytes. It is important to understand that the observed effects result from a combination of factors rather than from a single one. Thus, it may be difficult to ascribe given phenomena to change of a specific parameter. Interested readers may refer to the cited literature for in-depth description of the phenomena mentioned in this chapter.

Conditions necessary for complex coacervation to take place were first defined by Bungenberg de Jong for a system consisting of gum arabic and gelatin [12]. The process of complex coacervation is affected by polymer concentration, length of the polymer chain, mixing ratio of the polyelectrolytes, charge density, pH, salt concentration and type, and the temperature. Research shows that two oppositely charged polyelectrolytes are most likely to form complexes if they have long chains, high charge density with (close to) equal numbers of opposite charges, and if the complexation takes place at low salt concentration. It was argued that complexation resulting solely from electrostatic interactions can be driven by a decrease of the electric field energy or by increase of the translational entropy of counter-ions released upon complexation [29]. Spruijt *et al.* argued that surface charge as well as the curvature of the double layer determines which of these two contributions dominates the process: the complexation of highly charged surfaces with double layers of low curvature is entropy driven, and the complexation of less charged surfaces of highly-curved double layers is driven by the decrease of electric field energy.

4.2.1.1 Polyelectrolytes

Dobry *et al.* [30] studied experimentally phase separation in polyelectrolyte mixtures and discussed, among other aspects, the relationship between polymers concentration and molecular weight and the phase separation. The authors observed that for systems consisting of two polyelectrolytes, the phase separation range increases with increasing concentration of polyelectrolytes, and that for the longer polymer chains the limits of the phase separation are closer to low concentrations. In other words, in order to obtain phase separation of molecules with low molecular weight their concentration must be sufficiently high. Dobry's explanation of the phenomena was based on changes in the osmotic pressure of the respective polyelectrolytes in the mixture. He noted that the phase separation occurs at a point where the system deviates from van't Hoff's law. Overbeek *et al.* [14] concluded that for complex coacervation to occur, charge density or molecular weight, or both, have to be sufficiently high. Thus, phase separation in a given mixture of polyelectrolytes does not take place if length of (at least) one of the chains is below the critical chain length, for example, [31, 32], and if concentrations of the polyelectrolytes are too low.

Pawar *et al.* [32] suggested that the process of formation of the complex coacervate phase follows several steps, schematically shown in Fig. 4.3: (i) initially, soluble complexes are formed, (ii) these soluble complexes grow in time forming anisotropic complexes, followed by (iii) Ostwald ripening, and (iv) coacervation transition.

Complex coacervates have their "preferred composition" corresponding to 1:1 charge ratio. If one of the polyelectrolytes is in excess, the nature of the formed aggregates depends on the presence of salt in the system: in the presence of salt the formed aggregates tend to stay in the solvent-rich (diluted) phase, [33] whereas in salt-free systems the complex coacervate phase usually does not form. Priftis *et al.* [34] reported that in a system consisting of poly(L-lysine) and poly(L-glutamic acid), precipitation rather than complex coacervation was observed for mixing ratios different than 1:1. On the other hand, a system consisting of poly(acrylic acid) and poly(allylamide hydrochloride) was found to flocculate regardless of the mixing ratio at sufficiently low salt concentrations [35]. A different behavior was observed in systems consisting of bulky or branched polyelectrolytes [36]. In such systems, the apparent preferred composition of the complex coacervate phase may deviate from 1:1 charge ratio. Although the exact reason for this phenomenon is not clear, it may be partially related to the steric effects whereby the oppositely charged groups of the polyelectrolytes cannot be in close proximity.

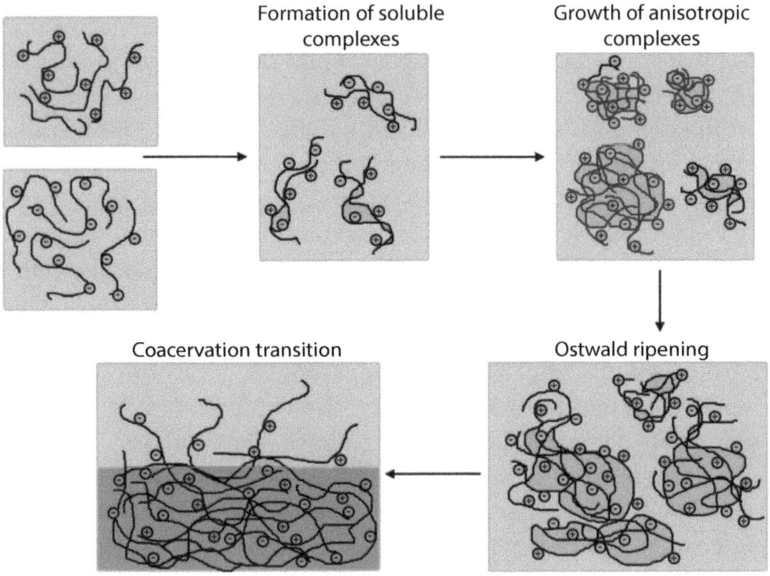

Figure 4.3 A schematic representation of steps leading to formation of the complex coacervate phase. Cartoon areas with darker colors correspond to regions of higher polymer concentration.

4.2.1.2 Ions

For a given pair of oppositely charged polyelectrolytes, there is a range of salt concentrations beyond which complex coacervation does not take place. Complex coacervation is replaced by flocculation or precipitation if the salt concentration is sufficiently low [34, 33]. At sufficiently high salt concentration, referred to as "critical salt concentration," complex coacervation is suppressed due to increased solubility of the polyelectrolytes [37]. The critical salt concentration depends on the length of polymer chains and is higher for longer chains. For sufficiently long chains, the critical salt concentration becomes independent of the length of the polymer chain. It strongly depends on the type of ions present. The presence of large ions with thin hydration shells seem to result in lower critical salt concentration. Van der Gucht *et al.* [33] argue that strongly hydrated ions (Na^+, Cl^-) have the ability to dehydrate the coacervate phase and thus, to strengthen the interactions between the oppositely charged groups. Less hydrated and highly polarizable ions may disrupt the complexes due to lower free energy of polymer–salt ion pairs and/or by lowering the energy of the ionic bond rapture process [33]. The authors noted that an increase in temperature has a similar effect on the composition of the coexisting phases as salt concentration, that is, the difference in polymer concentration in solvent-rich and polymer-rich phases, becomes greater. Formation of the complex coacervate phase is also affected by the valence of the ions present in the solution. Perry *et al.* [38] studied complex coacervation of poly(acrylic acid sodium salt) and poly(allylamine hydrochloride), and reported that the observed critical salt concentration was lower for divalent ions than for monovalent ions. However, the presence of monovalent ions promoted the formation of the complex coacervate phase as compared to divalent ions. The authors confirmed Hofmeister-like behavior in their system.

Molecular weights as well as concentration of polyelectrolytes determine the number and the concentration of the chargeable groups. Dissociated chargeable groups are points where ionic bonds between the oppositely charged polymer chains are formed. Thus, the more charges present the stronger the link between the polyelectrolytes becomes. The number of charges increases with increasing numbers of monomers forming a polyelectrolyte and, in case of weak polyelectrolytes, it strongly depends on pH. At pH most favorable for complex coacervation to occur, the concentration of charges is the highest and thus the electrostatic interactions between the oppositely charged chains are the strongest.

The degree of dissociation is not a simple function of pH. The charge depends also on other local conditions in a solution [39], such as local electrostatic potential, salt concentration, dielectric constant variation and the presence of other charged molecules. The effect of charge regulation must also be taken into account, particularly in systems including bioploymers. Charge regulation refers to a phenomenon where dissociation and association of ionizable groups is observed in response to the changed inter-particle distance [40, 41]. A first theoretical model describing complexation of proteins and polyelectrolytes with pH-dependent charges, taking charge regulation into account, was proposed by Biesheuvel *et al.* [42, 43]. A dissociation of the coacervate phase is expected close to the point of zero charge of (at least) one of the polyelectrolytes, assuming purely electrostatic interactions or at salt concentrations high enough to sufficiently suppress electrical double layers of the interacting polyelectrolytes.

4.2.1.3 Temperature

Interactions between polyelectrolytes are strengthened by the decline in temperature due to the increase of the Flory-Huggins interaction parameter [34]. In a system consisting of synthetic polyelectrolytes, the change in temperature affects the strength of hydrophobic interactions as well as the dissociation of chargeable groups. In case of biopolymers a window for temperature manipulation is restricted by potential denaturation and conformational changes, and is usually narrower than for synthetic polyelectrolytes. Prifits *et al.* report that a feasible range of temperature changes may not be sufficient to affect complex coacervation of biopolyelectrolytes in absence of salt [34]. The introduction of salt seems to make the system more responsive towards temperature changes. The authors observed that systems with higher total polymer concentrations are more sensitive to temperature variation. Anvari *et al.* studied the effect of temperature on complex coacervate formation by fish gelatin and gum arabic [44]. The authors reported an increase of volume fraction of the complex coacervate phase from 7.0% to 7.8%, and increased partition of the polyelectrolytes into the coacervate phase as temperature at which the complexation took place decreased from 40 °C to 10 °C, at pH 3.5. The partition of gum arabic was greater than the partition of gelatin. The authors also reported changes in the rheological behavior in the complex coacervate phase in response to temperature changes. Coacervates formed at lower temperatures were characterized by higher viscosity, more resistance towards structural deformation at low shear, and stronger shear thinning above the critical shear rate. Moreover, the complex coacervate phase formed at lower temperature had a denser structure and higher tendency for gelation. The resulting gels were reported to be more rigid and thermostable than the complex coacervate phase formed at higher temperatures [44].

4.2.1.4 Foreign molecules

In an idealized laboratory investigation of a model system, the complex coacervate phase is formed in solutions that contain only the polyelectrolytes that participate in the complex coacervation, solvent and salt. In applied systems the polyelectrolytes may be accompanied by large foreign molecules. These are the molecules that are not designed to constitute the complex coacervate phase. These molecules can be emulsifiers, stabilizers, or crosslinkers, constituents of the dispersed phase, or the compound to be encapsulated. Each of these foreign molecules has the potential to affect the process of formation of the complex coacervate phase.

During the process of encapsulation, the compound of interest must be dispersed in the aqueous phase to form emulsion (liquid compounds) or suspension (solid compounds). Wetting of the interface between the dispersed phase and aqueous solution by the complex coacervate phase is one of the most crucial steps during the encapsulation process (Fig. 4.4). Thus, any modification of this interface determines the process outcome. The surface-active molecules are commonly used as dispersants and stabilizers during the emulsification step. Their addition makes the wetting easier and usually enhances the formation of monodispersed emulsions. Mayya *et al.* [45] studied the encapsulation of paraffin oil using the gum arabic–gelatin complex coacervate in the presence of anionic surfactant (sodium dodecyl sulfate, SDS), at a concentration below critical micelle concentration. First, SDS and gelatin were mixed, followed by the addition of gum arabic. Under experimental conditions SDS and gelatin, used as a dispersant during the emulsification step, had opposite charges. The authors observed that in the presence of SDS, a yield increased from 35% to 70%, and the size distribution of oil drops changed as compared to the system without SDS. They also observed a difference in the morphology of the capsule walls. Gelatin and SDS formed a highly surface-active primary coacervate layer onto which a secondary complex coacervate layer, consisting of gum arabic and gelatin, was adsorbed. These two layers had different morphologies. Similar effects were observed in the presence of a cationic surfactant (dodecyl trimethyl ammonium bromide) [46]. However, it has also been observed that a too high concentration of the surface-active molecules can suppress formation of the complex coacervate phase [47, 48] altogether. Thus, there is only a specific range of concentrations within which the process benefits from the addition of a surfactant. Nakagawa *et al.* [47] noted that as far as the addition of the surfactants is concerned, there must be a compromise between the yield and the monodispersity of the emulsion as the concentrations of the surfactant required to improve the yield are usually too low to narrow down the drops size distribution.

Prata *et al.* [49] provided an interesting insight into the effect of the oil phase on formation of the complex coacervate phase. The authors studied the encapsulation of three different oils, namely vetiver oil, mineral oil and almond oil, in complex coacervates consisting of gum arabic and gelatin. The oils differed between each other with respect to hydrophobicity and hydrophilic–lipophilic balance. The type of oil influenced the process of complex coacervation between the two polyelectrolytes. At the most favorable complexation conditions determined in the absence of oil, the efficiency of complexation in the presence of almond oil and vetiver oil was above 90%, whereas in the presence of the most hydrophobic mineral oil, the efficiency decreased to approximately 46%. The coacervate formed in the presence of mineral oil had the highest water content (approximately 58%), followed by coacervates formed in the presence of vetiver oil (approximately 40%) and almond oil (approximately 33%). Coacervate formed in the presence of almond oil had the highest content of dry mass (approximately 20%).

4.2.2 Properties of Complex Coacervate Phase

Physicochemical properties of the complex coacervate phase determine its effectiveness as potential encapsulation material. In this section the relevant properties are discussed in more detail. We consider a system consisting of the solvent, the complex coacervate phase and the material to be encapsulated. The complex coacervate phase and the compound to be encapsulated may have a common solvent, that is, both are water soluble, or have incompatible solvents, that is, water soluble complex coacervate phase and oil soluble compound to be encapsulated. We discuss how such a system may respond to changes in salt concentration, pH, water content, and temperature, as well as its interfacial behavior.

4.2.2.1 Response to changes in external conditions

Once formed, the complex coacervate phase responds to changes in its environment. Thus, during a design of the capsules, the process parameters as well as subsequent storage and application parameters must be taken into consideration to ensure satisfactory encapsulation yield, stability, and functionality of the final product. As in many applications, the conditions may vary significantly at different stages and additional stabilization measures may become necessary, for instance crosslinking of the assembled complex coacervate carriers [50].

4.2.2.1.1 *Effect of salt*

As mentioned in Section 4.2.1, the salt concentration and type affect the formation of the complex coacervate phase. Changes in salt concentration have a pronounced effect on the existing complex coacervate phase, formed at low salt concentration as well. As the ionic strength of the solution increases, electrostatic interactions between the oppositely charged chains weaken due to suppression of the electrostatic double layers. Solubility of the polyelectrolytes increases and at sufficiently high ionic strength, the complex coacervate phase disintegrates. This has been observed experimentally in coacervates consisting of weak polyelectrolytes by, for instance, measurements of interfacial tension between the complex coacervate phase and its coexisting solvent phase. The interfacial tension was observed to decrease with an increasing concentration of NaCl. At the critical salt concentration, the interfacial tension was not observable any more [51].

The effect of ion valence must not be neglected either. Multivalent ions may reduce the critical salt concentration as compared to monovalent ions [38], and thus lead to unintended disintegration of the capsules. Moreover, some ions promote salting-in (e.g., I^-, Ca^{2+}) and some promote salting-out (e.g., NH_4^+, SO_4^{2-}) of the relevant solutes, thus making the process of encapsulation more challenging. Dautzenberg *et al.* [52] reported secondary aggregation and flocculation induced by the increased concentration of NaCl in the complex coacervate phase formed by polyelectrolytes with strong ionic groups. This effect depended on charge ratio in the complex coacervate phase. At fixed salt concentration, a stronger aggregation was observed for the higher mixing ratio of the polyelectrolytes. The authors noted that systems with strong ionic groups tend to separate into smaller particles at low salt concentrations. These observations are particularly important for a design of conditions for the encapsulation process, as certain compositions changes in salt concentration may prevent efficient formation and/or loading of the capsules. On the other hand, assuming favorable capsule composition, controlled change of salt concentration could be used to control the aggregation of capsules as a tool for separation of the final encapsulation product from the reaction solution.

4.2.2.1.2 *Hydration*

The complex coacervate phase is very strongly hydrated. In their experimental work various researchers found the water content to be between 65% and 85%, depending on the chemical composition of the coacervate phase [37, 53]. There seems to be only a weak dependence of water content in the complex coacervate phase and length of the polyelectrolyte chain. Water content was determined to increase with rising salt concentrations, with the increase becoming more rapid closer to the critical salt concentration where it attains the value characteristic for homogeneous solution (close to 99%). During dehydration, the thickness of the complex coacervate capsule walls decrease significantly. If the process of drying progresses rapidly, one may expect the formation of pores or the rupture of the capsules.

4.2.2.1.3 *Effect of pH*

Changes in pH particularly affect systems consisting of weak polyelectrolytes. In such systems, pH influences the degree of dissociation of the chargeable groups in constituting polyelectrolytes. Thus the change in pH may be used to trigger a release of the compound encapsulated in the complex coacervates. An example of such applications may be the delivery of compounds to a specific part of the digestive system, for example, [2]. The systems that were not subjected to crosslinking may completely disintegrate upon sufficient pH change, whereas crosslinked systems may respond by swelling and thus allowing the active compound to diffuse out from the capsule. The method of the capsule "opening" determines the rate of the compound release. In some applications, burst release may be beneficial whereas in others a prolonged, gradual release is preferred.

4.2.2.1.4 *Change of temperature*

Change of temperature affects the behavior of the complex coacervate phase in response to changes in its environment. As mentioned in Section 4.2.1, complex coacervation can be entropy-driven. Thus, the effect of temperature is expected to be significant. The general conclusion from studies of model systems is that the temperature changes result in shifts of phase transition boundaries. Chollakup *et al.* [35] observed a fall in the critical salt concentration as the temperature of the complex coacervate phase consisting of poly(acrylic acid) and poly(allylamide hydrochloride) was raised from 20°C to 50° C. Moreover, the secondary aggregation, observed at high salt concentration at 20 °C was not observed at 50 °C. The latter observation was explained by the greater degree of dissociation of the polyelectrolytes at higher temperatures, as secondary aggregation is attributed to coupled coil shrinkage and charge screening.

The temperature variations can make the process of encapsulation more challenging if, for instance, the gelation temperature of the capsule material is close to the limit of the thermal stability of the encapsulated compound. Alternatively, temperature changes may benefit the encapsulation process as, for instance, heating may affect size and improve the stability of the complex coacervate capsules comprising temperature responsive polyelectrolytes [54].

4.2.2.2 Wetting

Most commonly, the complex coacervate phase is used to encapsulate hydrophobic compounds. Such a process requires the coexistence of three different phases: oil (dispersed phase, with or without a compound to be encapsulated), complex coacervate phase (pre-formed or to-be-formed), and aqueous (continuous) phase. As shown in Fig. 4.4, for the capsule to be formed, a droplet or a particle of the dispersed phase must be coated by at least one of the components of the coacervate phase during or shortly after the process of emulsification is completed, depending on whether the components of the coacervate phase are used as emulsifiers. Subsequently, the second component of the coacervate phase is added, and the conditions (pH, temperature) are adjusted to induce complex coacervation as the capsule is formed. The capsules should remain dispersed in the aqueous phase. It is evident that for the process to occur, a sufficient wetting of the water–oil interface by the complex coacervate phase is crucial. For the encapsulation to be most efficient, the interfacial wetting should be spontaneous and homogenous. However, it was shown that the coacervate phase does not necessarily spread spontaneously at the water–oil interface [55] as the interactions between the aqueous phase and the coacervate phase affect the interfacial equilibrium. Interfacial tension between the coacervate phase and the coexisting aqueous phase, a parameter necessary for understanding drop formation and hydrodynamics, is relatively low. This poses a significant challenge with respect to the design of the experimental measurements as well as the process of encapsulation. The reported measured values range from 1 μN/m (measured by a capillary method) [56] to hundreds of μN/m (measured by atomic force microscopy) [51]. To understand and to control this process it is necessary to understand energy balance in the three-phase system consisting of two immiscible liquid drops surrounded by a third immiscible liquid [55]. The established equilibrium is determined by spreading coefficients and interfacial tensions [57]. A situation ideal from the perspective of encapsulation is the complete engulfing of the oil drop by the coacervate phase. This corresponds to the situation where the complex coacervate phase wets oil and aqueous interfaces equally well. If affinity of the coacervate phase to one of the interfaces is less, a partial engulfing is observed. In extreme cases, the coacervate phase does not have the affinity to the oil phase and the interface cannot be formed.

The application of sheer and/or electric field may help bring oil and coacervate droplets together, and eventually enhance engulfing [57]. However, in systems where the oil–coacervate interface does not form, the application of external force can bring the drops close together but cannot induce engulfing. In the most recent encapsulation techniques, the emulsion is formed in a microchannel whereby oil drops are forced to flow through the coacervate phase. As the drop of the oil phase flows through the coacervate phase, the phase forms a cup on this drop. The thickness of that cup as well as the degree of coverage depends on the viscosity of the continuous phase rather than the viscosity of the oil phase. It has also been argued that the coacervate shell is formed by a gradual build-up of the coacervate phase at the oil–aqueous interface rather than in a single step [55].

The balance between oil, water, and coacervate phases can be severely affected by the addition of surfactants. Rabiskova *et al.* [58] studied the encapsulation of soybean, olive, and peanut oils in gelatin–acacia complex coacervates in the presence of different surfactants. These surfactants had different hydrophile–lipophile balances (HLB). The research showed that oil encapsulation was most efficient when surfactants with HLB between 1.8 and 6.7 were used. Surfactants with higher HLB values lead to the formation of more stable emulsions but the resulting change of the interfacial properties made oil encapsulation impossible, that is, empty capsules were formed.

4.2.2.3 Rheological properties

Viscosity of the liquid phase(s) significantly influences the process of encapsulation. It has been reported that the more viscous coacervates, the lower the tendency of the droplets to coalesce [59]. Studies conducted to date focus on the relationship between viscosity of the coacervate phase and properties (pH, ionic strength, polymer concentration) and composition (ratio between the polyelectrolytes, molecular weights of the polyelectrolytes) of the coacervate phase. Maximum viscosity was shown to coincide with conditions of the strongest electrostatic interactions between the polyelectrolytes constituting the coacervate phase [60, 61]. The strength of these interactions is determined by the concentration and molecular ratio of the polyelectrolytes, as well as salt concentration, pH, and temperature. Spruijt *et al.* reported strong dependence of the response of the coacervate phase to the oscillatory strain on salt concentration [62]. The studied coacervate phase behaved like soft solids at low salt concentration and similar to liquid at high salt concentration. The deformations in the linear viscoelastic regime were hardly influenced by salt concentration. The effect of changes in salt concentration became significant in the non-linear regime as salt accelerates speed of the dynamic processes within polyelectrolyte assemblies, leading to faster equilibration upon disruption of these aggregates. Salt concentration has a much weaker effect on the modulus (G) than on the apparent relaxation time (τ) for various chain lengths. Polyelectrolyte chains constituting the coacervate phase interact with each other by means of interactions between oppositely charged groups (ion–ion pairs). Rubinstein *et al.* refer to these as "sticky points" that slow down the mobility of the chains and thus, the relaxation [63]. The authors have shown that the energy barrier for rearrangement of the ion–ion pairs depends on the square root of salt concentration, and exponential dependence of the relaxation time on salt concentration. The relaxation time is also a function of temperature and polymer concentration, that is, concentration of the "sticky points".

Similarly, pH determines the number of chargeable groups in weak polyelectrolytes that are dissociated and available to form ionic bonds between the polyelectrolytes. The more interactive the groups, the higher the energy needed to break them, and the more time required to have them reformed. Weinbreck *et al.* [61] determined that change of viscosity of the complex coacervate phase in response to pH change follows the same trend as the response to polyelectrolyte concentration. The maximum viscosity was measured under conditions of strongest electrostatic interactions. The relatively high viscosity of the complex coacervate phase seems to help in reducing the coalescence of the capsules [59]. Liberatore *et al.* [64] showed that complex coacervation can be induced by sheer, under specific temperature. At temperatures below liquid–liquid transition, thermodynamically stable complex coacervate behaves like Newtonian liquid under sheer. Above the temperature of liquid–liquid phase transition, the increase of sheer results in non-Newtonian behavior (sheer thinning) of the complex coacervate phase. As a result, a single, homogenous phase becomes a heterogeneous and anisotropic system with two distinct phases. Modulation of the sheer rate allows control over the size of the formed droplets of the complex coacervate phase [27, 65]. In a system consisting of poly(allyldimethylammonium chloride) and non-ionic surfactant Trion X-100, studied by Liberatore *et al.* [64] the size of the thus generated droplets ranged from 1 to 4 μm.

4.3 Process of Encapsulation

The complex coacervate phase is most commonly used for encapsulation of the lipophilic compounds. Encapsulation of such compounds usually requires formation of emulsion, with or without the addition of an emulsifier, followed by formation of the coacervate shell stabilized by means of gelation and/or crosslinking and drying. The process is schematically shown in Fig. 4.4. The first two steps can be realized in two different ways: a compound to be encapsulated is mixed with one of the polyelectrolytes to form emulsion, followed by mixing with the second polyelectrolyte and subsequently with the crosslinking agent [66]. Subsequently, the pH of the system is adjusted to initiate complex coacervation. Alternatively, both complex coacervate-forming components are pre-mixed, followed by addition of the oil phase and emulsification. Subsequently, the pH of the system is adjusted to trigger formation of the coacervate phase, followed by the addition of the cross linking agent [67]. The crosslinking agent can be a molecule (e.g., formaldehyde, glutaraldehyde) or a multivalent ion (e.g., Ca^{2+}). The thus formed capsules can be dried or mixed into application formulation directly after the washing step. An alternative way may be the formation of capsules by layer-by-layer deposition of the oppositely charged polyelectrolytes onto (usually) a spherical, inorganic template (e.g., silica, calcium carbonate), followed by dissolution of the template. Depending on the exact chemical composition, the thus formed capsules can be subsequently loaded with a water-soluble, active compound of interest. Subsequently, the capsules are "sealed" by shrinking in response to heating. The resulting loaded capsules are usually small in size, uniform in shape and are characterized by low polydispersity [7, 68, 69]. The LbL deposition technique can also be used to form capsules on lipid vesicles, and for encapsulation of oil-soluble compounds [70]. Attempts were made to encapsulate water soluble compounds in a coacervate phase restricted in growth to nanoscale. A good example may be encapsulation of lysozyme and lipase in so called complex coacervate core micelles (C3Ms) [71, 72]. C3Ms, also known as Ionomer Complexes, are formed upon mixing of oppositely charged polyelectrolytes at least one of which contains a neutral, water soluble block. This neutral block stabilizes the complex coacervate in solution, preventing macroscopic phase separation. An interested reader may refer to available reviews for further reading, for example, Ref. [73].

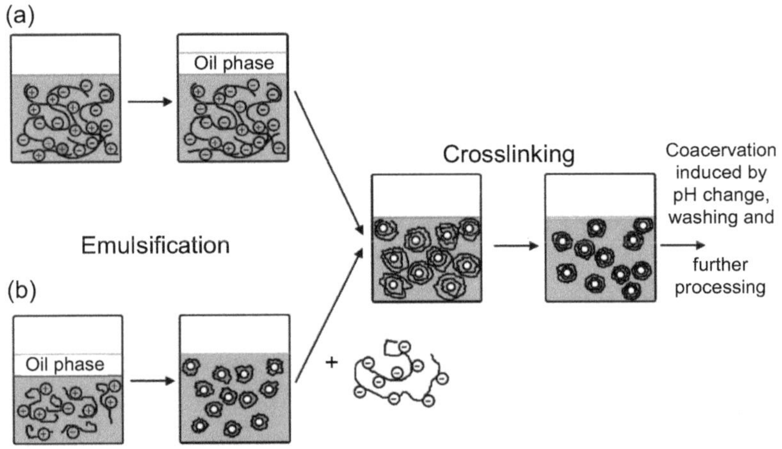

Figure 4.4 A simplified schematic representation of process of encapsulation of water insoluble material in complex coacervate phase, as described in the text. For clarity, charges of the polyelectrolytes are not indicated in every part of this schematic drawing.

Not every pair of oppositely charged polyelectrolytes forming the coacervate phase is suitable for encapsulation purposes. This is because the process of coacervation can be accompanied by other phenomena. A good example may be system consisting of gum arabic and poly(acrylic acid) investigated and described by Mathieu *et al.* [53]. In this system the formation of the complex coacervate phase is accompanied by flocculation. This leads to aggregation of the (micro) capsules. The flocculation process was shown to be suppressed by means of affecting the solubility of the formed complexes or complex composition. The solubility is affected by temperature, ionic strength, molecular mass and the total concentration of molecules. The complex composition is changed by adjustment of the mass ratio of the components. The authors emphasized that the changes in coacervate composition also result in lower coacervation yield. Another way to control an excess charge of the complex coacervates that may induce the secondary aggregation and, as a result flocculation, is through change of pH.

4.3.1 Emulsification

Emulsification is a process of dispersing one immiscible liquid (dispersed phase) in another immiscible liquid (dispersing phase). In the process of encapsulation in complex coacervates, the aqueous phase constitutes the dispersing phase, and the oil phase the dispersed phase. The droplets of the dispersed phase form a template on which a complex coacervate shell is built. Thus, the morphology of the emulsion determines the shape, the polydispersity, and the morphology (mono- versus polynucleated cores) of the capsules, and as such needs to be well controlled.

There are two main aspects of the emulsification process: mechanical and physicochemical [55, 74, 75]. The mechanical aspect is defined by the equipment used to form droplets, for example, mixers, homogenizers, or microfluidics devices, and their operational settings (homogenization rate, homogenization time). The physicochemical aspect refers to interfacial phenomena that govern the equilibria between the involved phases and thus define the morphology and the stability of the emulsion.

Commonly, emulsions are formed using rotor-stator systems, high-pressure systems, membrane systems, ultrasonic systems [76] or by extrusion using (arrays of) micro-channels [74, 75, 77, 78]. The former, relatively easy to scale-up, techniques result in droplets of same size distribution and diameters larger than 1 μm, whereas the latter, more costly technique, allow fabrication of monodisperse droplets with diameters smaller than 1 μm. The choice of the emulsification technique depends on the particular application in mind. Some applications, like paints and coatings, have fewer restrictions regarding capsule dimensions and polydispersity, whereas others like drug delivery require minimum polydispersity to ensure control over the delivered dose and the drug release. Moreover, the diameter of the capsules for biological applications should be small enough to ensure effective cellular uptake [79].

A broad overview of the available emulsification techniques, including discussions on the effect of the work parameters on structure and dimensions of the resulting emulsions, can be found in existing literature [75, 80–82]. In general, emulsions formed at higher sheer rates are characterized by smaller average droplets diameter and narrower polydispersity as compared to the emulsions formed at lower sheer rates. Similarly, higher sheer rates and more turbulent flow enhance the formation of single droplets. However, upon reaching a certain threshold, foaming, which can lead to droplets coalescence, takes place [83]. Lamprecht *et al.* [84] reported that short homogenization time resulted in the formation of typical core-wall structures whereas longer homogenization times resulted in the formation of matrix structures. The prolonged homogenization resulted in more efficient oil encapsulation and smaller droplet diameters. Similar observations were made by Prata *et al.* [49]. Thus, the mechanical parameters governing the process of emulsification are a compromise between acceptable levels of polydispersity and the desired morphology. The application of microfabricated devices seems to be a very promising alternative to the existing methods of emulsification, as they offer control over both parameters [74, 75]. By careful design of channel dimensions and control over hydrodynamic conditions, separated drops of desired sizes and polydispersity can be generated [55, 75]. The remaining issues are the efficiency of such systems considering industrial applications as well as the stability of the thus obtained emulsions. Alternatively, Piacentini *et al.* [85] proposed the use of membranes for emulsification and demonstrated their effectiveness by forming an emulsion of sunflower oil prior to its encapsulation in gelatin.

The tendency of the droplets to coalesce, and the droplet size strongly depend on interfacial interactions between components constituting the given emulsion. These can be modulated by the addition of emulsifiers [86] or by the choice of the oil constituting the dispersed phase. Moreover, a delicate balance between the phases that form the emulsion is changed upon the introduction of the complex coacervate phase. Emulsions formed in the presence of surfactants (emulsifiers) are more stable against changes of the process conditions. Nakagava *et al.* [47] attempted to encapsulate soybean oil in capsules consisting of gelatin and acacia, using microchannel emulsification. Gelatin solution was used as the continuous phase and no surfactant was added to the system. Decagrycerol monolaurate was used as an emulsifier in control samples. The authors reported that gelatin solution was used successfully to form emulsion but the emulsions were less stable than those prepared in the presence of decagrycerol monolaurate. Both gelatin and decagrycerol monolaurate produced monodispersed droplets, as defined by microchannel dimensions. The process was sensitive to pH changes. In her excellent review, Tcholakova *et al.* discussed in detail the stabilization of emulsions by solid particles, proteins and surfactants, and compared the effectiveness of various molecules as emulsifiers [86]. Not all of the described systems are applicable in the process of encapsulation. However, the general understanding of the stabilization mechanisms allows for more efficient process designs. It is worth mentioning that during the process of encapsulation, the emulsion is only a temporary "state" of a system. Therefore, long-term emulsion stability is not as important as emulsion morphology because the morphology defines the morphology of the capsule core as well as their size and polydispersity. Solid particles, proteins, and surfactants that can be used as emulsifiers differ in size, shape, and affinity to the interfaces. These differences correlate with different droplet sizes in the following manner: in general, particles of slower kinetics of adsorption, higher desorption energy, and smaller surface coverage per particle lead to the formation of larger droplets. On the other hand, particles with high desorption energies provide better long-term stability, and are reported to be able to completely suppress Ostwald ripening [86]. As the choice of the interface-stabilizing molecules is restricted by the final application of the capsules, one may consider controlling interfacial properties by changing the composition of the liquid phases. Several researchers studied the effect of oil used in a dispersed phase on encapsulation. Lamprecht *et al.* [84] found that the efficiency of encapsulation of omega-3 fatty acids in gelatin–acacia capsules strongly depended on the oil/polymer ratio as well as on total homogenization time. As the oil/polymer ratio increased, the amount of encapsulated oil increased as well but the encapsulation rate and total process yield decreased.

Prata *et al.* [49] considered various process variables and statistically determined that the effect of the polyelectrolytes ratio and process temperature had a more significant effect on emulsification and encapsulation yield than the rate and time of homogenization. However, the homogenization parameters seem to have greater effect than the type of the core material and polymer concentration. The polymer concentration and core material strongly affect the complex coacervation, which is the next crucial step in capsule formation. Thus, it becomes clear that it is very difficult to provide one recipe for the "perfect" emulsion. Physicochemical and mechanical aspects must be considered and addressed for each, specific system, and the resulting process protocol is a compromise between the effects of different variables.

4.3.2 Loading

The complex coacervate carriers can be of two distinct morphologies: capsules and particles. Capsules are characterized by a complex coacervate shell surrounding a coacervate-free core. Particles are characterized by the lack of a coacervate-free core. In capsules, the active compound occupies the core and it is separated from the environment by the complex coacervate shell. In particles, for example, hydrogels, the active compound occupies space within the polyelectrolytes network across the whole particle volume, and remains there due to favorable interactions with the particle-forming material.

Loading of the active compound can take place during or after formation of the carrier. The first method of loading, schematically shown in Fig. 4.4, yields the highest encapsulation efficiency. It requires formation of the emulsion as an intermediate step. In systems, in which polyelectrolytes constituting the shell and the active compound have immiscible solvents, the formation of a single emulsion is sufficient. In other systems in which polyelectrolytes constituting the shell and the active compound have one common or two miscible solvents, the formation of double emulsion is recommended. This was demonstrated by, for instance, Comunian *et al.* [87] for encapsulation of water soluble ascorbic acid, by Cardenas-Bailon *et al.* for encapsulation of insulin [88] and by Rocha-Selmi *et al.* for encapsulation of sucralose [89]. During encapsulation of the water-soluble compounds using double emulsion systems, oil creates a barrier, preventing diffusion of the compound out of the capsule and offers additional protection against molecules and solvents used for capsule processing. The capsule shell is created on the emulsion droplets containing the active compound during the coacervation step. As already mentioned, the composition of the dispersing and the dispersed phases as well as the type of additives used during emulsification determine the process outcome [90]. It needs to be emphasized that the use of surfactants does not only allow control over the morphology of the emulsion but also affects the loading. The loading is different if the same surfactant is used below and above its critical micellar concentration (CMC) [91].

Alternatively, the active compound itself may be a constituent of the complex coacervate phase. Such systems are commonly formulated into particles rather than capsules, provided that there is no additional active compound to be included.

The active compound can be present in a solid form. In such cases the compound is milled into a fine powder and dispersed in solvent compatible with the solvent of the polyelectrolytes [92]. The complex coacervate shell is formed at the solid–liquid interface.

A very interesting method of protein encapsulation in the complex coacervate phase was proposed by Jin *et al.* [93] for the fabrication of injectable gels for protein drug delivery. This method combines the benefits of complex coacervation between polyelectrolytes and proteins [27, 94] with thermoresponsivity of the material. The loading was done during formation of the complex coacervate phase consisting of gelatin A and a model therapeutic protein (fluorescein isothiocyanate conjugated bovine serum albumin), but the final encapsulation was to take place *in vitro* at the target location. This was achieved by addition of the thermoresponsive biopolymer (methylcellulose) co-aggregated into the complex coacervate phase. The gelling temperature of methylcellulose was adjusted to the body temperature by the addition of salting-out ammonium sulfate. The thus formed capsules allowed for stable release of the protein during a period of 25 days, with minimal initial burst release.

A post-loading can be applied to load hydrophilic active compounds to the hydrogel particles [95] or capsules made with thermoresponsive polyelectrolytes. Such carriers have relatively open structures that allow the compound to diffuse in. Subsequently, the compound is "locked" inside by interactions with the carrier material or due to narrowing of the pores of the particle in response to the thermal treatment [54]. The selection of carrier material is of crucial importance as it determines not only the conditions for complex coacervation, but also the type and strength of the interactions between the carrier material and the encapsulated compound, and thus potential losses during the washing step and the compound release profile. The thermal properties of the material must be considered so that the temperatures required to shrink the capsules do not cause denaturation or decomposition of the active compound [96].

Eventually, the final loading efficiency as well as the encapsulation yield is affected by each step of the encapsulation process. It begins with selection of the materials and process design and ends with the processing of the formed capsules that must be carefully controlled to avoid damage and unintentional release of the compound. A good example may be a choice of solvents used to wash and to harden the capsules. Application of some solvents may result in capsule cracking and leakage of the encapsulated material [91].

4.3.3 Crosslinking

In many systems crosslinking is required to further stabilize and harden the complex coacervate carriers. The choice of the crosslinking agent influences the structure, the properties, and the applicability of the final product. Traditionally, coacervates containing gelatin were stabilized with aldehydes. Alvim *et al.* [50] compared the effect of two molecules, namely glutaraldehyde and transglutaminase on the formation, processing and stability of the capsules. The authors reported the encapsulation efficiency at higher than 96% for both crosslinking molecules. However, the capsules crosslinked with glutaraldehyde retained the core material more effectively than those crosslinked with transglutaminase. The fabricated capsules were polydispersed, decreased in size during drying and returned to approximately the original size upon rehydration. The capsules were dried by freeze and spray drying. Capsules that were not crosslinked were successfully dried by freeze drying but disintegrated during spray drying. The capsules crosslinked with either molecule could be successfully dried using both techniques.

Muniyandy *et al.* [92] studied complex coacervates consisting of carboxymethyl sago pulp and gelatin for the encapsulation of ibuprofen. The capsules were double-crosslinked in a solution of toluene saturated with glutaraldehyde and aqueous aluminum chloride. The authors claim that the application of double crosslinking allowed for better control over the release profile. The toluene residues were removed on rinsing with isopropanol and drying. The use of such toxic organic crosslinkers demands a rigorous washing procedure to ensure their complete removal. Thus, there is a preference for more environmentally friendly and safe alternatives. Huang *et al.* studied the application of glycerol as a crosslinking agent in a gelatin–acacia complex coacervate system for the encapsulation of shikonin [97]. The authors showed that the encapsulation yield increased as the concentration of glycerol increased up to 6%. At higher concentrations the encapsulation efficiency decreased, most probably due to the inability to integrate glycerol into the complex coacervate network. Lamprech *et al.* [84] used dehydroascorbic acid to crosslink and harden microcapsules, and reported no major differences in capsules structure as compared with capsules hardened with ethanol. Genipin was successfully used to crosslink gelatin–carrageenan capsules containing neem seed oil [66]. More recent studies aim to eliminate crosslinkers completely. In order to increase the stability of the gelatin capsules Vandelli *et al.* [98] proposed their treatment with microwaves. The authors showed that irradiation for 10 min at a temperature above 220°C resulted in crosslinking of gelatin capsules, and hypothesized that the process relayed on condensation reactions between amino groups and carboxylic acid. Upon increase of the irradiation temperature from 220°C to 250°C, the morphology of the gelatin capsules changed: the cross section of pores changed from the shape of an ink bottle to a linear shape. The thus formed capsules could be post-loaded by soaking in drug (diclofenac) solution, and were reported to be biodegradable. Ifeduba *et al.* [99] took advantage of the Millard reaction to increase the stability of gelatin–gum arabic capsules for encapsulation of soybean oil. Moreover, the by-products of the reaction proved to possess antioxidant properties and thus added to the oxidation stability of the encapsulated oil.

The preferred crosslinking method must not affect the stability of the encapsulated compound. The heat treatment, even though it offers advantages with respect to capsules size distribution, compound retention and purity of the final product, may damage loaded drugs or proteins. On the other hand, the introduction of crosslinking molecules is limited by their potential toxicity. A search for effective, nontoxic, and environmentally friendly crosslinking agents, preferably of plant origin, is ongoing.

4.3.4 Separation and Further Processing

Once formed and loaded, the capsules need to be separated from the reaction solution by means of sedimentation, filtration or centrifugation. Subsequently, the capsules are washed to remove residues of polyelectrolytes, compounds, crosslinking and/or hardening agents, and dried. Capsules are usually rinsed in pure water, alcohol, or nitrogen [97] depending on the process and the nature of the crosslinker used. Capsules can be dried by means of dehydration using a suitable solvent, or by means of a physical process. Some solvents allow combining drying and hardening processes into a single process [84, 91]. However, their use may be restricted by potential damage to the capsules and leaking of the encapsulated compound [91, 67]. Thus, drying is usually done during a separate physical process. The drying techniques most commonly applied to capsules are air drying, spouted bed drying, spray drying, and freeze drying. Interested readers may refer to the excellent literature reviews available, for example, [100–103] on details regarding these processes. Each of the above mentioned methods has a potential to significantly affected the morphology of the capsules and, as a result, the core release profile [50]. Erate *et al.* [104] compared the effect of spray drying and freeze drying on capsules containing oil and live cells. Both methods rendered comparable residual moisture below 4%, and comparable water activity. Spray-dried capsules showed better oil encapsulation efficiency and higher stability against oil oxidation, related to lower surface oil content, than freeze-dried capsules. On the other hand, freeze-dried capsules were characterized by higher cell survival. The Spray-dried capsules were approximately spherical in shape, with concavities, whereas the freeze-dried capsules were flake-like and porous. Koupantsis *et al.* noted that capsules can be damaged at low temperatures required for freeze drying resulting in uncontrollable release of volatile compounds [105]. Bosio *et al.* claimed that the addition of carboxymethyl guar gum to capsule formulation protected gel structure during freeze drying, preventing the rapid release of the encapsulated compound. Jiang *et al.* proposed a combination of coacervation and spray drying, using three-fluid nozzle spray drying [106]. The thus fabricated capsules were not only smaller but also less porous and provided better enzyme stability than capsules formed in the classic manner. Oliveira *et al.* [107] applied the spout bed drying process to capsules formed by complex coacervation between pectin and casein, containing *B. lactis* and *L. acidophilus*. The authors reported that the process resulted in the rupture of capsules and exposure of the microorganisms to acid medium; whereas the same capsules dried by means of spray drying effectively shielded the microorganisms from the acidic environment. The presented examples clearly illustrate the importance of process design in correlation with the physical and mechanical properties of the designed encapsulation system.

4.4 Application of Complex Coacervates for Encapsulation

The first recorded use of the coacervate phase for encapsulation dates to 1957. Green and Schleicher described and patented microcapsules formed with gum arabic for use in carbonless copy paper [108]. Soon, it became evident that applicability of the thus formed capsules was much wider. Attempts were made to use coacervate-based microcapsules in the food industry. The very first capsules were formed using pigskin gelatin and gum arabic chemically crosslinked with glutaraldehyde. Due to rising health concerns, these components were subsequently replaced by fish gelatin and whey proteins. The ultimate goal was to replace them completely with biopolymers of plant origin. Some examples of such biopolymers can be found in literature: maltodextrins, gum acacia and other gums, sodium alginate, and whey proteins are reportedly used as flavor encapsulants. An excellent review of current developments in the application of coacervates for encapsulation in the food industry was presented by Thies [109]. Mc Clements *et al.* [2] have presented an interesting discussion over targeted delivery and controlled digestion of lipophilic food components. Encapsulation also becomes a preferred method of protection of sensitive compounds in other applications like pharmaceuticals, personal care products, coatings, textiles, diagnostics, and agriculture. Particularly interesting are drug delivery systems developed for proteins and small molecule drugs [3] as well as vitamins [110] as the choice of the encapsulation material allows for careful control over the release kinetics. Many of these applications require using nontoxic, biodegradable, and environmentally friendly compounds that are easily degradable under specific conditions (pH, exposure to light, shear, etc.). There is also the need to make encapsulation systems less complex by reducing the number of necessary components and fabrication steps. In view of these requirements, the application of complex coacervation for encapsulation becomes a method of choice. Table 4.1 provides several examples of application methods. Its purpose is to illustrate the versatility of the concept, schematically illustrated in Fig. 4.4, rather than to provide a complete overview of the existing systems. It is worth mentioning that the encapsulation methods as well as the preferred components of the coacervate phase remain the same in most of the examples despite the different properties of the encapsulated compounds, with gelatin being one of the most frequently used component. The most significant differences between the presented examples are observed in emulsification as well as in the crosslinking steps.

4.5 Concluding Remarks

The objective of this chapter was to discuss the properties of complex coacervates in the context of their applicability as an encapsulation material. The concept is not new, as the first application examples had emerged already in the 1950s. The relatively simple mechanism of coacervation, the large number of potential constituents of animal and plant origin, and quite well understood effects of varying environmental conditions on complex coacervates offer significant flexibility with respect to potential applications. Complex coacervates were successfully applied to encapsulate drugs, dyes, fertilizers, oils, and fragrances. Commonly used for encapsulation of hydrophobic compounds, they also allow encapsulation of hydrophilic compounds. The main advantage of encapsulation using complex coacervates over the other techniques is very high encapsulation yield. The process of encapsulation is realized under mild conditions contrary to, for instance, spray drying that requires high temperatures to form capsules. Microcapsules produced by complex coacervation are water insoluble, temperature resistant and have excellent characteristics for controlled release [131]. Coacervates are stable for long periods

Table 4.1 Application of complex coacervates for encapsulation of active compounds

Application	Composition of the complex coacervate phase	Encapsulated compound	Remarks on fabrication	Ref.
Food	Lactoferin–β-Lactoglobulin	Vitamin B9	Vitamin B9 was mixed with lactoferin, followed by addition of β-lactoglobulin. Coacervates formed spontaneously and were prone to coalescence and sedimentation in time. Particles size and shape depended on ratio between the components. Encapsulation ranged from 6 to 16 mg B9 / g protein, which was comparable to efficiency of other technologies for B9 encapsulation.	[110]
Food	Gelatin–Gum arabic	Palm oil	Warm gelatin solution and palm oil were emulsified, followed by addition of gum arabic solution and pH adjustment to induce coacervation. After cooling, sedimentation and decantation the capsules were frozen and lyophilized. Obtained capsules were spherical in shape with broad size distribution (average diameter 284.74 ± 207.26 μm).	[111]
Food	Gelatin–Sodium alginate	Olive oil	Olive oil and warm gelatin solution were emulsified, followed by addition of sodium alginate and pH adjustment to trigger coacervation. Subsequently, capsules were crosslinked with glutaraldehyde, filtered, washed with hexane and water, and freeze dried. Encapsulation efficiency increased with increase of oil, polymer, and glutaraldehyde concentrations.	[112]

Application	Composition of the complex coacervate phase	Encapsulated compound	Remarks on fabrication	Ref.
Food	Hydroxypropylmethyl cellulose–Sodium carboxymethyl cellulose–Sodium dodecylsulfate	Sunflower oil	Sunflower oil was homogenized with hydroxypropylmethyl cellulose-sodium dodecylsulfate mixture to form a primary emulsion. Subsequently, sodium carboxymethyl cellulose was added and coacervation took place. The capsules were rinsed and freeze dried. The content of SDS in a mixture determined viscoelastic properties of the coacervate phase deposited on the surface of the drop and, as a result, permeability of the capsules. Concentration of sodium dodecyl sulfate exceeding 1 % suppressed formation of coacervate phase.	[113, 114]
Food	Cold water fish gelatin–Gum arabic	Sunflower oil	Emulsion was formed by continuous membrane emulsification: oil was injected through the membrane into gelatin–gum arabic solution. Addition of sodium dodecyl sulfate promoted formation of smaller drops. Subsequently, pH was adjusted with lactic acid to induce coacervation. Formed capsules were crosslinked with glutaraldehyde and washed. Capsules had single cores. Optimal biopolymers concentration for successful encapsulation was below 4%.	[85]
Food	Whey protein isolate–Gum arabic	Omega-3 fatty acids, Tuna oil, and probiotic bacteria *Lactobacillus casei 431*	Emulsion of tuna oil and whey protein isolate was formed, followed by addition of cell solution and of gum arabic. Subsequently, pH was adjusted using citric acid to induce coacervation. The capsules were spray dried. Viability of *L. casei* was improved in presence of oil owed to the cells being separated from coacervate shell by oil phase, and in freeze-dried capsules. The stability of tuna oil against oxidation was improved by encapsulation. Produced capsules were amorphous.	[104]

(*Continued*)

Table 4.1 *(Continued)*

Application	Composition of the complex coacervate phase	Encapsulated compound	Remarks on fabrication	Ref.
Food	Gelatin–Acacia gum	Fish oil	Fish oil and gelatin were emulsified, followed by addition of acacia gum. Subsequently, the temperature was increased and pH was adjusted with acetic acid to induce coacervation. The capsules were lyophilized. Capsules were spherical in shape with broad size distribution (0.5–10 μm). Capsules were successfully incorporated into yoghurt.	[115]
Food	Soybean protein isolate–Gum arabic	Sweet orange oil	Sweet orange oil and soybean protein isolate were emulsified, followed by addition of gum arabic. Coacervation was induced by pH adjustment with hydrochloric acid. Encapsulation efficiency depended on effectiveness of the emulsifier (soybean protein isolate). Microencapsulation efficiency and yield increased upon addition of sucrose (reported to enhance the stability of protein matrixes).	[90]
Food	Gelatin–Gum arabic	Sucralose	Double-emulsion was prepared prior complex coacervation. Sucralose aqueous solution was emulsified with soybean oil with addition of soy lecithin. Subsequently, warm solution of gelatin was added followed by addition of gum arabic. Next, pH was adjusted with hydrochloric acid to induce coacervation. Capsules were freeze dried. Capsules were spherical in shape and contained multiple cores. Their size varied from 81 to 114 μm. The encapsulation efficiency was 43 to 89%. The solubility of encapsulates sucralose was 4–10 times less as compared with free sucralose.	[89]

Application	Composition of the complex coacervate phase	Encapsulated compound	Remarks on fabrication	Ref.
Food	Whey protein isolate–Gum acacia	β-carotene	Oil–water emulsion was created by mixing whey protein isolate and β-carotene dissolved in soy oil. Subsequently, solution of gum acacia was added, followed by pH adjustment with hydrochloric acid. Coacervate was crosslinked with glutaraldehyde. Capsules were lyophilized. The capsules were spherical in shape with smooth surfaces, with sizes ranging from 130 to 170 μm. The encapsulation efficiency was above 70%. Strong antioxidant activity was retained after 3 months.	[116]
Food	Gelatin–Gum arabic	Fucoxanthin	Fuxoxantin was dissolved in mixture of cetyl palmitate and canola oil, and mixed with gelatin solution to prepare emulsion. Subsequently, gum arabic and Tween 80 were added, and pH adjusted with acetic acid to induce coacervation. The cooled particles were crosslinked with tannic acid. After washing with water capsules were freeze-dried. Wet microcapsules aggregated into clusters of irregular shapes attributed to presence of tannic acid. Encapsulation efficiency was highest in absence of cetyl palmitate (91.5%) and decreased in presence of cetyl palmitate (85.5%). Encapsulation efficiency was lower in dry capsules.	[117]

(Continued)

Table 4.1 (*Continued*)

Application	Composition of the complex coacervate phase	Encapsulated compound	Remarks on fabrication	Ref.
Food	Gelatin–Carboxymethyl cellulose	Capsaicin	Capsaicin was added to warm gelatin solution and carboxymethyl cellulose, and emulsified. Subsequently, pH was adjusted to induce coacervation. Capsules were crosslinked with transglutaminase. After filtration and washing capsules were freeze-dried. For optimized conditions encapsulation yield was >90%. Dried capsules were approximately 100 μm in diameter. Capsules cracked as a result of freeze drying.	[118]
Food	Calcium (Ca) alginate–Ca-alginate + Chitosan	Amylase	Comparison of (i) one-step encapsulation and (ii) two-step encapsulation (ii): (i) calcium chloride solution mixed with chitosan and amylase was spray dried at 200°C (inlet) and 90°C (outlet), (ii) solution of sodium alginate with amylase was sprayed into calcium chloride solution, or into chitosan + calcium chloride solution, followed by freeze drying. One-step method rendered smaller, less porous particles and better enzymatic stability.	[106]
Food	Gelatin–Gum arabic	Allyl isothiocyanate	Warm gelatin solution and allyl isothiocyanate solution in olive oil were mixed to form emulsion. Subsequently solution of gum arabic, containing Tween 80, was added and pH was adjusted with acetic acid to induce coacervation. Capsules were crosslinked with tannic acid. Encapsulation efficiency depended on ratio between components, and was 89.7% for optimum conditions.	[119]

Application	Composition of the complex coacervate phase	Encapsulated compound	Remarks on fabrication	Ref.
Food	Gelatin–Gum acacia wall and medium-chain-triglyceride	Menthol or limonene	Solution of gum acacia and warm gelatin were emulsified with limonene solution (liquid core) or mixed with menthol crystals (solid core) at low pH. After cooling down, pH was adjusted with sodium hydroxide and the resulting capsules were crosslinked with glutaraldehyde. Capsules were washed and freeze dried. Capsules with liquid core (325±112 nm) were larger than capsules with solid core (219±132 nm) but had the same wall thickness. Encapsulation efficiency was higher for solid core-capsules (>90%).	[120, 121]
Food, Personal care	Gelatin–Sodium hexametaphosphate	Vitamins: A, D-3, D-, E, K-2, curcumin, coenzyme Q10; Tuna oil	Active compounds were dissolved in tuna oil. Initially, gelatin and tuna oil (under N_2 flow) were heated up and emulsified, followed by addition of sodium hexametaphosphate. pH was adjusted with phosphoric acid to induce coacervation. After cooling down the capsules were crosslinked with transglutaminase and freeze dried. Capsules contained multiple cores, were oval in shape, and their size varied from 50 to 80 μm prior to drying. No oil oxidation was observed.	[122], [123]
Food, Pharmaceutical	Isolated soy protein–Pectin	Propolis	Isolated soy protein was homogenized with propolis followed by slow addition of pectin. Subsequently, pH was adjusted with hydrochloric acid and system was cooled down. Capsules were allowed to sediment under refrigeration and freeze dried. Capsules were round but rather irregular in shape. Their size ranged from 10 to 90 μm. Propolis was shown to retain its antioxidant and antibacterial (*S. aureus*) activity.	[124]

(*Continued*)

Table 4.1 (*Continued*)

Application	Composition of the complex coacervate phase	Encapsulated compound	Remarks on fabrication	Ref.
Food, Pharmaceutical	Alginate–Chitosan	Lipid	Medium chain triglyceride was mixed with β-lactoglobulin solution to form a primary emulsion. Subsequently, alginate solution was added. The thus formed particles were entrapped within a hydrogel formed by complex coacervation between alginate and chitosan. Method of shell formation (order of mixing) affected the shell structure and thus, charges, porosity, stability, and release characteristics. The size of the microcapsules increased significantly on pH change from pH 1 to pH 7.	[125]
Food, Personal care, Pharmaceutical	Gelatin–Sodium alginate	Eugenol	Eugenol was mixed with warm gelatin and sodium alginate solutions to form emulsion. Coacervation was induced by pH adjustment with hydrochloric acid. Capsules were crosslinked with formaldehyde. Application of dehydration agent (isopropyl alcohol) resulted in microcapsules shrinkage and formation of cracks on the microcapsules resulting in eugenol leaking.	[67]
Pharmaceutical	Low molecular weight heparin–Polyaminomethacrylates	Low molecular weight heparin	Low molecular heparin was mixed with polyaminomethacrylates solution and polysorbate used as emulsifier. PEG derivatives were used as polymer solvents so that solvent removal was not required at the end of the encapsulation process. Formed particles were washed and air-dried. Particles were of irregular shape. Encapsulation efficiency was >80% and depended on PEG derivative used. Bioavailability of low molecular weight heparin in rabbits was significant.	[126]

Application	Composition of the complex coacervate phase	Encapsulated compound	Remarks on fabrication	Ref.
Pharmaceutical	N-acetylated chitosan–Methacrylic acid-hydroxyethyl methacrylate–Methyl methacrylate	Live cells	Drops of hepatocytes solution in solution of acetylated chitosan were extruded into solution of polycation solution, where coacervation took place. Capsule porosity depended on chitosan content. Live cells (hepatocytes) remained vital.	[127]
Pharmaceutical	Starch derivatives	IgG1, RNase A, insulin, and vancomycin	Starch nanoparticles were formed spontaneously upon mixing oppositely charged starch derivatives. Particle size varied from 138.9 to 346.9 nm and depended on ratio between two starch derivatives. The particles were spherical in shape with low polydispersity. The particles were post-loaded by mixing with aqueous protein solutions. Loading efficiency was highest for the smallest and most hydrophobic protein. Capsules were degradable by α-amylase.	[95]
Pharmaceutical	Whey protein isolate–Sodium carboxymethylcellulose or sodium alginate	Insulin	Insulin was mixed with glycerol and m-cresol to form inner aqueous phase of double emulsion. This solution was emulsified with soybean oil containing polyglycerolpolyrricinoleate. The thus obtained emulsion was re-emulsified with whey protein isolate. Subsequently, the complex coacervate shell was created by coating with carboxymethylcellulose or sodium alginate. The capsules were spray dried. Capsules were spherical in shape with relatively high polydispersity. Capsules were well soluble at pH 7 (>90%), capsules containing sodium alginate had lower solubility at pH 2 (<9%) offering better protection of insulin under stomach condition.	[88]

(*Continued*)

Table 4.1 *(Continued)*

Application	Composition of the complex coacervate phase	Encapsulated compound	Remarks on fabrication	Ref.
Pharmaceutical	Carboxymethyl sago pulp–Gelatin	Ibuprofen	Sago pulp and ibuprofen were homogenized, followed by addition of warm gelatin solution and pH adjustment with acetic acid to induce coacervation. The capsules were double-crosslinked: first using toluene saturated with glutaraldehyde quenched with sodium metabisulfite, and after rinsing with isopropanol and drying with aluminum chloride. Size distribution of the spherical particles was related to size of ibuprofen particles. Double crosslinking allowed better control over release.	[92]
Pharmaceutical, Cosmetics, Food, Printing	Gelatin–Acacia	Shikonin	Emulsion was formed by mixing shikonin in sesame oil with acacia solution in the presence of polysorbate 80. Subsequently, warm solution of gelatin was added and pH adjusted with acetic acid to induce coacervation. After cooling down capsules were separated from solution and crosslinked. Glycerol was used as nontoxic crosslinking agent. The gelatin shells formed with glycerol are reported to be more permeable than these formed by formaldehyde, and dissolve at pH above pH 6.	[97]
Botanical pesticide	Gelatin–κ-Carrageenan	Neem seed oil	Carrageenan was emulsified with neem seed oil. Subsequently, warm solution of gelatin was added and pH was adjusted with glacial acetic acid to induce coacervation. Capsules were crosslinked with genipin. Capsules were washed with Tween 80 and water, and freeze dried. Oil release rate decreased with increased polymer concentration, explained by formation of capsules with thicker walls. The release rate decreased as concentration of the crosslinker increased.	[66]

Application	Composition of the complex coacervate phase	Encapsulated compound	Remarks on fabrication	Ref.
Footwear	Gelatin–Sodium carboxymethyl-methyl cellulose	Tea tree oil	Tea tree oil and warm gelatin solution were emulsified, followed by addition of cellulose and pH adjustment with acetic acid. Capsules were crosslinked with 1:1 formaldehyde/glutaraldehyde solution and freeze dried. Capsules were oval in shape but shape changed upon change in gelatin to cellulose ratio. The highest loading efficiency was obtained for ratio 10 (63%). Capsules ruptured upon rubbing and heating to 160 °C (high iron setting).	[128]
Textiles	Gum arabic–gelatin	n-hexadecane, n-octadecane and n-nonadecane	Warm solution of gelatin was mixed with sodium dodecyl sulfate and $K_2S_2O_8$ and $Na_2S_2O_3$ solutions, followed by addition of gum arabic. Subsequently, the oil phase (phase changing materials for enhanced thermal performance) was added, followed by addition of sodium carbonate, cooling in ice bath and pH adjustment. Capsules were crosslinked with formaldehyde. Capsules containing molecules with longer chains were more prone to leakage under the experimental conditions.	[129]
Textiles	Silk fibroin–Chitosan	n-Eicosane	Melted n-eicosane (phase changing materials for enhanced thermal performance) was mixed with silk fibroin solution and emulsified with Span-20 as emulsifier. Subsequently, chitosan solution was added and pH was adjusted with sodium hydroxide to induce coacervation. Capsules were crosslinked with glutaraldehyde. Encapsulation of the content of the crosslinking agent has significant effect on encapsulation efficiency.	[130]

of time and resistant to bacteria. However, the full potential of this method has not been exploited yet, mainly due to difficulties with scaling-up. The limiting steps are formation of sufficiently stable emulsions with narrow distribution of droplets sizes, prevention of coalescence and aggregation. Another factor narrowing down practical applications of complex coacervate capsules is the limited selection of effective and nontoxic crosslinkers. Thus, there is an ongoing effort to make this process more efficient and scalable as well as to eliminate potentially hazardous compounds from complex coacervate formulations.

References

1. Gouin S., Microencapsulation: industrial appraisal of existing technologies and trends, *Trends in Food Science & Technology*, 157–8 (2004) 330–347.
2. McClements D. J., Li Y., Structured emulsion-based delivery systems: Controlling the digestion and release of lipophilic food components, *Advances in Colloid and Interface Science*, 1592 (2010) 213–228.
3. Johnson N. R., Wang Y., Coacervate delivery systems for proteins and small molecule drugs, *Expert Opinion on Drug Delivery*, 1112 (2014) 1829–1832.
4. Trojer M. A., Nordstierna L., Bergek J., Blanck H., Holmberg K., Nyden M., Use of microcapsules as controlled release devices for coatings, *Advances in Colloid and Interface Science*, 222 (2015) 18–43.
5. Timilsena Y. P., Adhikari R., Casey P., Muster T., Gill H., Adhikari B., Enhanced efficiency fertilisers: a review of formulation and nutrient release patterns, *Journal of the Science of Food and Agriculture*, 956 (2015) 1131–1142.
6. O'Donnell P. B., McGinity J. W., Preparation of microspheres by the solvent evaporation technique, *Advanced Drug Delivery Reviews*, 281 (1997) 25–42.
7. Walde P., Ichikawa S., Enzymes inside lipid vesicles: Preparation, reactivity and applications, *Biomolecular Engineering*, 184 (2001) 143–177.
8. Garti N., McClements D. J., *Encapsulation Technologies and Delivery Systems for Food Ingredients and Nutraceuticals*, Woodhead Publishing Series in Food Science, Technology and Nutrition, Elsevier Science (2012). Available from: https://books.google.com.sg/books?id=qHNHAgAAQBAJ

9. Mittal V., *Encapsulation Nanotechnologies,* Wiley (2013). Available from: https://books.google.com.sg/books?id=O0BL1dj_f1YC
10. Mishra M., *Handbook of Encapsulation and Controlled Release,* CRC Press (2015). Available from:

    https://books.google.com.sg/books?id=pY7wCgAAQBAJ
11. Tiebackx F. W., Gleichzeitige Ausflockung zweier Kolloide, *Zeitschrift für Chemie und Industrie der Kolloide,* 84 (1911) 198–201.
12. de Jong H. G. B., Kruyt H. R., Coacervation (Partial miscibility on colloid systems) (Preliminary communication), *Proceedings of the Koninklijke Akademie Van Wetenschappen Te Amsterdam,* 326/10 (1929) 849–856.
13. de Jong H. G. B., Kruyt H. R., Coacervation. (Decomposition in colloid systems). Preliminary announcement, *Kolloid-Zeitschrift,* 501 (1930) 41–52.
14. Overbeek J. T. G., Voorn M. J., Phase separation in polyelectrolyte solutions. Theory of complex coacervation, *Journal of Cellular and Comparative Physiology,* 49S1 (1957) 7–26.
15. Bamford C. H., Tompa H., The theory of coacervation, *Transactions of the Faraday Society,* 460 (1950) 310–316.
16. Broda A., Chodkowska M. B., The problem of polymer coacervation, *Journal of Polymer Science,* 30121 (1958) 639–644.
17. de Kruif C. G., Weinbreck F., de Vries R., Complex coacervation of proteins and anionic polysaccharides, *Current Opinion in Colloid & Interface Science,* 95 (2004) 340–349.
18. Lawrence A. S. C., Coacervation. 2. Introductory paper, *Discussions of the Faraday Society,* 18 (1954) 229–230.
19. Voorn M. J., Complex coacervation. 1. General theoretical considerations, *Recueil Des Travaux Chimiques Des Pays-Bas-Journal of the Royal Netherlands Chemical Society,* 753 (1956) 317–330.
20. Veis A., A review of the early development of the thermodynamics of the complex coacervation phase separation, *Advances in Colloid and Interface Science,* 1671–2 (2011) 2–11.
21. Bohidar H. B., Coacervates: A novel state of soft matter- an overview, *Journal of Surface Science and Technology,* 24 (2015), 1893.
22. Momeni A., Filiaggi M. J., Rheology of polyphosphate coacervates, *Journal of Rheology,* 601 (2016) 25–34.
23. Mohanty B., Bohidar H. B., Systematic of alcohol-induced simple coacervation in aqueous gelatin solutions, *Biomacromolecules,* 44 (2003) 1080–1086.

24. Lazko J., Popineau Y., Legrand J., Soy glycinin microcapsules by simple coacervation method, *Colloids and Surfaces B: Biointerfaces,* 371–2 (2004) 1–8.
25. Seweryn A., Wasilewski T., Bujak T., Effect of salt on the manufacturing and properties of hand dishwashing liquids in the coacervate form, *Industrial & Engineering Chemistry Research,* 554 (2016) 1134–1141.
26. de Ruiter L., de Jong H. B., The interfacial tension of gum arabic–gelatin complex coacervates and their equilibrium liquids, *Proceedings of the Koninklijke Nederlandse Akademie van Wetenschappen,* 50 (1947) 836–848.
27. Kizilay E., Kayitmazer A. B., Dubin P. L., Complexation and coacervation of polyelectrolytes with oppositely charged colloids, *Advances in Colloid and Interface Science,* 1671-2 (2011) 24–37.
28. Tavernier I., Wijaya W., Van der Meeren P., Dewettinck K., Patel A. R., Food-grade particles for emulsion stabilization, *Trends in Food Science & Technology,* 50 (2016) 159–174.
29. Spruijt E., *Strength, Structure and Stability of Polyelectrolyte Complex Coacervates.* PhD, (2012) Wageningen University, Wageningen.
30. Dobry A., Boyer-Kawenoki F., Phase separation in polymer solution, *Journal of Polymer Science,* 21 (1947) 90–100.
31. Abe K., Koide M., Tsuchida E., Selective complexation of macromolecules, *Macromolecules,* 106 (1977) 1259–1264.
32. Pawar N., Bohidar H. B., Anisotropic domain growth and complex coacervation in nanoclay-polyelectrolyte solutions, *Advances in Colloid and Interface Science,* 1671-2 (2011) 12–23.
33. van der Gucht J., Spruijt E., Lemmers M., Stuart M. A. C., Polyelectrolyte complexes: Bulk phases and colloidal systems, *Journal of Colloid and Interface Science,* 3612 (2011) 407–422.
34. Priftis D., Tirrell M., Phase behaviour and complex coacervation of aqueous polypeptide solutions, *Soft Matter,* 836 (2012) 9396–9405.
35. Chollakup R., Smitthipong W., Eisenbach C. D., Tirrell M., Phase behavior and coacervation of aqueous poly(acrylic acid)–poly(allylamine) solutions, *Macromolecules,* 435 (2010) 2518–2528.
36. Brzozowska A. M., Keesman K. J., de Keizer A., Leermakers F. A. M., Formation and structure of ionomer complexes from grafted polyelectrolytes, *Colloid and Polymer Science,* 2898 (2011) 889–902.
37. Spruijt E., Westphal A. H., Borst J. W., Stuart M. A. C., van der Gucht J., Binodal compositions of polyeleetrolyte complexes, *Macromolecules,* 4315 (2010) 6476–6484.

38. Perry S., Li Y., Priftis D., Leon L., Tirrell M., The effect of salt on the complex coacervation of vinyl polyelectrolytes, *Polymers*, 66 (2014) 1756.
39. Adzic N., Podgornik R., Field-theoretic description of charge regulation interaction, *European Physical Journal E*, 37 (2014) 49.
40. Markovich T., Andelman D., Podgornik R., Charge regulation: A generalized boundary condition? *Epl*, 1132. (2016)
41. Ninham B. W., Parsegian V. A., Electrostatic potential between surfaces bearing ionizable groups in ionic equilibrium with physiologic saline solution, *Journal of Theoretical Biology*, 313 (1971) 405–428.
42. Biesheuvel P. M., Lindhoud S., Stuart M. A. C., de Vries R., Phase behavior of mixtures of oppositely charged protein nanoparticles at asymmetric charge ratios, *Physical Review E*, 734. (2006)
43. Biesheuvel P. M., Stuart M. A. C., Cylindrical cell model for the electrostatic free energy of polyelectrolyte complexes, *Langmuir*, 2011 (2004) 4764–4770.
44. Anvari M., Pan C.-H., Yoon W.-B., Chung D., Characterization of fish gelatin–gum arabic complex coacervates as influenced by phase separation temperature, *International Journal of Biological Macromolecules*, 79 (2015) 894–902.
45. Mayya K. S., Bhattacharyya A., Argillier J. F., Micro-encapsulation by complex coacervation: influence of surfactant, *Polymer International*, 524 (2003) 644–647.
46. Bhattacharyya A., Argillier J., Microencapsulation by complex coacervation: effect of cationic surfactants, *Journal of Surface Science and Technology*, 213/4 (2005) 161.
47. Nakagawa K., Iwamoto S., Nakajima M., Shono A., Satoh K., Microchannel emulsification using gelatin and surfactant-free coacervate microencapsulation, *Journal of Colloid and Interface Science*, 2781 (2004) 198–205.
48. Duquemin S. J., Nixon J. R., The effect of sodium lauryl sulfate, cetrimide and polysorbate 20 surfactants on complex coacervate volume and droplet size, *Journal of Pharmacy and Pharmacology*, 3710 (1985) 698–702.
49. Prata A. S., Grosso C. R. F., Influence of the oil phase on the microencapsulation by complex coacervation, *Journal of the American Oil Chemists Society*, 927 (2015) 1063–1072.
50. Alvim I. D., Grosso C. R. F., Microparticles obtained by complex coacervation: influence of the type of reticulation and the drying

process on the release of the core material, *Food Science and Technology (Campinas)*, 30 (2010) 1069–1076.

51. Spruijt E., Sprakel J., Cohen Stuart M. A., van der Gucht J., Interfacial tension between a complex coacervate phase and its coexisting aqueous phase, *Soft Matter*, 61 (2010) 172–178.
52. Dautzenberg H., Karibyants N., Polyelectrolyte complex formation in highly aggregating systems. Effect of salt: response to subsequent addition of NaCl, *Macromolecular Chemistry and Physics*, 2001 (1999) 118–125.
53. Mathieu F., Ugazio S., Carnelle G., Ducini Y., Legrand J., Complex coacervation of the gelatin-poly(acrylic acid) system, *Journal of Applied Polymer Science*, 1011 (2006) 708–714.
54. Köhler K., Shchukin D. G., Möhwald H., Sukhorukov G. B., Thermal behavior of polyelectrolyte multilayer microcapsules. 1. The effect of odd and even layer number, *The Journal of Physical Chemistry B*, 10939 (2005) 18250–18259.
55. Dardelle G., Erni P., Three-phase interactions and interfacial transport phenomena in coacervate/oil/water systems, *Advances in Colloid and Interface Science*, 206 (2014) 79–91.
56. de Ruiter L., Bungenberg de Jong H.G., The interfacial tension of gum arabic-gelatine complex-coacervates and their equilibrium liquids, *Proceedings of the Koninklijke Nederlandse Akademie van Wetenschappen*, 50 (1947) 836–848.
57. Torza S., Mason S. G., Three-phase interactions in shear and electrical fields, *Journal of Colloid and Interface Science*, 331 (1970) 67–83.
58. Rabiskova M., Valaskova J., The influence of HLB on the encapsulation of oils by complex coacervation, *Journal of Microencapsulation*, 156 (1998) 747–751.
59. Burgess D. J., Complex coacervation: microcapsule formation, *Macromolecular Complexes in Chemistry and Biology*, Dubin P., Bock J., Davis R., Schulz D. N. & Thies C., (eds.), (1994), Berlin, Heidelberg: Springer Berlin Heidelberg, pp. 285–300.
60. Thomasin C., Merkle H. P., Gander B. A., Physico-chemical parameters governing protein microencapsulation into biodegradable polyesters by coacervation, *International Journal of Pharmaceutics*, 1472 (1997) 173–186.
61. Weinbreck F., Wientjes R. H. W., Rheological properties of whey protein/gum arabic coacervates, *Journal of Rheology*, 486 (2004) 1215–1228.

62. Spruijt E., Sprakel J., Lemmers M., Stuart M. A. C., van der Gucht J., Relaxation dynamics at different time scales in electrostatic complexes: time-salt superposition, *Physical Review Letters,* 10520 (2010) 4.
63. Rubinstein M., Semenov A. N., Dynamics of entangled solutions of associating polymers, *Macromolecules,* 344 (2001) 1058–1068.
64. Liberatore M. W., Wyatt N. B., Henry M., Dubin P. L., Foun E., Shear-induced phase separation in polyelectrolyte/mixed micelle coacervates, *Langmuir,* 2523 (2009) 13376–13383.
65. Dickinson E., Interfacial structure and stability of food emulsions as affected by protein-polysaccharide interactions, *Soft Matter,* 45 (2008) 932–942.
66. Devi N., Maji T., Genipin crosslinked microcapsules of gelatin A and kappa-carrageenan polyelectrolyte complex for encapsulation of Neem (Azadirachta Indica A.Juss.) seed oil, *Polymer Bulletin,* 654 (2010) 347–362.
67. Shinde U., Nagarsenker M., Microencapsulation of eugenol by gelatin-sodium alginate complex coacervation, *Indian Journal of Pharmaceutical Sciences,* 733 (2011) 311–315.
68. Ariga K., McShane M., Lvov Y. M., Ji Q., Hill J. P., Layer-by-layer assembly for drug delivery and related applications, *Expert Opinion on Drug Delivery,* 85 (2011) 633–644.
69. Wohl B. M., Engbersen J. F. J., Responsive layer-by-layer materials for drug delivery, *Journal of Controlled Release,* 1581 (2012) 2–14.
70. Haidar Z. S., Hamdy R. C., Tabrizian M., Protein release kinetics for core–shell hybrid nanoparticles based on the layer-by-layer assembly of alginate and chitosan on liposomes, *Biomaterials,* 299 (2008) 1207–1215.
71. Lindhoud S., de Vries R., Norde W., Cohen Stuart M. A., Structure and stability of complex coacervate core micelles with lysozyme, *Biomacromolecules,* 87 (2007) 2219–2227.
72. Lindhoud S., de Vries R., Schweins R., Stuart M. A. C., Norde W., Salt-induced release of lipase from polyelectrolyte complex micelles, *Soft Matter,* 51 (2009) 242–250.
73. Voets I. K., de Keizer A., Stuart M. A. C., Complex coacervate core micelles, *Advances in Colloid and Interface Science,* 147-48 (2009) 300–318.
74. Muijlwijk K., Berton-Carabin C., Schroen K., Cross-flow microfluidic emulsification from a food perspective, *Trends in Food Science & Technology,* 49 (2016) 51–63.

75. Vladisavljevic G. T., Recent advances in the production of controllable multiple emulsions using microfabricated devices, *Particuology*, 24 (2016) 1–17.

76. Schultz S., Wagner G., Urban K., Ulrich J., High-pressure homogenization as a process for emulsion formation, *Chemical Engineering & Technology*, 274 (2004) 361–368.

77. Maan A. A. L., Nazir A., Khan M. K. L., Boom R., Schroen K., Microfluidic emulsification in food processing, *Journal of Food Engineering*, 147 (2015) 1–7.

78. Ushikubo F. Y., Oliveira D. R. B., Michelon M., Cunha R. L., Designing food structure using microfluidics, *Food Engineering Reviews*, 74 (2015) 393–416.

79. Hoshyar N., Gray S., Han H. B., Bao G., The effect of nanoparticle size on in vivo pharmacokinetics and cellular interaction, *Nanomedicine*, 116 (2016) 673–692.

80. Jeong H. H., Issadore D., Lee D., Recent developments in scale-up of microfluidic emulsion generation via parallelization, *Korean Journal of Chemical Engineering*, 336 (2016) 1757–1766.

81. Lee T. Y., Choi T. M., Shim T. S., Frijns R. A. M., Kim S. H., Microfluidic production of multiple emulsions and functional microcapsules, *Lab on a Chip*, 1618 (2016) 3415–3440.

82. Zhang M. J., Wang W., Xie R., Ju X. J., Liu Z., Jiang L., Chen Q. M., Chu L. Y., Controllable microfluidic strategies for fabricating microparticles using emulsions as templates, *Particuology*, 24 (2016) 18–31.

83. Lemetter C. Y. G., Meeuse F. M., Zuidam N. J., Control of the morphology and the size of complex coacervate microcapsules during scale-up, *Aiche Journal*, 556 (2009) 1487–1496.

84. Lamprecht A., Schafer U., Lehr C. M., Influences of process parameters on preparation of microparticle used as a carrier system for Omega-3 unsaturated fatty acid ethyl esters used in supplementary nutrition, *Journal of Microencapsulation*, 183 (2001) 347–357.

85. Piacentini E., Giorno L., Dragosavac M. M., Vladisavljevic G. T., Holdich R. G., Microencapsulation of oil droplets using cold water fish gelatine/gum arabic complex coacervation by membrane emulsification, *Food Research International*, 531 (2013) 362–372.

86. Tcholakova S., Denkov N. D., Lips A., Comparison of solid particles, globular proteins and surfactants as emulsifiers, *Physical Chemistry Chemical Physics*, 1012 (2008) 1608–1627.

87. Comunian T. A., Thomazini M., Alves A. J. G., de Matos F. E., Balieiro J. C. D., Favaro-Trindade C. S., Microencapsulation of ascorbic acid

by complex coacervation: Protection and controlled release, *Food Research International,* 521 (2013) 373–379.

88. Cardenas-Bailon F., Osorio-Revilla G., Gallardo-Velazquez T., Microencapsulation of insulin using a W/O/W double emulsion followed by complex coacervation to provide protection in the gastrointestinal tract, *Journal of Microencapsulation,* 323 (2015) 308–316.
89. Rocha-Selmi G. A., Theodoro A. C., Thomazini M., Bolini H. M. A., Favaro-Trindade C. S., Double emulsion stage prior to complex coacervation process for microencapsulation of sweetener sucralose, *Journal of Food Engineering,* 1191 (2013) 28–32.
90. Xiao J. X., Yu H. Y., Yang J. A., Microencapsulation of sweet orange oil by complex coacervation with soybean protein isolate/gum arabic, *Food Chemistry,* 1254 (2011) 1267–1272.
91. Tirkkonen S., Turakka L., Paronen P., Microencapsulation of indomethacin by gelatin acacia complex coacervation in the presence of surfactants, *Journal of Microencapsulation,* 116 (1994) 615–626.
92. Muniyandy S., Sathasivam T., Veeramachineni A. K., Janarthanan P., Dual cross-linked carboxymethyl sago pulp-gelatine complex coacervates for sustained drug delivery, *Polymers,* 76 (2015) 1088–1105.
93. Jin K. M., Kim Y. H., Injectable, thermo-reversible and complex coacervate combination gels for protein drug delivery, *Journal of Controlled Release,* 1273 (2008) 249–256.
94. Turgeon S. L., Schmitt C., Sanchez C., Protein–polysaccharide complexes and coacervates, *Current Opinion in Colloid & Interface Science,* 124–5 (2007) 166–178.
95. Barthold S., Kletting S., Taffner J., Carvalho-Wodarz C. D., Lepeltier E., Loretz B., Lehr C. M., Preparation of nanosized coacervates of positive and negative starch derivatives intended for pulmonary delivery of proteins, *Journal of Materials Chemistry B,* 413 (2016) 2377–2386.
96. Ward M. A., Georgiou T. K., Thermoresponsive polymers for biomedical applications, *Polymers,* 33 (2011) 1215.
97. Huang Y.-I., Cheng Y.-H., Yu C.-C., Tsai T.-R., Cham T.-M., Microencapsulation of extract containing shikonin using gelatin–acacia coacervation method: A formaldehyde-free approach, *Colloids and Surfaces B: Biointerfaces,* 582 (2007) 290–297.
98. Vandelli M. A., Romagnoli M., Monti A., Gozzi M., Guerra P., Rivasi F., Forni F., Microwave-treated gelatin microspheres as drug delivery system, *Journal of Controlled Release,* 961 (2004) 67–84.

99. Ifeduba E. A., Akoh C. C., Microencapsulation of stearidonic acid soybean oil in complex coacervates modified for enhanced stability, *Food Hydrocolloids,* 51 (2015) 136–145.

100. Broeckx G., Vandenheuvel D., Claes I. J., Lebeer S., Kiekens F., Drying techniques of probiotic bacteria as an important step towards the development of novel pharmabiotics, *International Journal of Pharmaceutics,* 5051-2 (2016) 303–318.

101. Geankoplis C. J., *Transport Processes and Separation Process Principles: Includes Unit Operations,* Prentice Hall Professional Technical Reference (2003). Available from: https://books.google.com.sg/books?id=i9-TQgAACAAJ

102. Mujumdar A. S., *Drying Technology in Agriculture and Food Sciences,* Science Publishers (2000). Available from: https://books.google.com.sg/books?id=0plKAAAAYAAJ

103. Mujumdar A. S., *Handbook of Industrial Drying,* Taylor & Francis (2014). Available from: https://books.google.com.sg/books?id=oT36AwAAQBAJ

104. Eratte D., McKnight S., Gengenbach T. R., Dowling K., Barrow C. J., Adhikari B. P., Co-encapsulation and characterisation of omega-3 fatty acids and probiotic bacteria in whey protein isolate-gum arabic complex coacervates, *Journal of Functional Foods,* 19 (2015) 882–892.

105. Koupantsis T., Pavlidou E., Paraskevopoulou A., Flavour encapsulation in milk proteins - CMC coacervate-type complexes, *Food Hydrocolloids,* 37 (2014) 134–142.

106. Jiang H., Zhang M., McKnight S., Adhikari B., Microencapsulation of alpha-Amylase by carrying out complex coacervation and drying in a single step using a novel three-fluid nozzle spray drying, *Drying Technology,* 3116 (2013) 1901–1910.

107. Oliveira A. C., Moretti T. S., Boschini C., Baliero J. C. C., Freitas L. A. P., Freitas O., Favaro-Trindade C. S., Microencapsulation of B-lactis (BI 01) and L-acidophilus (LAC 4) by complex coacervation followed by spouted-bed drying, *Drying Technology,* 2510 (2007) 1687–1693.

108. Green B. K., Schleicher L., 1957. Oil-containing microscopic capsules and method of making them. Google Patents.

109. Thies C., Encapsulation by complex coacervation, In: *Encapsulation and Controlled Release Technologies in Food Systems,* (2016), John Wiley & Sons, Ltd, pp. 41–77.

110. Chapeau A. L., Tavares G. M., Hamon P., Croguennec T., Poncelet D., Bouhallab S., Spontaneous co-assembly of lactoferrin and

beta-lactoglobulin as a promising biocarrier for vitamin B9, *Food Hydrocolloids,* 57 (2016) 280–290.

111. Marfil P. H. M., Vasconcelos F. H. T., Pontieri M. H., Telis V. R. N., Development and validation of analytical method for palm oil determination in microcapsules produced by complex coacervation, *Quimica Nova,* 391 (2016) 94–99.

112. Devi N., Hazarika D., Deka C., Kakati D. K., Study of complex coacervation of gelatin A and sodium alginate for microencapsulation of olive oil, *Journal of Macromolecular Science Part A: Pure and Applied Chemistry,* 4911 (2012) 936–945.

113. Katona J. M., Sovilj V. J., Petrovic L. B., Microencapsulation of oil by polymer mixture-ionic surfactant interaction induced coacervation, *Carbohydrate Polymers,* 793 (2010) 563–570.

114. Katona J. M., Sovilj V. J., Petrovic L. B., Milanovic J. L., Preparation and characterization of oil containing microcapsules obtained by an interaction induced coacervation, *Journal of Dispersion Science and Technology,* 3112 (2010) 1679–1684.

115. Tamjidi F., Nasirpour A., Shahedi M., Rheological characteristics of yogurt enriched with microencapsulated fish oil, *Journal of Agricultural Science and Technology,* 165 (2014) 1073–1082.

116. Jain A., Thakur D., Ghoshal G., Katare O. P., Shivhare U. S., Microencapsulation by complex coacervation using whey protein isolates and gum acacia: an approach to preserve the functionality and controlled release of beta-carotene, *Food and Bioprocess Technology,* 88 (2015) 1635–1644.

117. Quan J., Kim S. M., Pan C. H., Chung D., Characterization of fucoxanthin-loaded microspheres composed of cetyl palmitate-based solid lipid core and fish gelatin-gum arabic coacervate shell, *Food Research International,* 501 (2013) 31–37.

118. Gao Y., Xie D., Han D. N., Ouyan J. X., Liu Y. L., Preparation and characterization of gelatin/CMC microcapsule by complex coacervation, In: *Advances in Chemical Engineering Ii, Pts 1-4 Advanced Materials Research*. Liu Z., Peng F., Liu X., (eds.), (2012), Stafa-Zurich: Trans Tech Publications Ltd, pp. 1005–1009.

119. Zhang Z. Q., Pan C. H., Chung D., Tannic acid cross-linked gelatin-gum arabic coacervate microspheres for sustained release of allyl isothiocyanate: Characterization and in vitro release study, *Food Research International,* 444 (2011) 1000–1007.

120. Leclercq S., Harlander K. R., Reineccius G. A., Formation and characterization of microcapsules by complex coacervation with

liquid or solid aroma cores, *Flavour and Fragrance Journal,* 241 (2009) 17–24.

121. Leclercq S., Milo C., Reineccius G. A., Effects of cross-linking, capsule wall thickness, and compound hydrophobicity on aroma release from complex coacervate microcapsules, *Journal of Agricultural and Food Chemistry,* 574 (2009) 1426–1432.

122. Wang B., Vongsuiuut J., Adhikari B., Barrow C. J., Microencapsulation of tuna oil fortified with the multiple lipophilic ingredients vitamins A, D-3,D- E, K-2, curcumin and coenzyme Q(10), *Journal of Functional Foods,* 19 (2015) 893–901.

123. Wang B., Adhikari B., Barrow C. J., Optimisation of the microencapsulation of tuna oil in gelatin-sodium hexametaphosphate using complex coacervation, *Food Chemistry,* 158 (2014) 358–365.

124. Nori M. P., Favaro-Trindade C. S., Matias de Alencar S., Thomazini M., de Camargo Balieiro J. C., Contreras Castillo C. J., Microencapsulation of propolis extract by complex coacervation, *LWT - Food Science and Technology,* 442 (2011) 429–435.

125. Li Y., McClements D. J., Controlling lipid digestion by encapsulation of protein-stabilized lipid droplets within alginate-chitosan complex coacervates, *Food Hydrocolloids,* 255 (2011) 1025–1033.

126. Viehof A., Lamprecht A., Oral delivery of low molecular weight heparin by polyaminomethacrylate coacervates, *Pharmaceutical Research,* 308 (2013) 1990–1998.

127. Zhu J. H., Wang X. W., Ng S., Quek C. H., Ho H. T., Lao X. J., Yu H., Encapsulating live cells with water-soluble chitosan in physiological conditions, *Journal of Biotechnology,* 1174 (2005) 355–365.

128. Perez-Liminana M. A., Paya-Nohales F. J., Aran-Ais F., Orgiles-Barcelo C., Effect of the shell-forming polymer ratio on the encapsulation of tea tree oil by complex coacervation as a natural biocide, *Journal of Microencapsulation,* 312 (2014) 176–183.

129. Onder E., Sarier N., Cimen E., Encapsulation of phase change materials by complex coacervation to improve thermal performances of woven fabrics, *Thermochimica Acta,* 4671-2 (2008) 63–72.

130. Deveci S. S., Basal G., Preparation of PCM microcapsules by complex coacervation of silk fibroin and chitosan, *Colloid and Polymer Science,* 28712 (2009) 1455–1467.

131. Dong Z., Ma Y., Hayat K., Jia C., Xia S., Zhang X., Morphology and release profile of microcapsules encapsulating peppermint oil by complex coacervation, *Journal of Food Engineering,* 1043 (2011) 455–460.

Chapter 5

Improving Drug Biological Effects by Encapsulation into Polymeric Nanocapsules

Jieyi Peh[a] and Xian Jun Loh[a,b,c]
[a]*Department of Materials Science and Engineering, National University of Singapore, 9 Engineering Drive 1, Singapore 117576*
[b]*Institute of Materials Research and Engineering (IMRE), 3 Research Link, Singapore 117602*
[c]*Singapore Eye Research Institute, 11 Third Hospital Avenue, Singapore 168751*
lohxj@imre.a-star.edu.sg

Nanoscale materials are of great interest because of their promising potential in the biomedical field for targeted drug delivery. In the research ground of nanostructures, much focus has been placed on polymeric nanocapsules as drug carriers because of the superior advantageous qualities that they offer in comparison with other nanoscale materials. The small particle size of nanocarriers have a high surface area to volume ratio which aids in the permeation of cell barriers. The polymeric shells of nanocapsules are important determinants in offering high protection against chemical- and

Polymer Capsules
Edited by Ye Liu and Xian Jun Loh

ISBN 978-981-4774-60-4 (Hardcover), 978-0-429-42873-9 (eBook)
www.jennystanford.com

photodegradation of drugs susceptible to external factors. As such, it improves the chemical stability of encapsulated drugs and increases the interaction with cells and tissues. Enhanced stability increases the bioavailability of drugs and thus allows nanocapsules to stay in the blood circulation for a prolonged period which subsequently enhances cellular uptake at targeted diseased cells. The permeation and retention effect will therefore be greatly enhanced. In addition, the surface property of the polymeric shell can be covalently bonded with specific ligands for active targeting of drug delivery. This will help to increase the amount of drugs at the targeted cancerous cells which are more receptor-responsive than other normal healthy cells and also prevent pre-mature release of drugs before reaching the targeted sites. In this way, polymeric nanocapsules help to ensure a good control of drug release at an optimum therapeutic range while at the same time, minimizing unwanted toxic side-effects. Of all the different types of polymeric nanocapsules, lipid-core nanocapsules have shown to have the highest effective level of encapsulated drugs delivered. Further optimization of the thickness, size, rigidity, and surface properties of the polymeric wall could be done to improve the efficiency of nanocapsules. The effectiveness and safety concerns of various nanocapsules in *in vitro* and *in vivo* tests are discussed in this paper as well.

5.1 Introduction

In recent years, much attention has been focused on drug delivery research with the use of nanoscale biomaterial because of the promising abilities of nanostructured materials could offer. The delivery of drugs in the nanoscale range helps to regulate the drug dosage in its therapeutic index level, thus keeping the toxic side effects to a minimum with better control of the drug release rate and distribution in the physiological environment. Nanoscale drug delivery systems can be tuned to achieve desirable biological properties for a wide variety of biomedical applications. The biocompatibility, solubility, size, shape, distribution of molecular weight, functionality, degradation rate, and interaction between encapsulated drugs and the polymer shell of carriers are factors which should be considered in nanoscale drug delivery systems.

The intake of drugs can be delivered orally, by inhaling, through transdermal patches, implantation, topical medication, and injection.

Nanoscale drug delivery systems work towards addressing the intrinsic challenges of conventional medicine with its insufficient stability in the body, short plasma half-life, and the induced immune response. The particle size and surface properties of the nanostructures are important elements of its functionality. Particle size is an important contributor in determining the properties and thus behavior of the particles in the drug delivery system. In the range of the nanometer size, biomaterials are small enough to pass through cell membranes. This, in turn, reduces the possibilities of immunogenicity and decreases the likelihood of undesirable uptakes by the reticuloendothelial system. Due to its small size, nanoscale materials have a large surface area to volume ratio which helps to speed up its rate of dissolution and thus aid in the adsorption process. In the physiological environment, the optimal size of the nanoparticle in drug delivery systems is about 10 to 100 nm [1], where the particles can be encircled in vesicles by endocytosis. The circulation time and bioavailability depends greatly on the size of the particles, albeit in the nanometer scale. Particles with a diameter smaller than 10 nm are usually promptly cleared via extravasation and renal plasma clearance; while larger particles of diameter size between 10 to 70 nm have a slightly longer circulation time and are able to pass through small capillaries. Particles with a diameter of larger than 70 nm but smaller than 200 nm usually circulate in blood for the longest period of time relative to those in other particle size ranges. Beyond 200 nm, particles are usually isolated by the spleen and removed via phagocytosis [1].

In addition, the surface property of nanostructure material is a crucial property that ensures active targeting and hence determines the effectiveness of the drug delivery system. The surface can be bonded to a ligand that will enable the nanostructures to possess receptor-responsive sites. Nanostructures with specific surface-targeting molecules will then be able to bind to targeted diseased cells or biomolecules and get adsorbed through physical interaction, electrostatic binding, covalent bonds, and complementary bonds [1]. This will, in turn, help to achieve effective targeted drug delivery to specific organelles in the cells where drugs can be released locally in a controlled manner at targeted sites to provide an optimum dosage of drugs within the therapeutic range. An example of active targeting involves the modification in the surface charge of nanostructures, such that negatively charged particles are targeted to stay in lysosomes and positively charged particles get attracted and attached to mitochondria in cytoplasm [1].

Working towards nanoscale biomaterial structures for medical applications, various agents such as drugs, enzymes, nuclei acids, and magnetic crystals could be loaded in the inner cavity of nanostructures. In this paper, the focus will be on drugs as the cargo. Nanostructures research has placed great emphasis on the ability of a drug carrier to transport and deliver drugs with zero premature release until it reaches the targeted site. In addition, the surface of the polymer nanoparticles is modified so that the polymer wall is responsive to the signals produced by stimuli in diseased areas and thus able to release an appropriate drug dosage level in these areas which require treatment. In this way, the release of drugs can be controlled effectively and efficiently. There are a varieties of existing nanostructure morphologies for drug delivery systems. These comprise nanoemulsion and nanoparticles (including nanocapsules, nanospheres, nanotubes, nanogels, and dendrimers) [2]. Each system has its own benefits and drawbacks.

5.2 Nanostructures

5.2.1 Nanoemulsion

Nanoemulsion involves two immiscible liquids where one phase is dispersed into another with the help of an emulsifier known as surfactant and/or co-surfactant to stabilize the mixtures thermodynamically. The dispersed phase refers to the discontinuous phase distributed in the external phase (i.e., continuous phase). The size of the nanoemulsion droplet is about 20 to 200 nm [3]. There are three typical kinds of nanoemulsion systems, namely oil (dispersed phase) in water (continuous phase) nanoscale emulsion, water dispersed in oil nanoemulsion, and bicontinuous (where the dispersed phase is interconnected or globular).

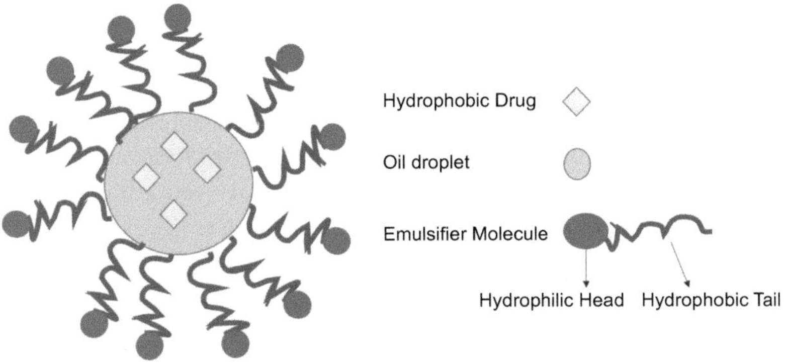

Figure 5.1 Schematic drawing of oil/water nanoemulsion.

5.2.2 Nanospheres

Nanospheres are solid polymeric nanoparticles with drugs dispersed within the network of the polymeric phase.

Figure 5.2 An illustration of scanning electron micrograph of conducting-polymer PEDOT nanotube. Reprinted with permission from Ref. [4], Copyright 2006, John Wiley and Sons.

5.2.3 Nanotubes

The structure of nanotubes (which resembles the shape of a straw) has several advantages. Nanotubes have a large inner capacity to hold a few therapeutic drug agents. Nanotubes also have open ends which makes the drug loading process simpler [1] and help to improve the efficiency of the drug loaded. The divided inner and exterior surfaces of nanotubes are advantageous because both sides could be differentially altered to increase the compatibility in the inner cavity to hold the encapsulated drug cargo within; while at the same time, chemically modify the outer surface for targeted-site drug delivery and prevent immunogenicity *in vivo* [2].

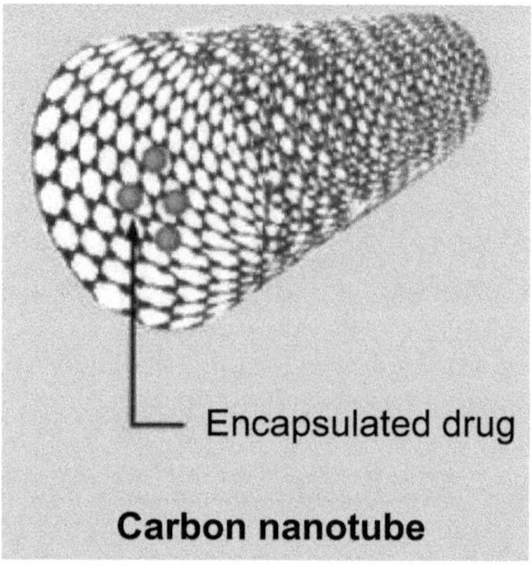

Figure 5.3 An illustration of carbon nanotube with encapsulated drugs. Reprinted with permission from Ref. [5], Copyright 2006, Future Medicine Ltd.

5.2.4 Nanogels

Nanogels are nanoscale hydrogels which swell and form a network of covalent crosslinks when placed in aqueous solution. When reacting to stimuli, nanogels swell, allow the flow of water inwards, thereby releasing the drugs. Improvised and more advanced forms of nanogels can be deliberately modified to be stimuli-responsive to pH, temperature and ionic charge [1], When there is a change in the environmental parameters, polymer will response accordingly. At targeted sites, polymer would become a gel and release the drugs. This would help to control the release of drugs.

Besides being biocompatible and possessing surface-tunable properties, one of the key advantages of the nanoscale hydrogel for specific-site drug delivery is having the option to exclude the drug cargo in its synthesis process. This helps to prevent the possibility of drug aggregation during the synthesis of hydrogels. Drugs are loaded subsequently after nanogels are produced, via self-assembly processes without the involvement of covalent bonds. The synthesis process of nanogels is relatively simple compared to other techniques and allows the loading of large amounts of drugs in it. In addition, polymeric nanogels are able to demonstrate sustained release of drugs, good stability *in vivo* and can be made to be less susceptible to enzymatic degradation attacks [1]. However, the use of lipids as a nanogel outer layer coating to enhance colloidal stability poses a serious challenge on the bonding efficiency of its surface conjugates.

5.2.5 Dendrimers

Dendrimers are greatly branched polymers with closely packed surface groups. The size and shape of dendrimers can be specifically regulated to tailor to its functionality [7]. In dendrimers, drugs are confined in the core via different kinds of bonding. The bonding includes physical interactions (such as induced dipole, permanent dipole, van der Waals interactions, and hydrogen bonding), electrostatic interactions or attachment to the surface via chemical interactions. With more surface groups incorporated into dendrimer, the core–shell structural design of dendrimers will then extend greatly out in radius size. The branches entangle with one another as the surface groups grow, thus constructing a highly dense structure that surrounds the drug cargo in the core. The outer surface of dendrimers has closely-packed and dense functional groups tuned to enhance surface properties such as lower cytotoxicity, biocompatibility, and improved molecular recognition for targeted drug delivery systems. For instance, dendrimers with polyethylene glycol (PEG) on its surface have shown to have lower cytotoxicity [1].

5.2.6 Nanocapsules

Nanocapsules usually consist of drugs in the inner cavity, oil in the internal core, surrounded by polymer as the outer shell, and water [8].

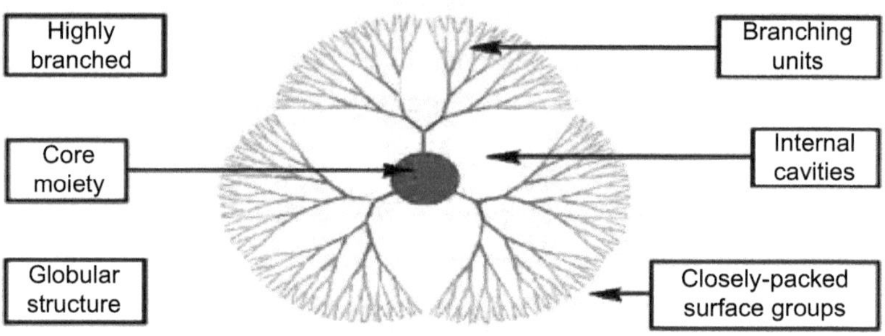

Figure 5.4 The dendritic structure. Reproduced from Ref. [6], Copyright Somwanshi et al., JPSI (2013).

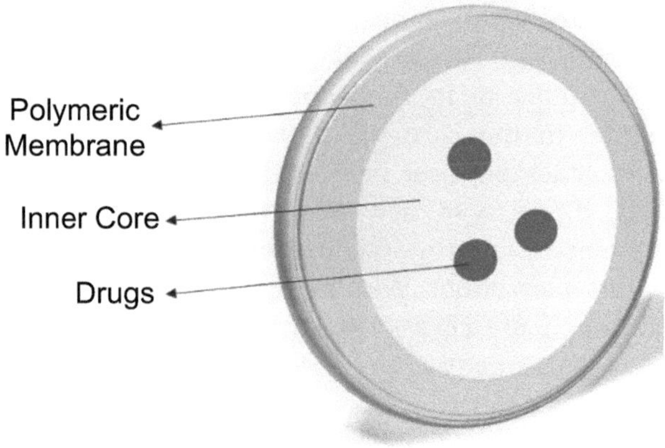

Figure 5.5 An illustration of nanocapsule structure with drug in the inner core and outer polymeric membrane.

Antitumor drugs are entrapped in the cavity of polymeric nanocapsules to treat cancerous or malignant cells. This is beneficial for the delivery of drugs which are not able to permeate well through the cell membrane on their own. The polymeric shell of nanocapsules protects the incorporated drugs from the external environment. The drugs encapsulated in nanocapsules are usually in a dissolved or dispersed state. It is important to prevent the drug cargo in nanocapsules from saturating as it would eventually form into a precipitate [9]. Therefore, in order to successfully synthesize nanocapsules, there is a need to eliminate other unwanted impurities such as nanoscale crystals, spheres, and emulsions which could be formed during the process.

The inner phase of nanocapsules is usually oil-based material or in some uncommon cases, an aqueous core. There are 3 different types of an oil-based core, namely a liquid lipid, a combination of several liquid lipids or a lipid-core nanocapsule which is an amalgam of organogel (lipid in solid phase) and lipid in liquid phase.

Nanocapsules consist of polymeric coating entrapping the incorporated drug within it. The outer polymer shell coating controls the stability of dispersion in a solution and its physiological reaction to stimulus. The polymeric shell of nanocapsules are usually biodegradable. The modification and degradation rate of surface moieties of the polymer wall are important determinants of a good control of drug release and thus an effective drug delivery system. The polymer surface ligands ensure molecular recognition at its targeted sites. At specific sites in the physiological environment, the polymer wall would degrade and dissolve away, thereby releasing the drug. For most of the nanocapsules, the polymer wall would be a crucial factor in determining the diffusion rate of drugs. However, in the case of lipid-core nanocapsules, the diffusional release of drug depends on not just the polymer wall, but also on the organogel core in its inner oil-phase [8]. The degradation rate has to be optimized to limit the unintended release of drugs to a minimal during circulation, before reaching the targeted sites. The common polymers used for nanocapsules are poly(ε-caprolactone) (PCL), poly(lactide) (PLA) and poly(lactide-*co*-glycolide) (PLGA) [8].

5.3 Nanocapsule and Its Advantages over Other Nanostructures

As drug carriers, nanocapsules possess more superior qualities than other nanoscale structures.

Compared to nanoemulsion droplets, nanocapsules have an additional layer of polymeric wall to protect encapsulated drug molecules from the external environment. This will help to reduce the exposure and hence the interaction of drugs with the physiological environment [10]. This, in turn, prevents irritation at the administration spot, ensures a decrease in the degradation of drug molecules and also minimizes the premature release of drugs during blood circulation. Toxic side effects in the physiological environment will then be greatly reduced.

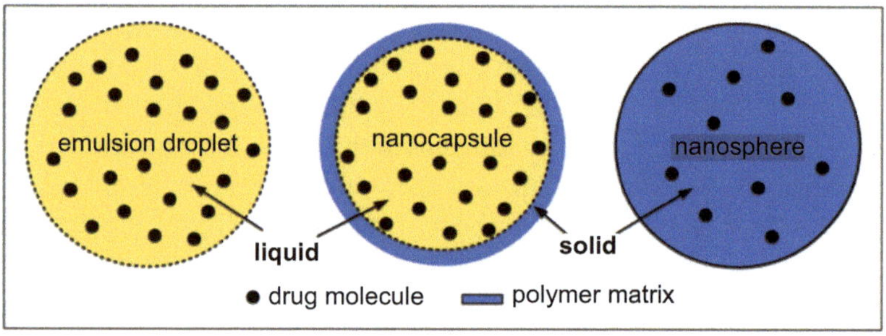

Figure 5.6 Comparison of nanoemulsion droplet, nanocapsule, and nanosphere [10].

In addition, nanocapsules are able to have a much higher drug loading capacity than nanospheres. The ratio of drug to polymer in nanospheres is about 1:10 while in nanocapsules, the ratio is approximately 5:1 [10]. This characteristic makes nanocapsules exceptionally valuable as drug carriers since they are able to carry a high capacity of incorporated drugs using a less amount of polymer in its assembly as compared to nanospheres.

Generally, nanocapsules have 3 main advantages compared to other nanoscale structures. These advantages include an increased drug stability against chemical- or photodegradation, better interaction with cells and tissues *in vivo* and being able to tune its surface properties for targeted drug delivery. This, in turn, ensure amore controlled and sustained release of the drug and thus an effective drug delivery system at targeted sites. These advantages will be elaborated on later in Section 5.3.

5.3.1 Nanocapsules

5.3.1.1 Efficiency parameters of nanocapsules

Some significant parameters such as the size of the nanocapsules, the distribution and diameter of the radius, the thickness of the coating, the rate of membrane decomposition and kind of surfactant used are all important elements of their efficiency in drug delivery systems. The size and distribution of the radius affect the stability and bioavailability of drugs in the physiological environment. It determines the period of time that nanocapsules are able to remain in the blood circulation before being recognized by the immune system, cleared via extravasation and renal plasma clearance or before degradation [1]. The thickness and type of surfactant used for the polymeric shell of nanocapsules determines the mechanical properties (including the rigidity and mechanical stability) and chemical stability (the type of surface ligands which can be covalently bonded to the polymeric wall). This is further discussed in Sections 3.2.3 and 3.2.4.

5.3.1.2 Fabrication techniques

There are several fabrication techniques used to synthesize nanocapsules [8]:

1. Interfacial deposition
2. Emulsion–diffusion
3. Double emulsification
4. Layer-by-layer deposition
5. Emulsion–coacervation or precipitate

5.3.1.2.1 *Interfacial deposition*

Interfacial deposition is the most common method of synthesizing nanocapsules due to its ease of handling and reproducibility. This technique involves organic solvent, a phospholipid mixture, polymer, and a surfactant in aqueous solution. The organic solvent diffuses rapidly in the aqueous solution, thus causing the polymer to precipitate around the phospholipid droplets and form nanocapsules. Subsequently, organic solvent is removed by reducing the pressure of the experiment. For this method, the drugs have to be highly soluble in the organic-oily phase for them to be incorporated in the inner-lipid phase which otherwise would not be contained within the lipid phase, resulting in the development of precipitate in the aqueous phase during formation of the nanocapsules [11].

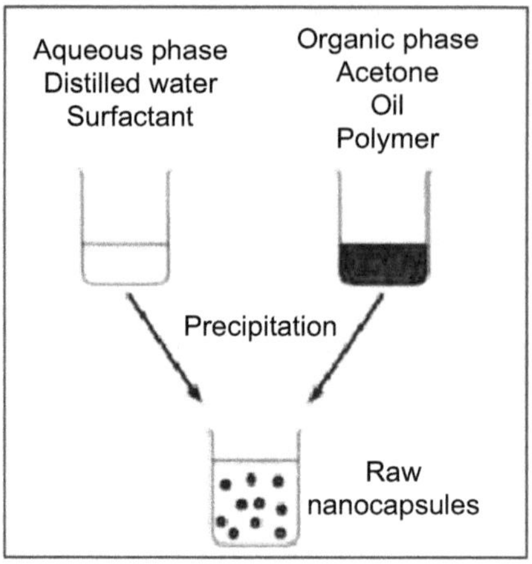

Figure 5.7 Illustration of the preparation of nanocapsules by interfacial deposition technique [11].

5.3.1.2.2 *Emulsion–diffusion*

Emulsion–diffusion comprises drug and polymer in organic oil solvent dispersed in an aqueous phase (continuous phase). After stirring and adding a stabilizer solution, oil solvent diffuses into the polymer phase, thus, forming polymer precipitate with the inner oil phase [12].

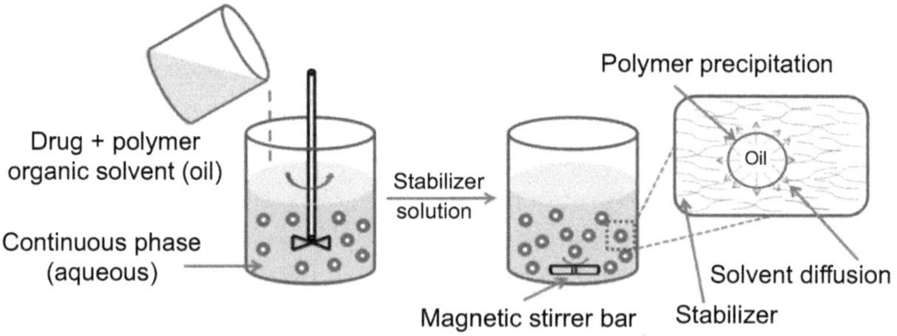

Figure 5.8 The emulsion–diffusion technique. Reproduced from Ref. [12] under Creative Commons Attribution License.

5.3.1.2.3 *Double emulsification*

Double emulsification involves an aqueous solution of polymer homogenized and dispersed in the oil phase to form the first emulsion water/oil (w/o emulsion). Consequently, the second emulsion (w/o/w) is formed by adding an aqueous solution (w) into the first emulsion (w/o). Nanocapsules are then produced after separation, washing, and drying [11].

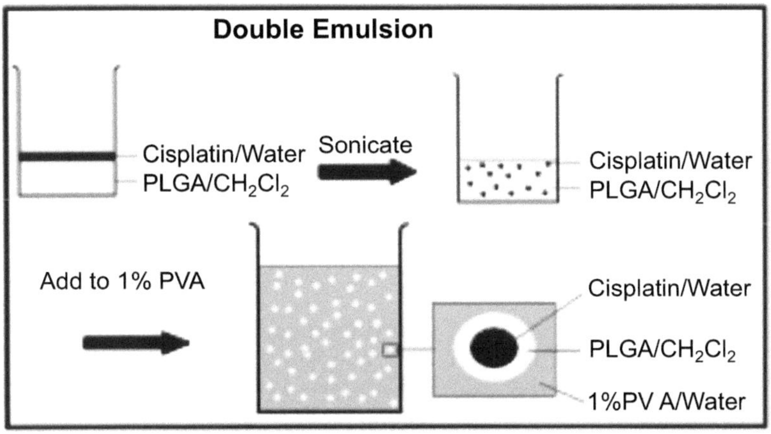

Figure 5.9 A schematic representation of double emulsion method used to prepare encapsulated cisplatin nanocapsules. Reproduced from Ref. [13] under Creative Commons Attribution License.

5.3.1.2.4 *Layer-by-layer deposition*

Layer-by-layer deposition starts with a vesicular polymeric template and subsequent additional layers of oppositely charged polymer deposition to form a core–shell complex. The vesicles are then transited to micelles upon addition of a surfactant such as Triton X. Micelles are then used to form nanocapsules [14].

5.3.1.2.5 *Emulsion–coacervation*

Emulsion–coacervation involves the preparation of nanocapsules by adding polymeric solution such as chitosan into a basic organic solvent like sodium hydroxide, sodium hydroxide methanol or ethanediamine, through a compressed air nozzle. Polymer starts to precipitate at pH value of more than 6.5. Nanocapsules are then produced after filtering, centrifuging, and regulating the air pressure [12].

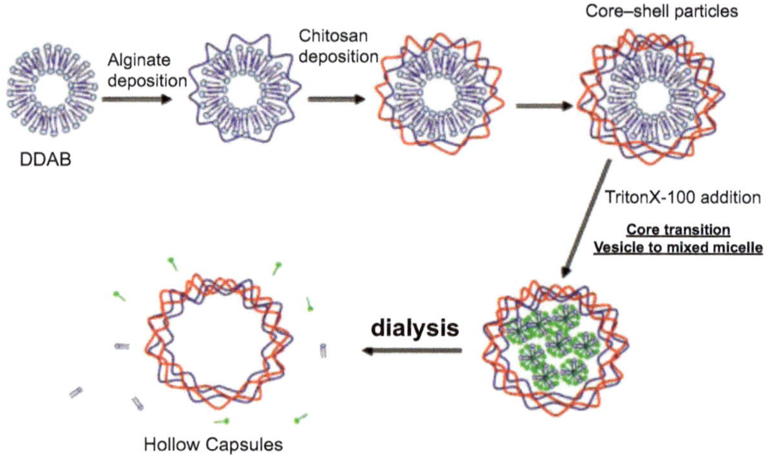

Figure 5.10 An illustration of layer-by-layer deposition of core–shell nanoparticles, where Triton X is used to dissolve the core and allow vesicles to micelles transition. Reprinted with permission from Ref. [14], Copyright 2010, American Chemical Society.

5.4 Benefits of Polymeric Nanocapsules

Nanocapsules ensure a controlled release of drugs and hence an effective drug delivery system by offering an increase in stability of drugs in nanocarriers due to better resistance against degradation, enhanced interactions with cells and tissues *in vivo* and surface-functionalized properties to deliver drugs to specific sites. These three advantages are elaborated on in Sections 5.4.1 and 5.4.2 as follows.

5.4.1 Increased Drug Stability Against Chemical- and Photodegradation

Nanocapsules can stabilize the drug cargo at a much greater efficiency compared to free drugs which are not encapsulated within a nanocapsule because of the additional layer of a polymeric shell. The encapsulation of drugs in the lipid-core of polymeric nanocapsules address the instability issues faced by non-encapsulated drugs. This instability challenge is prominently observed especially in the case when lipids, solely, are used to deliver drugs. The addition coating of the polymer wall helps to protect and improve the stability of drug carriers in the physiological media, where external factors such as pH, exposure to light, temperature, and osmotic pressure could be a great concern.

The encapsulation of the drug in nanocapsules aids in the prevention of degradation at unintended sites *in vivo* and thus enhances the chemical stability of the drug. This was observed in the case of anticancer drugs such as cisplatin in polymeric nanocapsules which have demonstrated a high level of cytotoxicity against tumors *in vivo* tests as compared to non-encapsuled drugs [1]. Different types of polymer were used as the shell or coating to help in protecting the drugs from different degradation stimuli. For instance, PLA has shown to inhibit the chemical degradation (hydrolysis) process of drug cargo such as curcumin in varying pH levels [8]. Different polymers are used to protect light sensitive drugs such as E-resveratrol, clobetasol propionate, and rutin against photodegradation, usually triggered by UVA light [8]. Polymer such as PCL prevents drugs in nanocapsules from oxidizing by shielding them against heat and light [8].

Polymeric nanocapsules enhance drug stability against chemical and photodegradation with lipid-core nanocapsules exhibiting the best observed results [8]. The degradation rate of nanocapsules in the physiological environment is better controlled with the presence of both lipid and polymer in lipid-core nanocapsules, which as a result, ensures an increased stability of such drugs. Lipid-core nanocapsules have established very high stability as drug carriers in studies for cargo, such as Tretinoin, which is typically used to treat acne and other skin conditions, protecting it against photodegradation under UVA and UVC lights [8]. Lipid-core nanocapsules stand out from other nanocapsules because of their unique ability to scatter UV radiation [8]. This, in turn, helps to significantly slow down the photodegradation of light-sensitive drugs cargo such as benzophenone-3, a common compound found in UV sunscreen.

As such, the polymeric membrane of nanocapsules play an important role in protecting the encapsulated drugs from chemical- and photodegradation. Nanocapsules are able to regulate their degradation rate and therefore, balance their stability *in vivo*. This ensures a good drug bioavailability and also a longer plasma half-life for cellular intake.

5.4.2 Increased Interaction with Cells and Tissues and Drug Targeting

5.4.2.1 High Specific Surface Area to Volume Ratio

Nanocapsules, which have a large specific surface area to volume ratio, were shown to be able to demonstrate higher interaction with cells and tissues. Nanocapsules are small enough to penetrate through cell barriers and thus increase the uptake of nanoparticles by the cells. In addition, in the range of 10 to 100 nm [1], where nanocapsules are able to be encircled in vesicles by endocytosis, they reduce the unwanted uptake by the reticuloendothelial system. Hence, nanocapsules enhance the permeation and retention effect.

5.4.2.2 Polymeric Shell

Another factor which boosts the interaction of nanocarriers and cells is largely dependent on the kind of polymeric membrane used in the nanocapsules. The type of polymer wall used determines the cells-nanocapsules interaction in the physiological environment. A good selection of polymeric membrane used in nanocapsules will help to safeguard better cells-nanocapsules interaction *in vivo*. For instance, nanocapsules with a flexible polymer wall such as PLA allow deeper penetration into the cells as compared to rigid polymers.

5.4.2.3 Surface Modifications

In addition, compared to other nanocarriers, the polymeric wall of nanocapsules allows the modification made to their surface properties so that nanocapsules are able to stay longer *in vivo* with a slower degradation rate and to prevent premature discharge of drugs before reaching the targeted sites. Better designed compatible surface moieties on nanocapsules helps to facilitate better interaction with cells and tissues and hence, increase penetration through cell barriers. The polymeric surface shell of nanocapsules could be covalently bonded with PEG, as illustrated in Fig. 5.11, to mask off the identification by opsonins (molecules that induce immune response and phagocytosis via introduction of antigens) so as to allow polymeric nanocapsules to stay in the blood circulation for a prolonged period of time [15]. This will therefore help to enhance permeability and retention in targeted tumor cells than in normal healthy cells.

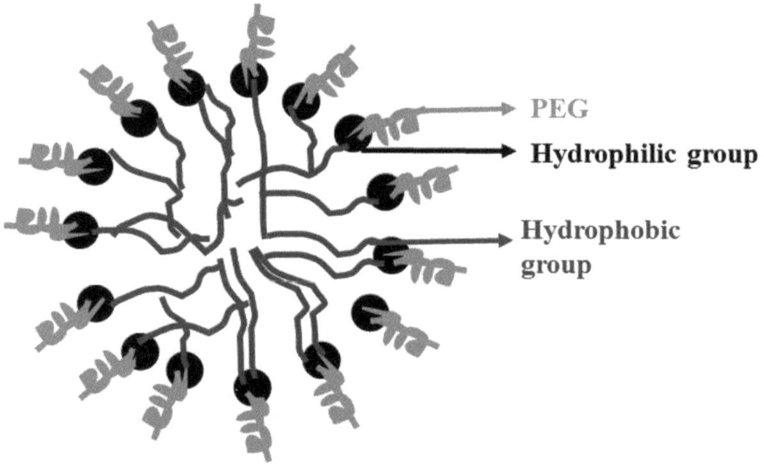

Figure 5.11 Schematic illustration of a PEGylated polymeric micelle.

Furthermore, modified surfaces can be optimized to improve active targeting of diseased cells which require treatment. Specific ligands can be attached to the polymer surface of nanocapsules through various means, usually via chemical bonding, to increase the surface interaction of nanocapsules and the cells and tissues at targeted diseased sites [15]. The active targeting exploits the fact that abnormal or cancerous cells are distinguished from the normal cells because of their differences in parameters such as pH, ionic charge, and chemical bonds. Aiming towards these differences between cancerous and healthy cells, the ligands attached to the polymer surface of a polymer-drug conjugate can be made bio-responsive to various stimuli such as pH, certain enzymes, temperature, or net charge [2], in order to target cancerous cells. This would ensure that the encapsulated drug remains in the cavity of the polymer and not released in the circulation until it reaches the targeted sites where the polymer degrade or dissolve in the physiological environment and release the drug.

The pH value of cancerous cells or tumors is slightly lower than that of normal cells. The pH value of cancerous cells is about 5.5, while that of normal cells is approximately 7.4 [16]. The diseased cells receptors respond differently to chemical ligands as compared to normal cells. Unhealthy and healthy cells also possess different charges on their surfaces. In the presence of abnormal cells which require therapeutic treatment where pH could be lower than normal cells or when ionic charge differs from that or other healthy cells, the polymer would then degrade and release the encapsulated drug within it at targeted sites. This, in turn, reduces the chances of the premature release of the drug and hence a controlled release of the drug at the specific targeted sites. A representative of surface-active nanocapsules would be polyelectrolyte shell which can be produced by layer-by-layer deposition. The modification in the polymer surface helps to control the thickness of the membrane and also the drug release mechanism. Polyelectrolyte shells were more receptive to the physiological environment; they were able to remain inactive until they reached targeted sites where the presence of stimuli such as pH and ionic strength caused the polymer shell to open up and dissolve, thereby releasing the drug.

Besides the differences in pH, temperature, and ionic strength, tumor cells usually express more of certain enzymes and antigens than that of normal cells. Based on this reason, passive targeting can be achieved. In addition, active site-specific activation at targeted sites can be achieved by modifying the polymeric surface of nanocapsules so that the surface becomes more responsive to the enzymes and antigens at locations of cancerous cells than at other normal cells. The site-specific interaction will therefore ensure the effectiveness of drug delivery. Specific antibodies or ligands can be conjugated to the polymeric surface of nanocapsules so as to increase the site-specific interaction with tumor cells. The common surface targets of cancerous cells are folate, transferrin, and glycoproteins. For instance, antibody anti-CD 33 can be covalently bonded to the surface of polymeric nanocapsules loaded with calicheamicin to target acute myelogenous leukemia cells which express a high level of antigen CD33 on its surface [18]. Also, antibody anti-MM46 and anti-EL4 are attached to polymeric shells to target melanoma cells [19]. With site-specific activation, more encapsulated drugs loaded polymeric nanocapsules will be able to target cancerous cells and thus have a high localized concentration of drugs intake by the cancerous cells.

5.4.2.4 Representative Examples of Polymeric Nanocapsules with Enhanced Interaction with Cells and Tissues

Studies have shown that nanocapsules were able to have a higher cellular uptake than non-encapsulated drugs in the transport of anticancer drugs, for instance doxorubicin and dexamethasone for the therapeutic treatment of breast and liver cancer respectively [8]. Poly(lactic-*co*-glycolic acid) nanocapsules with PEG attached to its surface have been demonstrated to exhibit a much longer circulation time in blood and also, enhanced active targeting of tumor cells [15]. Surface ligands covalently attached to the polymer wall of nanocapsules could be designed to be specifically recognized by cancerous cells or tumors. For instance, folic acid was conjugated on the shell of polymeric nanocapsules to improve the targeting of tumor cells since cancerous cells usually express much higher folate receptors on its surface than healthy cells [15]. Some other examples of modified surface ligands also include metal–chitosan–lecithin complex such as scFv-anti-LDL(−) to target LDL (low density lipoprotein) used for the treatment of atherosclerosis patients and positively-charged surface functional groups used to increase adhesion to negatively charged mucosa linings [8]. Active targeting of drug delivery would, therefore, enable the increased intracellular uptake and penetration of drugs in targeted areas.

As a result, with appropriate specific surface ligands on polymeric walls of nanocapsules, premature release of encapsulated drugs would be minimized until the nanocapsules reach the diseased cell sites and get absorbed. As such, unwanted toxic side effects will be reduced and the bioavailability of the drugs will be increased with a longer plasma half-life. Encapsulated drugs will then have enough time to be able to penetrate into and be internalized by the cells. Hence, the cellular uptake by cells and tissues will be greatly enhanced. In this way, nanocapsules will have a better controlled and sustained release of drugs in the body via diffusion, enzymatic degradation, or oxidative degradation of the polymer. An effective therapeutic dosage of drugs can be achieved at targeted sites.

5.5 Other Ways to Enhance Efficiency of Polymeric Nanocapsules

Polymeric nanocapsules improve the efficiency of targeted drugs delivery with lower vulnerability against chemical- and photodegradation, better stability in a psychological environment, increased interaction with cells and tissues, and through active targeting mechanism. Polymeric nanocapsules were shown to be able to stay in the blood circulation for a long time extension, thus increasing the availability of drugs in diseased sites for cellular uptake. In addition to the above-mentioned factors, there are several ways to improve the polymeric shell properties. These involve optimizing its thickness, size, and stiffness [10]. This serves to improve the mechanical and chemical stability of nanocapsules in the external environment during administration of drug carriers into the body and in the physiological environment. A thicker shell would usually increase the rigidity and mechanical resistance of nanocapsules. Polymeric shells of nanocapsules with optimal thickness and rigidity offer higher protection for the incorporated drugs within it. This would also ensure a prolonged period time for the encapsulated drugs to reach targeted sites in the body.

5.6 Evaluation Tests on Efficiency of Polymeric Nanocapsules

The presence of surface ligands such as PEG and folic acid can be characterized using Fourier transform infrared spectroscopy (FTIR) and nuclear magnetic resonance (NMR) tests, and can be quantified by using the $BaCl_2/I_2$ colorimetric test [15]. The cytotoxicity and uptake of drugs by the cells of polymeric nanocapsules conjugated with folic acid and without folic acid can be compared using MTT assay (a colorimetric test for evaluating cell metabolic rate) and confocal laser scanning microscopy respectively.

In this paper, the proficiency test of polymeric nanocapsules with encapsulated antioxidant drugs, anti-inflammatory drugs, and most importantly, antitumor drugs, on top of photo-activity, toxicity, and protection against photodegradation mechanisms, are discussed in both *in vitro* and *in vivo*, in Sections 4.1 and 4.2 respectively.

In vitro refers to studies done in a controlled setting externally, outside of living organisms. *In vivo* refers to tests conducted in living organisms. *In vitro* tests are conducted before *in vivo* tests because they are cheaper and do not involve ethical issues. *In vivo* studies are conducted to further affirm the predictability of *in vitro* tests and confirm the efficiency of polymeric nanocapsules in drug delivery.

5.6.1 *In Vitro* Research Test of Polymeric Nanocapsules

There are many tests to assess the effectiveness of polymeric nanocapsules in the therapeutic treatment of killing cancerous cells. Some of the ways employed to quantify the efficacy of polymeric nanocapsules involved, but not limited to, the evaluation of the antioxidant, anti-inflammatory, anti-proliferative, and anti-microbial drugs and photodynamic effects of drugs in diseased cells.

5.6.1.1 Antioxidative Effects of Drugs

The efficacy of antioxidative drugs in polymeric nanocapsules could be evaluated by comparing the amount of free radical hydroxyl groups. A higher antioxidant property will show a lower quantity of reactive free radical hydroxyl species. Studies have shown that encapsulated antioxidants drugs in polymeric nanocapsules have exhibited a lower amount of reactive free radical species than drugs which are not encapsulated or in nanoemulsion arrangement [8]. This, therefore, supported the fact that the polymeric membrane of nanocapsules were more effective in drug delivery by contributing to the factor that such polymeric nanocapsules could more effectively diminish the effect of photodegradation than non-encapsulated drugs. This consequently ensured the higher stability of these drugs, thus maximising their efficacy.

5.6.1.2 Anti-inflammatory Effects of Drugs

The anti-inflammatory effectiveness in polymeric nanocapsules could be assessed by measuring their capability in reducing the production of cytokines, which are a group of proteins released by the cells for cell signaling in organisms. Studies conducted by employing *in vivo* tests have shown that using encapsulated anti-inflammatory drugs such as dexamethasone and indomethacin in polymeric nanocapsules have produced less reactive oxygen species and more anti-inflammatory cytokine, interleukin-10 [8], than in non-encapsulated drugs. The results were in accordance with the fact that encapsulated drugs in polymeric nanocapsules could be much better regulators in immunogenicity and are able to maximise the anti-inflammatory drug effect. Table 5.1 summarizes the various functions of macrophages in the human body.

5.6.1.3 Anti-proliferative Effects of Drugs

The anti-proliferative proficiency of encapsulated antitumor drugs in polymeric nanocapsules used in the treatment of various cancers such as prostate, breast, leukaemia, and colon cancer cells, were evaluated too. Polymeric nanocapsules have shown a greater anti-proliferative outcome compared to drugs which were not encapsulated. This could be attributed to the increased interaction of cells and nanocapsules and consequently, the enhanced uptake of polymeric nanocapsules by the cells. As such, the treatment by antitumor drugs at diseased sites were greatly enhanced. Similarly, this trend was also observed in polymeric nanocapsules with the high concentration of an additional encapsulated anti-inflammatory drug, ketoprofen [8]. However, the efficacy under *in vitro* experimental conditions was limited to only a range of concentrations and time intervals because of the different procedures and plans in the experiments conducted.

Table 5.1 The function of macrophages in the physiological environment

Macrophage Functions	
Inflammatory function	Chemokines IL–1, IL–8, etc. Metabolites, acid, etc.
Anti-inflammatory function	IL-10, IL-1 Ra
Accessory function	Antigen Co-stimulatory cytokines
Effector function	Phagocytosis

5.6.1.4 Anti-microbial Effects

The results of the proficiency of anti-microbial drugs such as the antiseptic compound chlorhexidine in polymeric nanocapsules have demonstrated them to be better than gel without any encapsulation of drugs in *in vitro* tests [8]. This also supported the effectiveness of polymeric nanocapsules in maximising the effect of anti-microbial drugs.

5.6.1.5 Photodynamic Therapy

Furthermore, *in vitro* experiments involving polymeric nanocapsules were conducted for a more advanced cancer treatment, photodynamic therapy (PDT). PDT comprises the use of a photosensitive dye which produces a reactive oxygen species that helps destroy cells and proteins along with the exposure to a specific form of light (illustrated in Fig. 5.12). The activity of PDT is measured by examining the amount of reactive oxygen species and photochemical alteration of the drug caused by broken covalent bonds between the drugs and surrounding molecules (photobleaching). Polymeric nanocapsules in the studies have shown to reduce the photobleaching effect of photosensitive dyes such as cyanine, which therefore intensified the oxidation process, thereby killing more diseased cells [8]. Hence the research on the photodynamic effect in *in vitro* studies were in accordance with the effectiveness of polymeric nanocapsules in being able to boost the photodynamic effect of the photosensitive dye.

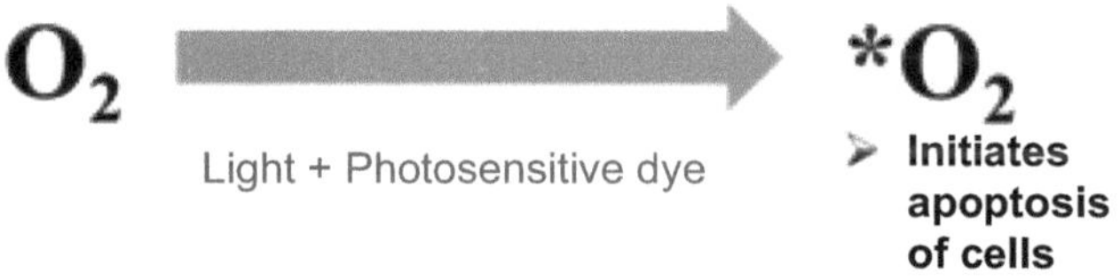

Figure 5.12 Photodynamic therapy mechanism.

5.6.2 *In Vivo* Research Test of Polymeric Nanocapsules

In vivo tests were conducted in living organisms after *in vitro* experiments to affirm the benefits and efficiency of numerous polymeric nanocapsules as drug delivery systems. Most of the *in vivo* experiments were done using animals as representation testing, from small to large size, to confirm the results obtained in *in vitro* testing. Some *in vitro* tests are discussed in the following paragraphs.

5.6.2.1 Anti-proliferative Effects of Drugs

In vivo onion root-tip cell tests were conducted to evaluate the anti-proliferative effects of polymeric nanocapsules which were loaded with the drug tretinoin used to treat acne. Studies have shown these tretinoin loaded polymeric nanocapsules had a lower mitotic index observed compared to drugs which were not encapsulated within a polymeric membrane. A lower mitotic index represents a lower number of cells going through mitosis, which therefore reflected the enhanced effect of the anti-proliferative action of the encapsulated drug. In addition, the efficiency of anti-proliferative effects can be measured by the body weight of the living organisms, their behavior and muscular activity. In *in vivo* experiments tumor-cell embedded mice, polymeric nanocapsules with cargo such as the anti-inflammatory drugs ketoprofen, curcumin, and indomethacin, have demonstrated higher anti-proliferative properties as compared to other morphologies of mobile drugs [8]. The effectiveness of the anti-proliferative properties in mice representatives can be assessed by quantifying the amount of antigens, such as proteins in the cells, by immunohistochemistry. While the anti-proliferative effects of polymeric nanocapsules observed for the above-mentioned anti-inflammatory drugs were significant, the effects were not prominent in polymeric nanocapsules with the encapsulated antibiotic drug, gentamicin, used to treat infection [8]. As such, this could be attributed to the limitations of the effectiveness of polymeric nanocapsules which could only reflect in some encapsulated drugs.

5.6.2.2 Surface Active Targeting Effects of Drugs

The effectiveness of active targeting in a modified surface of polymeric nanocapsules were also tested. *In vivo* tests concerning the specific functional groups, such as folate, attached onto the polymeric membrane surface were observed to have a higher localized amount on tumors or cancerous cells than in other parts of healthy cells [8]. This, therefore, supported the fact that surface-responsive ligands on the polymeric membrane of nanocapsules are effective in ensuring a targeted drug delivery system.

5.6.2.3 Photodynamic Therapy

The proficiency of anti-inflammatory drugs such as indomethacin in lipid-core nanocapsules were tested in acute and chronic inflammation mice as a comparison against non-encapsulated drugs in drug solutions. The results of anti-inflammatory drugs in lipid-core nanocapsules used to treat acute inflammation did not vary much from that of non-encapsulated drugs as both cases did not prevent edema swelling induced by inflammation [8]. However, in mice which experienced chronic inflammation, lipid-core nanocapsules with an incorporated anti-inflammatory drug were shown to be much more effective than non-encapsulated drugs. In addition, several tests were carried on PLGA nanocapsules on mice which suffered from periodontitis (an inflammation condition of the cells around the teeth and usually triggering the contraction of gums and loosening of teeth). The studies have obtained positive results which have shown a great reduction in inflammatory reactions and a more regulated immune response [8]. This was indicated by a decline in the amount of cytokines, which thus reflected on the decrease in the immune response *in vivo*, and also, an increase in anti-inflammatory factors such as interleukin-10. Studies have accounted these results to the controlled and continual release of drugs from the nanocapsules at the localized inflamed sites, thus supporting the effectiveness of polymeric nanocapsules as targeted drug delivery systems.

5.6.2.4 Medical Applications of Drug Delivery

Various types of polymeric membrane loaded with encapsulated drugs such as benzocaine [8], were compared against each other in their effectiveness in anaesthetic influence. Among PLGA, PLA, and PCL, PLA nanocapsules have shown to possess the greatest effectiveness in expressing the anesthetic effect. The anti-psychotic effect of encapsulated drugs and the extent of its toxic side effects such as muscular rigidity and orofacial disability were assessed *in vivo*. The effectiveness of the anti-psychotic effect of polymeric nanocapsules were prominent in the experiments in which nanocapsules with PCL polymeric membrane and lipid-core nanocapsules were shown to have the highest efficiency as carriers to deliver drugs, such as olanzapine [20], in comparison with other forms of delivery where drugs were free and mobile and not encapsulated.

5.6.2.5 Efficacy of Lipid-Core Nanocapsules

A number of polymeric nanocapsules were studied to test their in antimalarial treatment. Lipid-core nanocapsules with PCL as its polymeric wall, loaded with the antimalarial drug, pyrimethamine, were shown to have the best effective results compared to other forms of drug carriers with non-encapsulated drugs [8]. The efficiency of lipid-core nanocapsules could be attributed to the increased availability of drugs in the physiological environment and longer plasma half-life of drugs in nanocapsules. The high protection against chemical- and photodegradation, increased instability in the physiological environment, the enhanced interaction with cells and tissues offered by polymeric nanocapsules increases the bioavailability of the drugs over a longer period of time.

Lipid-core nanocapsules have been demonstrated to be the most effective form of drug carriers for Alzheimer's diseases as well. Compared to other methods where drugs are mobile, lipid-core nanocapsules prevented the drugs from oxidizing and degrading before it reaches the targeted sites. This, in turn, helped to ensure the prior release of drugs such as meloxicam and curcumin, before reaching the diseased cells or tissues. In addition, lipid-core nanocapsules loaded with the drug curcumin reduced the drug dosage by 20 times [8], thus preventing the sudden outbreak of drugs into the body. Hence, it helped to minimize the unwanted toxic effects in cases of premature release of drugs.

5.6.2.6 Polymeric Nanocapsules Efficiency

Polymeric nanocapsules have shown to have significantly reduced the amount of toxic side effects because of better control in the release of encapsulated drugs. This was supported by the observations in the oral delivery of the drugs where nanocapsules, especially lipid-core nanocapsules, helped to reduce the amount of pre-mature release of drugs in oral barriers [8]. The significant drop in lymphoctyes detected also indicated the effectiveness of polymeric nanocapsules in the reduction of unwanted toxicity. The high absorption of drugs through stomach and intestine barriers also contributed to the increased interaction of cells and nanocapsules.

Increased anti-proliferative, anti-inflammatory, antioxidant, anaesthetic, and anti-psychotic effects and the effectiveness of the drugs in polymeric nanocapsules support its success as drug carriers. These qualities contributed to better protection against degradation, higher stability, and the increased interaction of polymeric nanocapsules with cells and tissues. These advantages allowed the encapsulated drugs to remain in the physiological environment for a longer period of time, thus increasing the bioavailability of the drugs. As such, polymeric nanocapsules would be able to maintain a sustained release of drugs, thereby maximising the effectiveness of the encapsulated drugs. *In vivo* tests of the encapsulated drug cyclosporine (used to prevent organ rejection after transplantation) in nanocapsules have supported the fact that polymeric nanocapsules were effective in increasing the bioavailability of drugs [8]. The encapsulated drugs were able to stay in the blood for a period of time almost twice that of a non-encapsulated drug. Conversely, for some drugs such as quinine (used to treat malaria), the pharmacology results obtained did not show a significant difference between encapsulated drugs and free drugs [8]. More tests have to be conducted to confirm the efficacy of polymeric nanocapsules. However, the negative results obtained in the study could most probably be due to the unpredictability of the body response to a drug in different individuals even if the dosage drug level is specified [8].

5.7 Safety Concerns Over Polymeric Nanocapsules

There has been significant amount of research and attention focused on polymeric nanocapsules because of their promising potential in targeted drug delivery. Therefore, the safety concerns over polymeric nanocapsules will be an important issue to examine and discuss.

5.7.1 *In Vitro* Tests

Safety tests conducted under *in vitro* conditions include cytotoxicity, photo-irritation [21], sensitisation, hemocompatibility, and genotoxicity analysis. Cytotoxicity evaluates the toxicity to cells. Photo-irritation assesses the toxic reaction that is caused either after chemicals and continuous light exposure or by skin irradiation after the intake of chemicals. Sensitization measures the possibility of contact sensitization. Hemocompatibility is a measurement of the biocompatibility of chemicals with blood. Genotoxicity considers the destruction of genetic information by chemical substances which may therefore lead to mutations and cancer eventually.

Various tests of different polymeric nanocapsules were examined by conducting studies on different cancerous cells. This paper will discuss some of the safety tests conducted on polymeric nanocapsules. The safety of PLGA polymeric nanocapsules were assessed based on experiments conducted on cancerous breast cells and tissues. Positively charged surface-functionalized PLA were evaluated using malignant breast tumors, fibroblasts (cells which produces collagen and other kinds of fibers), and prostate cancerous cells. The safety trials of PCL nanocapsules could be examined based on the tests carried out on acute non-lymphocytic leukaemia cells and epidermal cells which produce keratin. All the above-mentioned polymeric nanocapsules did not reflect significant cytotoxicity, with PLGA nanocapsules being the only exception. The cytotoxicity case detected in PLGA nanocapsules were hypothesized to be due to the process of producing cell membranes through the use of surfactants or higher zeta potential (scale of electrostatic repulsion and attraction between particles) between the surface charge of nanocapsules and cells [8].

In addition, the cytotoxicity and photo-irritation of PLA polymeric nanocapsules were evaluated based on tests conducted on a human's predominant epidermis cell type and a mouse's fibroblasts. The results obtained showed that high cytotoxicity and photo-irritation levels were observed at higher concentrations. A possible explanation for this observation could be attributed to the existence of retinol palmitate [8]. However, more experiments will have to be carried out to further test this suggested explanation with the use of non-encapsulated retinol palmitate or nanocapsules which are deprived of retinol palmitate as controls for comparison.

Hemocompatibility tests of lipid-core nanocapsules and poly(*n*-butyl cyanoacrylate) nanocapsules were conducted using human blood to test for applications such as the administration of drugs intravenously. Studies have shown that these nanocapsules were found to be safe in these experiments with no toxic hemolysis observed. Additionally, the genotoxicity of PCL nanocapsules were assessed and have shown favorable results.

5.7.2 *In Vivo* Tests

In vivo safety can be assessed by various international guidelines, for instance, "Safety of Manufactured Nanomaterials Guidelines" from the Organization for Economic Cooperation and Development (OECD) [8]. Some examples of *in vivo* tests are, but not limited to, hemocompatibility, intracutaneous reactivity and acute systemic toxicity.

Some studies have shown that nanocapsules were significantly effective in treatments on rats with an acceptable range of hepatic toxicity (liver damage from toxic chemicals) and nephrotoxicity (toxic effect on kidneys) for drug dosage of 7.21×1013 nanocapsules/kg [8]. Above this drug dosage, the body's immune system started to react against the drug loaded nanocapsules and inflammation was evident in the spleen. Therefore, the results highlighted the importance of conducting more tests to find the optimum range of dose for the next stage of clinical trials.

5.8 Conclusion

Polymeric nanocapsules possess superior qualities as compared to other nanostructures in targeted drug delivery. Drugs loaded polymeric nanocapsules can act as effective drug carriers with high protection of drugs against chemical- and photodegradation, besides enhanced interaction with cells and tissues. Polymeric nanocapsules can provide higher stability for the encapsulated drugs and increase the plasma half-life of drugs. As such, this provides polymeric nanocapsules with the advantages of being able to stay in the blood for a prolonged time extension and also increases the bioavailability of drugs at targeted sites for drug intake by the cells. Nanocapsules have a great surface area to volume ratio which allows them to permeate through cells barriers and get adsorbed quickly. The optimum size for drug delivery is about 10 to 100 nm where nanocapsules can be encircled in vesicles by endocytosis. The surface property of the polymeric wall of nanocapsules is a very crucial element in determining the effectiveness of receptor-responsive sites. Specific ligands can be attached to the surface of the polymeric wall to ensure an effectual targeted drug delivery system.

The surface tunability of polymeric nanocapsules helps to ensure a controlled release of drugs and keep the drug dosage level at an optimum therapeutic range with reduced toxic side effects. Studies have shown that lipid-core nanocapsules were able to demonstrate a high level of effective therapeutic treatments as compared to other types of nanocapsules. The evaluation of polymeric nanocapsules have to be carried out under *in vitro* and *in vivo* conditions to evaluate their effectiveness. In addition, the safety concerns of polymeric nanocapsules have to be addressed as well.

References

1. Goldberg M., Langer R., Jia X., *Journal of Biomaterials Science, Polymer Edition*, **18** (2007) 241–268.
2. Chen G., *Pharmaceutica Analytica Acta*, **03** (2012).
3. Jaiswal M., Dudhe R., Sharma P. K., *3 Biotech*, **5** (2014) 123–127.
4. Abidian M. R., Kim D. H., Martin D. C., *Advanced Materials*, **18** (2006) 405–409.
5. Bamrungsap S., Zhao Z., Chen T., Wang L., Li C., Fu T., Tan W., *Nanomedicine (Lond).*, **7** (2012) 1253–1271.
6. Somwanshi S., Dolas R., Siddheshwar S., Merekar A., *Journal of Pharmaceutical and Scientific Innovation*, 4 (2013) 18–23.
7. Hughes G. A., *Nanomedicine*, **1** (2005) 22–30.
8. Frank L. A., Contri R. V., Beck R. C., Pohlmann A. R., Guterres S. S., *Wiley Interdisciplinary Reviews: Nanomedicine and Nanobiotechnology*, **7** (2015) 623–639.
9. Pohlmann A. R., Fonseca F., Paese K., Detoni C., Coradini K., Beck R. C. and Guterres S. S., *Expert Opinion on Drug Delivery*, **10** (2013) 623–638.
10. Nyakas C. Polyelectrolyte nanocapsules as modern drug delivery system for lipophilic drug candidates. Dissertation (Dr. rer. nat.), Martin Luther University Halle-Wittenberg (2012), 2.
11. Swetha S., Microspheres, http://www.slideshare.net/shikhaswetha/microspheres).
12. Wang Y., Li P., Thao T.-D. T., Zhang J., Kong L., *Nanomaterials*, **6** (2016) E26.
13. Jayasuriya A. C., Darr A. J., *Journal of Biomedical Science and Engineering*, **06** (2013) 586–592.

14. Cuomo F., Lopez F., Miguel M. G., Lindman B., *Langmuir*, **26** (2010) 10555–10560.

15. El-Gogary R. I., Rubio N., Wang J. T.-W., Al-Jamal W. T., Bourgognon M., Kafa H., Naeem M., Klippstein R., Abbate V., Leroux F., Bals S., Van Tendeloo G., Kamel A. O., Awad G. A. S., Mortada N. D., Al-Jamal K. T., *ACS Nano*, **8** (2014) 1384–1401.

16. Zhang Y., Hsu B. Y., Ren C., Li X., Wang J., *Chemical Society Reviews*, **44** (2015) 315–335.

17. Prabhu R. H., Patravale V. B., Joshi M. D., *International Journal of Nanomedicine*, **10** (2015) 1001–1018.

18. De Propris M. S., Raponi S., Diverio D., Milani M. L., Meloni G., Falini B., Foa R., Guarini A., *Haematologica*, **96** (2011) 1548–1551.

19. Kumar C. S., Mohammad F., *Advanced Drug Delivery Reviews*, **63** (2011) 789–808.

20. Dimer F., Ortiz M., Pase C., Roversi K., Friedrich R., Pohlmann A. R., Burger M., Guterres S. S., *Journal of Biomedical Nanotechnology*, **10** (2014) 1137–1145.

21. Contri R. V., Frank L. A., Kaiser M., Pohlmann A. R., Guterres S. S., *International Journal of Nanomedicine*, **9** (2014) 951–962.

Chapter 6

Drug and Protein Encapsulation by Emulsification: Technology Enhancement Using Foam Formulations

Suun Weng Cheong Jeremy[a] **and Xian Jun Loh**[a,b,c]
[a]*Department of Materials Science and Engineering, National University of Singapore, 9 Engineering Drive 1, Singapore 117576*
[b]*Institute of Materials Research and Engineering (IMRE), 3 Research Link, Singapore 117602*
[c]*Singapore Eye Research Institute, 11 Third Hospital Avenue, Singapore 168751*
lohxj@imre.a-star.edu.sg

6.1 Introduction

Drug and protein delivery has long been a key area for the biomedical industry. With traditional delivery methods such as injection or ingestion, the drugs or protein concentration in the body fluctuates severely. A large spike in concentration occurs when the drug or protein first enters the body, followed by a gradual drop in concentration as the drug or protein is removed from the body by biological processes—following a logarithmic progression known as the biological half-life. Both the large spike and eventual fall of

Polymer Capsules
Edited by Ye Liu and Xian Jun Loh

ISBN 978-981-4774-60-4 (Hardcover), 978-0-429-42873-9 (eBook)
www.jennystanford.com

the drug concentration is undesirable, as they can easily result in drug overdose or underdose, respectively. This can be mitigated by delivering the drug over multiple doses (see diagram below), even if the drug concentration still fluctuates over time, but within a smaller region. With more frequent but lighter doses, the fluctuations in concentration can be made even smaller, with the obvious end step being a fully controlled release of the drugs.

However, this is impractical with traditional delivery methods. Therefore, methods of controlled release of drugs and proteins are desirable. Encapsulation is currently one of the most promising types of drug delivery being researched.

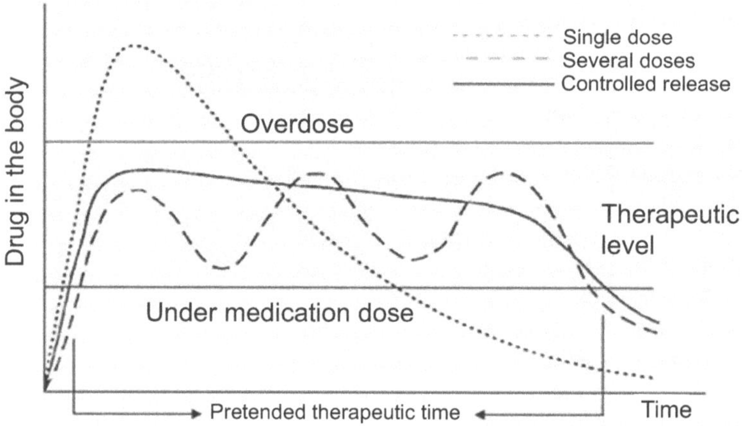

Figure 6.1 Drug Concentration in the body from various drug delivery schedules: single dose, multiple doses, and controlled release. Reprinted from Ref. [1], Copyright 2013, with permission from Springer Nature.

With encapsulation, drugs or protein are encapsulated inside a carrier material, typically a biodegradable polymer, to form microscopic particles, which can be placed within the body where they will gradually degrade. As the polymer degrades, the drugs or protein are released into the body. The drug delivery schedule is thus tied to the polymer degradation process, which provides a means of controlling the drug delivery schedule. Furthermore, the carrier material also protects the drugs or proteins from degrading within the body by preventing exposure with the rest of the body. This allows a greater measure of control over the dose. In some cases, the carrier material may degrade more quickly in specific parts of the body, allowing a localized dose to be delivered. Double emulsification is the typical method with which these microparticles are created. However, alternative methods have arisen for the encapsulation of drugs or proteins in which the microparticles are formed not through emulsification, but through the use of supercritical carbon dioxide gas. This review will investigate the development of supercritical carbon dioxide–based encapsulation and the potential benefits or consequences of adopting supercritical carbon dioxide encapsulation over traditional double emulsification techniques.

6.1.1 Particle Parameters

Several parameters are important when describing encapsulated particles. ***Particle size*** is of special concern because of its complex relationship with the drug delivery schedule. Larger particles are associated with a larger initial release (which is generally undesirable [2]). Particle size distribution is also important. After emerging from the encapsulation process, the size of the particles will fall into a bell curve-like distribution of sizes. A narrow size distribution would produce a sharper drug release as compared to a wider size distribution. The best size distribution is dependent on the release schedule required of the pharmaceutical product. ***Particle morphology*** also affects the drug release. Particles can possess various shapes. In general, particles arising from emulsification-based techniques are spherical or agglomerations of spheres. Particles arising from the supercritical carbon dioxide–based techniques described later, however, may come in a wider variety of shapes. Due to their rapid precipitation, the particle morphology may be spherical, needle-shaped, or simply irregular. The morphology of the particles affects the rate at which the carrier material degrades, and thus the drug release schedule.

Encapsulation efficiency is the percentage of drug or protein encapsulated over the total mass used in the particle manufacture. Loss of efficiency with double emulsification generally occurs as drug molecules disperse out into the continuous phase during the precipitation process [3]. A loss of efficiency is, of course, undesirable as it increases the cost of the process. In many cases, the cost of the drug or protein comprises a significant amount of the manufacturing cost, so wastage of the drug or protein is highly undesirable. However, if the drug or protein is inexpensive compared to the processing cost then other factors may be of greater concern. ***The drug or protein to polymer ratio*** of the particles affects the relative thickness of the polymer coating, which in turn affects the time taken for the initial release. Together with particle size, this ratio determines the total amount of drug or protein inside the particle, also known as the loading. The loading can affect other parameters, for example, particles with heavier loading are associated with more distorted particle morphologies [4]. However, there is no convenient formula for how each of these variables affect the drug release, or indeed, each other. They may be affected by the drug or protein, or carrier material's identity. For example, certain drugs may prefer to crystallize in needle-like conformations resulting in the encapsulated particles favoring needle-like morphologies as well [4].

6.2 Double Emulsification–Based Techniques

6.2.1 Water in Oil in Water Emulsification (W/O/W)

W/O/W emulsification techniques represent the most common type of double emulsification–based encapsulation mechanics. An aqueous drug or protein is mixed in an organic solvent containing the carrier material, forming a preliminary W/O emulsion. This emulsion is then dispersed into an aqueous phase containing an emulsifier: an amphiphilic substance such as poly(vinyl alcohol), forming micelles. This forms a W/O/W emulsion. From here, the micelles are typically solidified via solvent evaporation. The temperature is elevated until the volatile organic solvent evaporates. Alternatively, there is solvent extraction, in which a large amount of an extraction agent such as water or methanol is added for the organic solvent to diffuse out into. With both these methods, the solidification process generally presents the most time-consuming step.

Other than increasing the time taken, this can also affect the process in other ways. The drug or protein may diffuse out into the aqueous bulk during this step, decreasing encapsulation efficiency [3]. In some cases, the drug or protein may also have more time to react with the solvent or polymer. Several variables might be controlled in order to reduce the loss in encapsulation efficiency from this step. By decreasing the solubility of the carrier material in the organic solvent, the carrier material can be made to solidify more quickly, thus giving less time for the drug or protein to diffuse out. This can be achieved through several methods, including changing the solvent or increasing the molecular weight of the polymer to reduce solubility [5].

The concentration of the carrier material in the organic solvent is also another major factor which may be manipulated. This improves efficiency in two ways: the solidification process is faster if the carrier material is already concentrated, and increasing the concentration of carrier material makes the outer layer of the micelle more viscous, preventing diffusion of the drug or protein out of the micelle during solidification [5]. Using an organic solvent with a higher solubility in water (or other extraction agent) will also improve the speed of the solidification step (for solvent extraction) [5], as the time taken for the solvent to diffuse out is decreased. Methylene chloride is an example of an organic solvent with a high solubility in water.

6.2.2 Water in Oil in Oil Emulsification (W/O/O)

W/O/O emulsification is one method of avoiding the drug or protein leeching problem. In this case, after forming the W/O emulsion, it is added to another organic phase in which the polymer is insoluble in, such as silicon oil. This causes the polymer to coacervate into droplets containing the drug or polymer, analogous to the formation of micelles. Another hardening agent is then added which hardens the polymer.

Since there is no bulk aqueous phase in W/O/O emulsification, the leeching of aqueous drugs or proteins into the bulk is reduced, as the hydrophilic drugs tend not to migrate out of the micelle [3].

6.2.3 Solid in Oil in Water (S/O/W) or Solid in Oil in Oil (S/O/O) Emulsification

Another method of preventing leeching is to use the drug/protein in solid state, rather than liquid. The solid particles are mixed into the polymer solution instead of an aqueous solution. Due to their solid state, the drugs or proteins have reduced mobility, and the diffusion of the substance out of the micelle is reduced, improving encapsulation efficiency [3].

6.3 Supercritical Carbon Dioxide–Based Techniques

The following methods present a vastly different method of encapsulation. Common to each of the methods is the use of supercritical fluids. Supercritical fluids are gases which have been brought over a certain critical temperature and pressure until they enter a supercritical phase, a phase which possesses properties of both liquids and gases. The gas used is almost always carbon dioxide for reasons which will be described later. There are three major supercritical carbon dioxide–based techniques: the particles from gas saturated solutions (PGSS) technique, the rapid expansion of saturated solutions (RESS) technique, and the supercritical anti-solvent (SAS) technique. Many other techniques exist, but can generally be classified as variants of these three techniques. In some cases, a small number of the techniques may still involve emulsions. However, most of these are simple single emulsions, and are not reliant on the emulsification; rather they are a simple consequence of the mixing of two phases. Consequently, the term "double emulsification–based techniques" in this paper does not include these techniques.

6.3.1 Particles from Gas Saturated Solutions (PGSS)

PGSS and derived techniques comprise a large fraction of foam-related encapsulation methods. Patented by Weidner *et al.* in 1994 [6], PGSS produces microparticles through the introduction of carbon dioxide which is above its critical temperature and pressure and such that it is a supercritical fluid which bears properties of both gases and liquids. PGSS was originally intended for the food industry for the purpose of creating nanoparticles as an alternative for other micronization techniques like ball milling. The original particles made with PGSS were comprised of a single substance, however, PGSS can also be used to create encapsulated drugs and proteins with minimal modifications to the process. The typical process involves first mixing the drug or protein with the molten polymer. From here, the mixture is introduced into a heated highpressure chamber in which compressed carbon dioxide gas is rapidly stirred in till saturation is achieved. The gas may comprise up to 50% by weight of the mixture. The mixture is then allowed to depressurize through a nozzle, and expands rapidly, forming into tiny droplets. Carbon dioxide trapped within these droplets expands into a gas, further breaking up these droplets into even smaller particles. The particles solidify as the gas expands due to the rapid temperature decreases brought about by the Joule–Thomson effect [7].

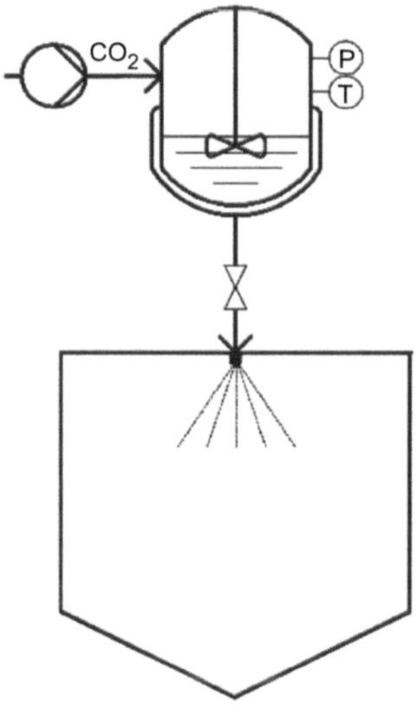

Figure 6.2 Basic PGSS setup [8].

Figure 6.2 shows the basic PGSS setup. The setup appears very simple, which is indicative of how simple the process actually is. The mixture is placed in the top tank and supercritical carbon dioxide is added, before the pressure is subsequently released rapidly through the nozzle into the second chamber. Carbon dioxide is used over other candidate gases such as chlorofluorocarbons and carbon tetrachloride, due to a number of factors. Carbon dioxide is relatively abundant in air, making it cheap and readily available. It is also relatively non-toxic. It also reaches its supercritical state at relatively low temperatures and pressures (31.10 °C, 7.39 MPa) [8] as compared to other gases, and is low even compared to many other processes. The primary benefit of PGSS over traditional double emulsification–based encapsulation lies in the use of carbon dioxide instead of organic solvents [8], for reasons similar to the reasons for choosing carbon dioxide over other gases. The use of organic solvents in double emulsification–based encapsulation may contaminate the product. Carbon dioxide, on the other hand, is relatively non-toxic. This non-toxicity also makes the process more environmentally friendly. Organic solvents are also more expensive than carbon dioxide which is abundantly available in the air. The PGSS process is also generally faster than traditional double emulsification techniques. No solidification process is needed to turn micelles into solid particles, which is usually one of the lengthy processes in encapsulation by double emulsification. The formation of particles in PGSS happens extremely rapidly after the mixture is expelled. Other than simply saving time, the quick solidification process is also useful for a number of factors: the solidification process in double emulsification techniques is the time when a significant amount of the encapsulate substance diffuses out; the PGSS process avoids this readily. One of the main drawbacks of PGSS over double emulsification–based and other supercritical gas related techniques is the temperatures required [8]. The polymer–drug or protein mixture must be in a molten state, which may be problematic for certain drugs or proteins which may degrade or decompose at elevated temperatures.

Despite the relative lack of research into PGSS encapsulation compared to traditional double emulsification techniques, PGSS is likely to be applied to a large number of drugs, proteins, and other useful substances due to several reasons. Firstly, unlike the rapid expansion from saturated solutions technique (see the following section), PGSS does not rely on the solubility of compounds in carbon dioxide. Even though only 1% of lipids can be dissolved in carbon dioxide, for example, lipids can contain up to 30 wt% of supercritical carbon dioxide at a pressure of 100 bars [7]. This greatly increases the variety of materials usable with PGSS. Secondly, high pressure of supercritical carbon dioxide causes a melting point depression effect in many substances, allowing many of these substances to be sprayed which could not originally be sprayed, such as highly viscous fluids. The exact extent of the melting point depression is based on the substance's interactions with the carbon dioxide [9].

Particle parameters can be altered by manipulating several process variables, such as pre-expansion pressure and temperature, and gas-to-product ratio. Increasing gas-to-product ratio tends to produce smaller particles for example. Particle morphology is also greatly influenced by these parameters. However, there seems to be no simple way of determining optimal variables. Many report no simple linear trends between the variables [10], and research papers employ trial and error in order to determine the best set of process variables. The following table is a list of various research papers involving encapsulation via PGSS, listing the drug or protein encapsulated, the carrier material, and the peak encapsulation efficiency reported by the paper. It should be noted that many papers claim that the lack of efficiency stems from problem with equipment, rather than inherent issues with the PGSS process [10, 11].

Table 6.1 Summary of research papers regarding PGSS, listing various combinations of drugs and proteins, carrier materials, and the peak encapsulation efficiency achieved

Encapsulate substance	Carrier material	Peak encapsulation efficiency achieved	Reference
β-carotene	Soybean lecithin	60%	[10]
Linalool	Poly-(ε-caprolactone)	50%	[11]
Lavandin oil	Poly-(ε-caprolactone)	45%	[11]
Solid-lipid pea particle	Glyceryl dipalmitosterate and olive oil	96%	[12]
Limonene	Modified starch	86%	[13]
S-(+)-Ibuprofen	Pluronic® F-127	85%	[14]
S-(+)-Ibuprofen	Pluronic® F-127 and Pluronic® L-64	90%	[14]
S-(+)-Ibuprofen	Pluronic® F-127 and Gelucire®	38%	[14]
S-(+)-Ibuprofen	Pluronic® L-64, glycerol monostearate	85%	[14]
Ketoprofen	Gelucire® and glyceryl monooleate	97%	[15]

From this table, we may observe that in many cases, PGSS has a high encapsulation efficiency. In many cases the efficiency is close to or on par with double emulsification-based techniques. The supercritical assisted atomization (SAA) technique is a variant of PGSS. Instead of melting the drug or polymer and carrier material together, the materials are instead dissolved in a solvent. After being mixed with the supercritical carbon dioxide and being sprayed through the nozzle, the stream of particles passes through another chamber through which heated nitrogen gas is passed. The heated nitrogen gas has a similar purpose as in spray drying, and indeed, the SAA process may be considered a combination of PGSS and SAA. The solvent rapidly evaporates and travels into the bulk of the nitrogen filled chamber, leaving the solidified particle [8, 16].

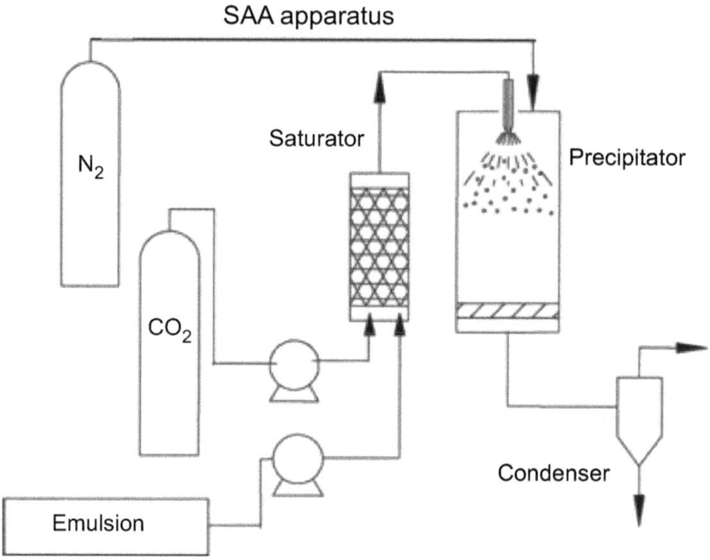

Figure 6.3 Basic SAA setup. Reprinted from Ref. [16], Copyright 2015, with permission from Elsevier.

The main advantage of this is the ability to produce encapsulated microparticles from drug/protein–carrier material pairs which cannot properly mix [16]. The spray drying process also allows a greater control over the already rapid solidification process, which allows fine control over particle size. Many variations of the PGSS process exist. It is possible to adapt the PGSS process to a continuous processing mode with only slight modifications. The polymer melt and drug or protein mixture is fed continuously through a pipe, while a second pipe feeds the supercritical carbon dioxide. The two streams mix in a static mixer before being expanded through the nozzle.

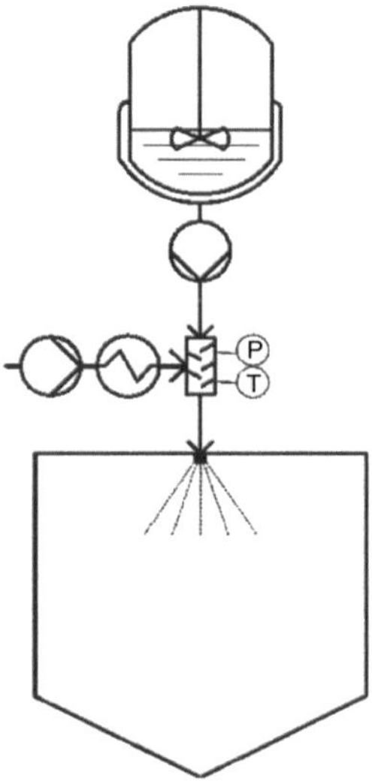

Figure 6.4 Continuous PGSS setup [8].

The PGSS process has also been adapted to the production of microbubbles by Rodriguez-Rojo *et al.* [17]. Microbubbles are essentially encapsulated microparticles containing a gas (i.e., perfluorocarbon gas) instead of a drug. The carrier material in this case prevents the bubble from bursting prematurely. The microbubbles may be used for ultrasonic imaging in order to create high contrasting areas. Alternatively, drugs or proteins may be incorporated into the polymer carrier itself. Such microparticles can then be used for targeted delivery by rupturing the particles with ultrasound. In Rodriguez-Rojo *et al.*'s setup, the perfluorocarbon gas was mixed with the carbon dioxide instead of with the carrier material (Gelucire®).

6.3.2 Rapid Expansion from Saturated Solutions (RESS)

RESS bears many similarities to PGSS. However, RESS processes rely on the ability of supercritical carbon dioxide to dissolve many organic compounds, while PGSS simply uses the Joule–Thomson effect of the rapidly expanding gas.

In RESS, the drug or protein must be soluble in supercritical carbon dioxide. The substance, carrier material, and optionally, a co-solvent [18] are placed in a pressure vessel, whereupon supercritical carbon dioxide is pumped in at an elevated temperature and pressure. The purpose of the co-solvent is to enhance the solubility of the polymer encapsulant in the carbon dioxide. The substance to be encapsulated, along with the carrier material, dissolves into the supercritical carbon dioxide. The solution is then sprayed out through a nozzle. Just as in PGSS, the rapid expansion divides the mixture into fine droplets, with the expansion of trapped carbon dioxide further subdividing the droplets. The sudden drop in pressure makes the carbon dioxide lose its solvent properties and causes encapsulated particles to rapidly precipitate, much like the way regular saturated aqueous solutions may precipitate their salts upon cooling.

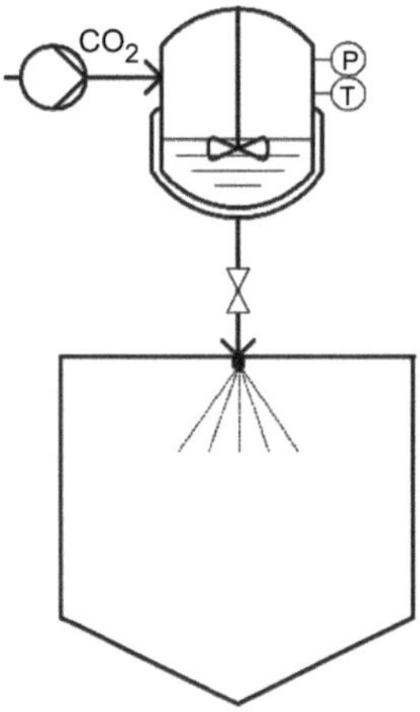

Figure 6.5 Basic RESS setup [8].

The basic setup for RESS is identical to that of PGSS, except for the larger holding and reaction tanks required due to the larger amount of supercritical carbon dioxide required in RESS as compared to PGSS. Like PGSS, RESS does not require the use of organic solvents, relying instead on supercritical carbon dioxide which is relatively harmless. RESS processes also share the speed of PGSS processes, with the same benefits such as cost-savings, and avoiding the loss of encapsulation efficiency during the emulsification solidification process. Another benefit of RESS versus PGSS is that the encapsulated material and carrier material need not be in the molten state, since it dissolves into the supercritical carbon dioxide. Consequently, materials at risk of decomposing or reacting at temperatures around or below their melting point are at less risk with RESS [19].

The obvious downside to RESS is the requirement of the materials to be soluble in the supercritical carbon dioxide. This restriction excludes RESS as a possibility for the encapsulation of many drugs, proteins, and carrier materials. It is possible to overcome this via the use of a co-solvent in order to allow the materials to dissolve. However, co-solvents negate the advantage of not having to contaminate the particles with organic solvents. Alternatively, other gases may be used in place of carbon dioxide. However, most other gases are not as easy or safe to use as carbon dioxide. Furthermore, since many of the possible gases are also non-polar, the solubility of materials in other candidate gases may be similar to that of carbon dioxide. Utilizing RESS also requires an intimate understanding of multi-phase interactions between each of the materials and carbon dioxide. The drug or protein, as well as the carrier material must both be soluble in supercritical carbon dioxide, then precipitate out after the rapid depressurization. The precise temperatures and pressures are typically obtained experimentally via trial and error; thus, research and expertise is required to determine if a given process is viable. Consequently, research on the use of RESS is limited [7].

Another downside is the low solubility ratio of most materials in carbon dioxide. This means a large volume of supercritical carbon dioxide is required to dissolve most substances, far greater than the amount used in PGSS. Although carbon dioxide is relatively cheap, the increase in costs may still be significant. Larger storage chambers and pressure chambers would also be required, which will of course increase setup costs [7]. However, given the relative costs of carbon dioxide versus pharmaceuticals, the cost increase may be insignificant. Additionally, Weidner [7] suggests that much of the carbon dioxide may be recovered after use, which results in large cost savings. The following table is a list of various research papers involving encapsulation via RESS, listing the drug or protein encapsulated, the carrier material, and the peak encapsulation efficiency reported by the paper. Again, some of the papers claim that their lack of efficiency stems from problems with equipment, rather than inherent issues with the PGSS process [20].

Table 6.2 Summary of research papers regarding RESS, listing various combinations of drugs and proteins, carrier materials, and the peak encapsulation efficiency achieved

Encapsulated substance	Carrier material	Peak encapsulation efficiency achieved	Reference
Anthocyanin	Poly(ethylene glycol)	80%	[18]
Rutin	Poly(ethylene glycol)	44%	[20]
Anthocyanin	Poly(ethylene glycol)	49%	[20]
Essential oil from *Baizhu*	Phosphatidylcholine and cholesterol	82%	[21]

The low number of papers is indicative of the low amount of research into RESS-based encapsulation. Other papers regarding RESS are not listed here because they are still examining the possibility of RESS rather than its industrial feasibility. It is speculated that the unpopularity of RESS is because it bears so much similarity to PGSS, yet bears few advantages over it; most substance and carrier material combinations which can be encapsulated via RESS can also be encapsulated by PGSS.

The situation of the effect of process parameters on particle parameters in RESS is much like that of PGSS. Although trends between factors such as drug/protein ratios and particle size are observable, there is no clear linear relationship, necessitating the experimental determination of process variables [7]. However, in the case of RESS, the spray distance has a relatively linear effect on particle size [22], which provides a simple method of controlling the particle size.

There are several variants of RESS possible which have also undergone some research. The rapid expansion of supercritical solutions into a liquid solvent process (RESOLV) is very similar to RESS, except that the mixture of drug or protein, carrier material, and supercritical carbon dioxide is expanded out of the nozzle into a tank full of a liquid solvent [23]. The term solvent in the name of the process is a bit confusing, as the solvent used should not dissolve the carrier material.

The purpose of the liquid solvent is to halt the growth of precipitating particles. This leads to a better particle size as the average particle size is lower and the size distribution is more narrow as compared to RESS [23]. RESOLV has a problem with a tendency for the microparticles to agglomerate while in the liquid solvent. However, this may be solved via the introduction of stabilizing agents into the liquid solvent. The use of the liquid medium also presents an opportunity to incorporate such additives inside the liquid so that they may take effect shortly after the encapsulated particles have formed.

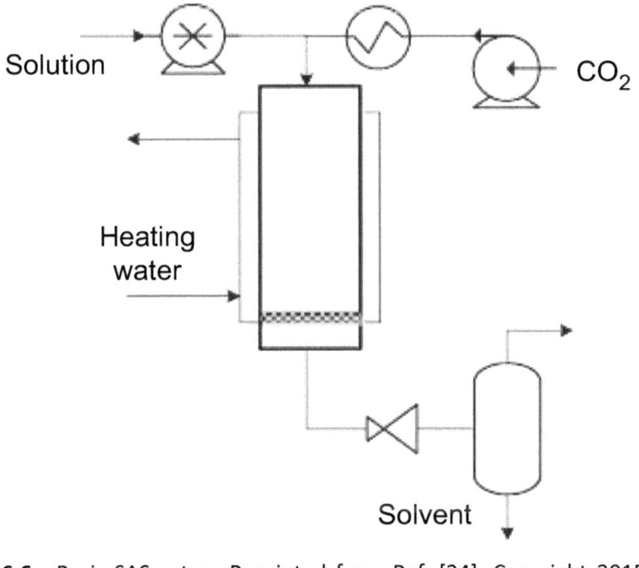

Figure 6.6 Basic SAS setup. Reprinted from Ref. [24], Copyright 2015, with permission from Elsevier.

6.3.3 Supercritical Anti-solvent (SAS)

The SAS technique takes advantage of the *lack* of solubility of some carrier materials in supercritical carbon dioxide. Adding supercritical carbon dioxide to a solution of carrier material in organic solvent causes the solubility of the carrier material to fall and precipitate out on the surface of the drug or protein. SAS requires an organic solvent which is soluble in supercritical carbon dioxide, a carrier material soluble in that solvent but insoluble or sparingly soluble in the mixture of that solvent and supercritical carbon dioxide (the temperature and pressure may be altered for this purpose), and microparticles of the desired drug/protein which must be *in*soluble in the organic solvent and supercritical carbon dioxide. The carrier material is dissolved in the organic solvent, and the microparticles of the drug/protein are added. The mixture must be stirred to ensure a proper dispersion of the materials. The mixture is then sprayed into a pressure and temperature controlled chamber containing supercritical carbon dioxide. The solvent rapidly mixes with the bulk of supercritical carbon dioxide, reducing the solubility of the carrier material dramatically, whereupon it begins to precipitate rapidly around the drug or protein to be encapsulated.

The basic setup for SAS is shown in Fig. 6.6, courtesy of Salgado *et al.* [24]. The organic solvent, drug or protein, as well as carrier material form the solution. Carbon dioxide is pumped into the chamber until it reaches the desired temperature and pressure; then the solution of carrier material and the drug or protein in the organic solvent is added. Immediately after the addition of the solution to the supercritical carbon dioxide, the chamber is depressurized via expansion through the nozzle. At this point the microparticles form in a second chamber whereupon they are collected. Like RESS, SAS does not generally require intensive temperatures and pressures. It is also relatively fast and provides less time for undesirable reactions to occur between the drug or protein and the solvated polymer. Since SAS requires an insolubility of the carrier material, it may often be employed where RESS might not. This partially makes up for the long checklist of requirements needed. The use of the drug or protein in a solid state is generally an inconvenience. However, it may be taken advantage of: the particles can be micronized via any method and the increased control over its parameters might be of use. Difficult to dissolve substances might also be encapsulated this way. Unfortunately, SAS uses an organic solvent, negating one of the main advantages of supercritical gas–based techniques. The solvent must be chosen based on the multi-phase interactions with the other materials. Dichloromethane is a common choice as it is commonly used with well-studied effects and is suitable for a large number of materials. Dichloromethane also dissolves into supercritical carbon dioxide more rapidly than other common solvents, making the precipitation time even more rapid [25]. Furthermore, SAS requires an even more in-depth knowledge of multi-phase interactions between the various substances compared to RESS. This is due to the four distinct materials and reagents which need to perform in a specific manner. These requirements may be overcome to some extent through clever selection of the organic solvent and carrier material, but even in such cases the multi-phase behavior can only be predicted and must still be experimentally confirmed. Due to the long checklist of requirements for the process to be used, employing SAS requires intensive research and expertise [7]. The following table is a list of various research papers involving encapsulation via SAS, noting the drug or protein encapsulated, the carrier material, the organic solvent used, and the peak encapsulation efficiency reported by the paper.

As can be seen from the table, high encapsulation efficiencies are possible with SAS, on par with double emulsification–based processes. This establishes it as a viable option for consideration against the double emulsion–based processes. One variant of SAS is the supercritical fluid extraction of emulsions (SFEE) process. Just like SAS, SFEE uses supercritical carbon dioxide as an anti-solvent in order to induce precipitation. However, the process also bears similarities to the solvent extraction process in double emulsification processes, and might indeed also be considered a double emulsification variant. In SFEE, the drug or protein and carrier material are emulsified using an organic solvent, then double emulsified into an aqueous phase (W/O/W). The mixture is sprayed into another tank, with the energy of the spraying promoting the formation of especially small micelles. The nozzle used is a double nozzle, with the outer nozzle spraying supercritical carbon dioxide simultaneously. The solubility of the organic solvent in supercritical carbon dioxide results in the rapid extraction of the organic solvent, causing the micelle to solidify [34].Though the process is very similar to solvent extraction in double emulsification, the SFEE precipitation process is extremely rapid, unlike typical solvent extraction. Thus SFEE, like other supercritical carbon dioxide–based processes, avoids long residence times which decrease encapsulation efficiency in normal double emulsification [34]. This makes SFEE an interesting alternative to double emulsification.

Table 6.3 Summary of research papers regarding SAS, listing various combinations of drugs and proteins, carrier materials, and the peak encapsulation efficiency achieved

Encapsulated substance	Carrier material	Solvent	Peak encapsulation efficiency achieved	Reference
Bixin	Poly(ethylene glycol)	Dichloromethane	62%	[20]
Asthaxanthin	Poly(ethylene glycol)	Dichloromethane	51%	[26]
Quercetin	Ethyl cellulose	Ethyl acetate	99%	[27]
Nano-curcumin	Poly(lactic-*co*-glycolic acid)	Acetone and ethanol	38%	[28]
Nano-curcumin	Poly(lactic-*co*-glycolic acid)	Acetone	33%	[28]
Hydroxycamptothecin	Soy lecithin and cholesterol	Dichloromethane	85%	[29]
Bixin	Poly(3-hydroxybutirate-*co*-hydroxyvalerate)	Dichloromethane	92%	[30]
Asthaxanthin	Poly(3-hydroxybutirate-*co*-hydroxyvalerate)	Dichloromethane	48%	[31]
Lutein	Hydrogenated phosphatidylcholine	Dichloromethane and ethanol	90%	[32]
Insulin	Hydroxypropyl methyl cellulose phthalate	Dimethyl sulfoxide and acetone	99%	[33]

6.4 Conclusion

Supercritical carbon dioxide–based techniques for microencapsulation are a promising method for encapsulating pharmaceuticals for drug delivery purposes. Although the concept is far younger than double emulsification techniques, the three processes of PGSS, RESS, and SAS, as well as their variants have achieved a level of viability comparative to double emulsification techniques. Supercritical carbon dioxide processes are more rapid than double emulsification techniques, saving valuable time, as well as preventing loss of encapsulation efficiency through diffusion. The main processing chemical is carbon dioxide, which is easily available, non-toxic, environmentally friendly, and most importantly, inexpensive. PGSS and RESS, particularly, may avoid the use of potentially toxic organic solvents altogether, preventing the potential for trace amounts of remaining solvent to contaminate the product. Perhaps most importantly, the encapsulation efficiencies of the three techniques are comparable to emulsification-based techniques. Encapsulation efficiency is a major factor in processing costs, indicating the economic viability of such processes.

References

1. Fonseca A. C., Ferreira P., Cordeiro R. A., Mendonça P. V., Góis J. R., Gil M. H., Coelho J. F. J., Drug delivery systems for predictive medicine: Polymers as tools for advanced applications, In *New Strategies to Advance Pre/Diabetes Care: Integrative Approach by PPPM*, Mozaffari, M. S., Ed. Springer Netherlands: Dordrecht (2013) 399–455. Doi: 10.1007/978-94-007-5971-8_16.
2. Costantino H. R., Johnson O. L., Zale S. E., Relationship between encapsulated drug particle size and initial release of recombinant human growth hormone from biodegradable microspheres, *Journal of Pharmaceutical Sciences*, **93** (2004) 2624–2634.
3. Tan M. X. L., Danquah M. K., Drug and protein encapsulation by emulsification: technology enhancement using foam formulations, *Chemical Engineering & Technology*, **35** (2012) 618–626.
4. Kishida A., Dressman J. B., Yoshioka S., Aso Y., Takeda Y., Some determinants of morphology and release rate from poly(L)lactic acid microspheres, *Journal of Controlled Release*, **13** (1990) 83–89.

5. Jyothi N. V., Prasanna P. M., Sakarkar S. N., Prabha K. S., Ramaiah P. S., Srawan G. Y., Microencapsulation techniques, factors influencing encapsulation efficiency, Journal of Microencapsulation, **27** (2010) 187–197.
6. Weidner E., Knez Z., Novak Z., PGSS (particles from gas saturated solutions): a new process for powder generation: Proceedings of the 3rd *International symposium on supercritical fluids,* Strasbourg, France (1994), 229–235.
7. Weidner E., High pressure micronization for food applications, *The Journal of Supercritical Fluids*, **47** (2009) 556–565.
8. Nunes A. V. M., Duarte C. M. M., Dense CO_2 as a solute, co-solute or co-solvent in particle formation processes: a review, *Materials*, **4** (2011) 2017–2041.
9. Pasquali I., Bettini R., Are pharmaceutics really going supercritical? *International Journal of Pharmaceutics*, **364** (2008) 176–187.
10. de Paz E., Martín Á., Cocero M. J., Formulation of β-carotene with soybean lecithin by PGSS (Particles from Gas Saturated Solutions)-drying, *The Journal of Supercritical Fluids*, **72** (2012) 125–133.
11. Varona S., Martín Á., Cocero M. J., Duarte C. M. M., Encapsulation of lavandin essential oil in poly-(ε-caprolactones) by PGSS Process, *Chemical Engineering & Technology*, **36** (2013) 1187–1192.
12. Saldanha do Carmo C., Nunes A. N., Silva I., Maia C., Poejo J., Ferreira-Dias S., Nogueira I., Bronze R., Duarte C. M. M., Formulation of pea protein for increased satiety and improved foaming properties, RSC Advances, **6** (2016) 6048–6057.
13. Machado L. C., Pelegati V. B., Oliveira A. L., Study of simple microparticles formation of limonene in modified starch using PGSS – Particles from gas-saturated suspensions, *The Journal of Supercritical Fluids*, **107** (2016) 260–269.
14. Fraile M., Martín, ÿ., Deodato D., Rodriguez-Rojo S., Nogueira I. D., Simplício A. L., Cocero M. J., Duarte C. M. M., Production of new hybrid systems for drug delivery by PGSS (Particles from Gas Saturated Solutions) process, *The Journal of Supercritical Fluids*, **81** (2013) 226–235.
15. García-González C. A., Argemí, A., Sousa A. R. S. d., Duarte C. M. M., Saurina J., Domingo C., Encapsulation efficiency of solid lipid hybrid particles prepared using the PGSS® technique and loaded with different polarity active agents, *The Journal of Supercritical Fluids*, **54** (2010) 342–347.

16. Petra Š., Renata A., Věra K., Ernesto R., Tomáš, S., Miroslav P., Supercritical assisted atomization of emulsions for encapsulation of 1-monoacylglycerols in an hydrophilic carrier, *The Journal of Supercritical Fluids*, **97** (2015) 183–191.

17. Rodríguez-Rojo S., Lopes D. D., Alexandre A. M. R. C., Pereira H., Nogueira I. D., Duarte C. M. M., Encapsulation of perfluorocarbon gases into lipid-based carrier by PGSS, *The Journal of Supercritical Fluids*, **82** (2013) 206–212.

18. Santos D. T., Albarelli J. Q., Beppu M. M., Meireles M. A. A., Stabilization of anthocyanin extract from jabuticaba skins by encapsulation using supercritical CO_2 as solvent, *Food Research International*, **50** (2013) 617–624.

19. Sergei G. K., Supercritical fluid impregnation of polymers for drug delivery. in *Supercritical Fluid Technology for Drug Product Development*, Marcel & Dekker Inc, New York (2004) 322–343.

20. Santos D. T., Meireles M. A. A., Micronization and encapsulation of functional pigments using supercritcal carbon dioxide. *Journal of Food Process Engineering*, **36** (2013) 36–49.

21. Wen Z., Liu B., Zheng Z. K., You X. K., Pu Y. T., Li Q. O., Preparation of liposomes entrapping essential oil from Atractylodes macrocephala Koidz by modified RESS technique. *Chemical Engineering Research & Design*, **88** (2010) 1102–1107.

22. Esfandiari N., Production of micro and nano particles of pharmaceutical by supercritical carbon dioxide. *The Journal of Supercritical Fluids*, **100** (2015) 129–141.

23. Dalvi S. V., Azad M. A., Dave R., Precipitation and stabilization of ultrafine particles of Fenofibrate in aqueous suspensions by RESOLV. *Powder Technology*, **236** (2013) 75–84.

24. Salgado M., Rodríguez-Rojo S., Alves-Santos F. M., Cocero M. J., Encapsulation of resveratrol on lecithin and β-glucans to enhance its action against Botrytis cinerea. *Journal of Food Engineering*, **165** (2015) 13–21.

25. Kumar R., Mahalingam H., Tiwari K. K., Selection of solvent in supercritical antisolvent process. *APCBEE Procedia*, **9** (2014) 181–186.

26. Machado Jr F. R. S., Trevisol T. C., Boschetto D. L., Burkert J. F. M., Ferreira S. R. S., Oliveira J. V., Burkert C. A. V., Technological process

for cell disruption, extraction and encapsulation of astaxanthin from Haematococcus pluvialis. *Journal of Biotechnology*, **218** (2016) 108–114.

27. Fernández-Ponce M. T., Masmoudi Y., Djerafi R., Casas L., Mantell C., Ossa E. M. d. l., Badens E., Particle design applied to quercetin using supercritical anti-solvent techniques. *The Journal of Supercritical Fluids*, **105** (2015) 119–127.
28. Zabihi F., Xin N., Li S., Jia J., Cheng T., Zhao Y., Polymeric coating of fluidizing nano-curcumin via anti-solvent supercritical method for sustained release. *The Journal of Supercritical Fluids*, **89** (2014) 99–105.
29. Liu G., Wang W., Wang H., Jiang Y., Preparation of 10-hydroxycamptothecin proliposomes by the supercritical CO_2 anti-solvent process. *Chemical Engineering Journal*, **243** (2014) 289–296.
30. Boschetto D. L., Aranha E. M., de Souza A. A. U., Souza S. M. A., G. U., Ferreira S. R. S., Priamo W. L., Oliveira J. V., Encapsulation of bixin in PHBV using SEDS technique and in vitro release evaluation. *Industrial Crops and Products*, **60** (2014) 22–29.
31. Machado Jr F. R. S., Reis D. F., Boschetto D. L., Burkert J. F. M., Ferreira S. R. S., Oliveira J. V., Burkert C. A. V., Encapsulation of astaxanthin from Haematococcus pluvialis in PHBV by means of SEDS technique using supercritical CO_2. *Industrial Crops and Products*, **54** (2014) 17–21.
32. Xia F., Hu D., Jin H., Zhao Y., Liang J., Preparation of lutein proliposomes by supercritical anti-solvent technique. *Food Hydrocolloids*, **26** (2012) 456–463.
33. Jin H. Y., Xia F., Zhao Y. P., Preparation of hydroxypropyl methyl cellulose phthalate nanoparticles with mixed solvent using supercritical antisolvent process and its application in co-precipitation of insulin. *Advanced Powder Technology*, **23** (2012) 157–163.
34. Kluge J., Fusaro F., Casas N., Mazzotti M., Muhrer G., Production of PLGA micro- and nanocomposites by supercritical fluid extraction of emulsions: I. Encapsulation of lysozyme. *The Journal of Supercritical Fluids*, **50** (2009) 327–335.

Chapter 7

Drug Delivery Vehicles with Improved Encapsulation Efficiency: Taking Advantage of Specific Drug–Carrier Interactions

Siti Amalina Binte Mohamed Kotharisa[a] and Xian Jun Loh[a,b,c]
[a]*Department of Materials Science and Engineering, National University of Singapore, 9 Engineering Drive 1, Singapore 117576*
[b]*Institute of Materials Research and Engineering (IMRE), 3 Research Link, Singapore 117602*
[c]*Singapore Eye Research Institute, 11 Third Hospital Avenue, Singapore 168751*
lohxj@imre.a-star.edu.sg

This chapter reviews the different specific drug–carrier interactions that have been studied in recent years to improve the encapsulation efficiency of the carrier. These interactions are specific to each type of carrier and drug. This chapter focuses on the encapsulation of various anticancer drugs such as doxorubicin (DOX) and paclitaxel (PAC) in various drug delivery carriers such as in dendrimers, solid

Polymer Capsules
Edited by Ye Liu and Xian Jun Loh

ISBN 978-981-4774-60-4 (Hardcover), 978-0-429-42873-9 (eBook)
www.jennystanford.com

lipid nanoparticles (SLNs), polymeric micelles, and liposomes. The strategies that were employed for different types of carriers differs according to the properties of the carrier. Thus, it is important to understand the properties of the different classes of carriers. It is also important to note that these strategies are not interchangeable for all types of drugs. Most of the studies discussed in this chapter focuses on the anticancer drug, DOX. However, this chapter also discusses the encapsulation of hydrophilic drugs using liposomes.

7.1 Introduction

Drug delivery refers to the process of introducing drugs into the body to prevent, cure or slow down any disease or conditions. This can be done by the direct insertion of drugs or by using drug delivery vehicles. Direct insertion of drugs into the body is generally not a feasible method as most drugs are made up of very small molecules of about 500 da [1] and this results in the rapid removal and suboptimal biodistribution of the drugs which may even lead to toxic side effects [2] which limits their clinical applications. Direct insertion of drugs also often leads to the biodistribution of drugs into other parts of the body which thus adversely affects the efficacy of the drugs [3]. These factors would lead to increased drug administration which is patient-unfriendly.

As such, studies have been carried out to use a vehicle to transport the drugs to eliminate these limitations. The use of drug delivery vehicles would allow targeted drug delivery, delivery of drugs with low solubility using micelles, increased lifetime and controlled drug release rate. Thus, the use of drug carriers leads to a more efficient use of drugs and a more patient friendly dosing regimen. This would also lead to reduced expenses for the patient as lesser dosage and drugs are required. Some examples of drug carriers are soluble polymers, insoluble natural and synthetic polymers, micro particles, cells, cell ghosts, lipoproteins, liposomes, and amphiphilic polymer-based micellar systems [4]. These drug carriers are also known as a prodrug in some cases and it is defined as a biologically inactive derivative of a parent drug molecule that is generally activated by enzymatic transformation within the body in order to release the actual drug and has better delivery properties than the parent drug [5]. One of the factors that affect the rate of release of drugs is the nature of the polymer–drug interaction. Thus it is important to explore and choose the right polymer so that the drugs can be released at the desired rate. We can also use this information to select the right polymer that would mediate the drug release when it is subjected to pH or enzymatic degradation to achieve an optimal drug release rate in the body. The mechanism of the drug release will be discussed further in the following section.

In addition, the use of nanoparticles in drug delivery has been gaining attention as (1) they are able to pass through the smallest capillary vessels due to their small size and volume and can escape from phagocytes which allows them to have a longer half-life in the bloodstream; (2) they are able to easily penetrate through veins and tissues to get to the targeted area; (3) they can control the drug release rate by modifying the properties of the polymers; and (4) they are able to reduce the toxic side effects of drugs. Most of the drug carriers mentioned above lie in the nanometer range and thus they all have these advantages as a drug delivery carrier.

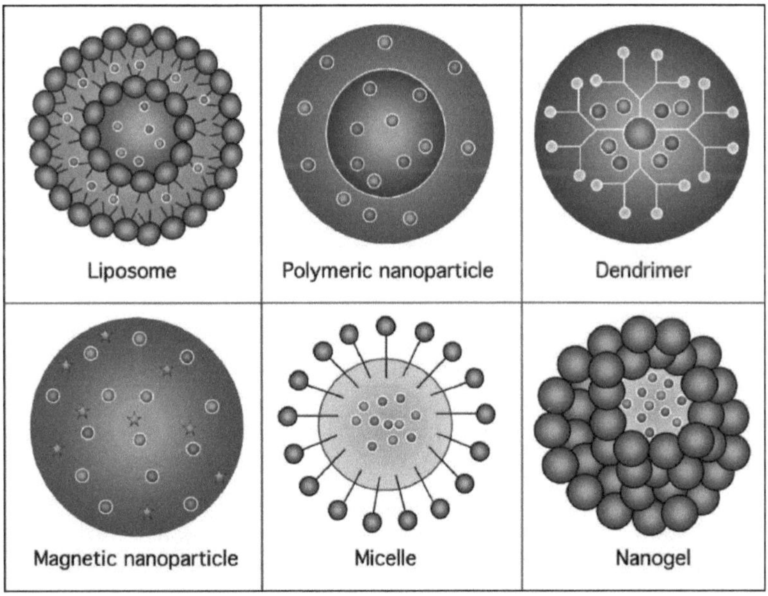

Figure 7.1 Types of polymeric nanocarriers. Reprinted from Ref. [6], Copyright 2013, Gowda et al.

7.1.1 Drug Delivery Mechanism

It is important to know how drugs work in order to understand what the requirements of a carrier are. Generally, drugs bind with the receptors on the cells to transmit/transfer specific signals so that the cells can respond accordingly. For example, DOX, an anticancer drug, is believed to control cancer by the apoptosis of cancerous cells by DNA damage through topoisomerase II inhibition and free radical generation by redox reaction [7]. Thus in this case it is important to design a carrier with the ability to optimize the cellular uptake mechanism so that these drugs can be delivered into the cells efficiently as per Fig. 7.2.

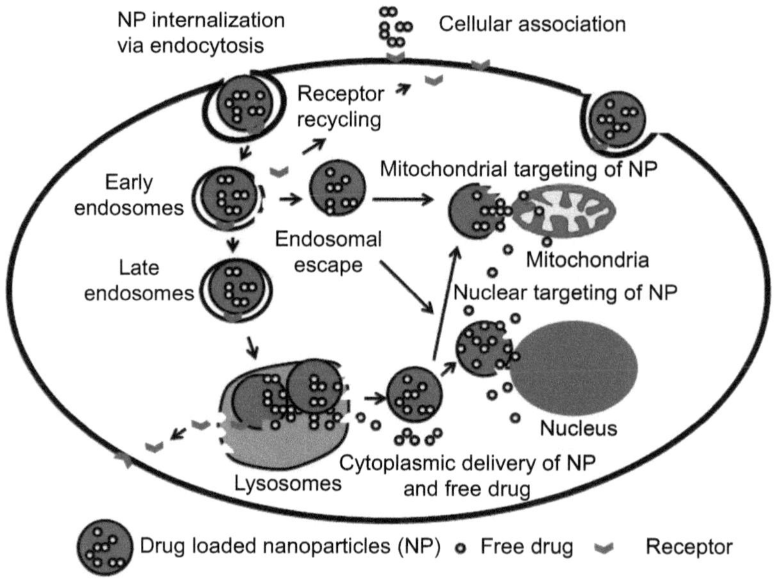

Figure 7.2 Drug delivery mechanism of nanoparticles. Reproduced from Ref. [8] under Creative Commons Attribution License.

As seen from the figure above, the cellular uptake mechanism involves several steps before the drug can be delivered successfully into the cell. Firstly, it is important to note that cell membranes are negatively charged. Thus, positively charged nanoparticles would increase its probability of attaching itself onto the cell membrane via electrostatic interactions. However, it should not be too positively charged either as it will result in the death of cells.

Upon the attachment of the nanoparticle onto the cell membrane, endocytosis occurs. Endocytosis refers to the process whereby the nanoparticle is internalized by the cell. Once it is internalized, it has to overcome various intracellular barriers in order to successfully deliver the drugs. Once in the endocytic pathway, the nanoparticle will have to escape from the endosome into the cytoplasm. There is also a probability where it would be expelled from the cell if it fails to escape from the endosome. Most of the nanocarriers, unfortunately, are unable to cross this barrier and thus the delivery of the drugs becomes limited. Once the nanoparticles are free from the endocytic pathway, they would continue to make their way to the nucleus via diffusion which is size dependent. The release of drugs in the cells is dependent on the type of carrier–drug interactions. Thus it is important to study these interactions to maximize both the encapsulation efficiency as well as the rate of release of drugs so that they will only be released when they are in the cells.

As mentioned above, the encapsulation efficiency of the polymer is an important parameter that needs to be considered when choosing a drug carrier. Entrapment/encapsulating efficiency (EE, in percent) represents the ratio of the amount of drug entrapped within the carrier to the drug amount initially introduced in the formulation [2]. High EE is desirable as it ensures the protection of drugs from degradative enzymes and prevents any leakage in undesirable areas [9]. The EE of the carriers are generally affected by the nature of the carrier and the type of drug that is being encapsulated [10]. Thus it is important to look at both the properties of the carrier and the drugs in order to obtain a high EE for each drug carrier system. In this paper, we will be focusing on improving the EE by specific drug–carrier interactions.

7.2 Commonly Used Anticancer Drug: Doxorubicin

DOX is one of the most widely used drugs in cancer therapy. It is a lipophilic, hydrophobic, and cationic compound. Its application as an anticancer drug is however limited due to its lack of solubility which would lead to acute toxicity. Thus it is important to find a suitable drug delivery system for DOX in order to reap its pharmaceutical benefits. DOX is less hydrophobic than PAC as it contains hydroxyl and amino groups which are polar. Thus it is harder to encapsulate DOX in a hydrophobic micelle core. DOX is commonly encapsulated using anionic or neutral copolymers [3] due to its cationic nature. However, there have been attempts to encapsulate DOX using cationic polymers as cationic polymers can mediate drug delivery into the cell since the cell membranes are anionic in nature. Consequently there is a need to search for the correct carrier with high EE and high cellular uptake to increase the efficacy of the drug delivery system.

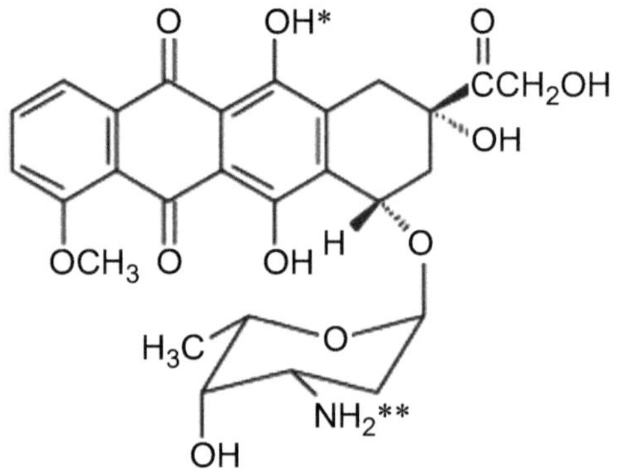

Figure 7.3 Molecular structure of DOX.

7.3 Types of Carriers

7.3.1 Dendrimers

Dendrimers are highly branched macromolecules that have a specific size, shape, and chemical functionality [11]. It is one of the most promising polymeric nanocarriers for different therapeutic categories of bioactives due to their unique properties [12], namely the ability to produce monodispersed dendrimers, specific shapes and sizes, and the ability to control its chemical functionality by adding specific end groups. For example, Hawker *et al.* reported that poly(aryl ether) dendrimers that have carboxylic group chain ends are able to solubilize pyrene in water [13]. The use of drug–dendrimer systems have also been known to improve the solubility of the drug and acts as a platform for dug targeting and Meiji and his team [14] successfully showed the entrapment of large molecules such as rose bengal (Fig. 7.4) is possible. There are two main ways in which a drug can be loaded onto the dendrimer: (1) drugs are entrapped within the dendritic architecture by physical interactions (electrostatic, hydrophobic, and hydrogen bond interactions) and (2) surface interactions of dendrimers and drugs by electrostatic and covalent interactions.

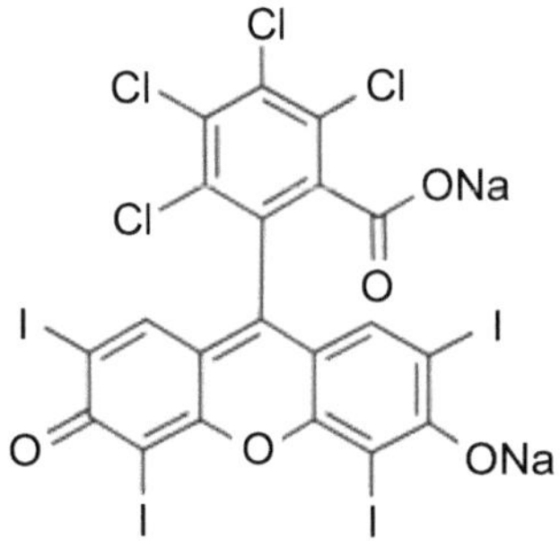

Figure 7.4 Chemical structure of rose bengal, a type of dye.

7.3.1.1 Properties of dendrimers

As discussed in the previous section, dendrimers have several unique properties which exhibit their potential as successful drug carriers.

7.3.1.1.1 *Monodispersity*

Generally, a monodispersed set of dendrimers can by synthesized by convergent methods [15, 16]. The monodispersity of the dendrimers is affected by factors such as dendrimer bridging and incomplete removal of ethylenediamine at each of the generation sequences which can be eliminated easily by proper purification at each step of the synthesis. The advantage of having the ability to produce a monodispersed set of dendrimers offers researchers the possibility to work with a tool of a well-defined and reproducible scalable size, which is a key advantage for commercial production [17].

7.3.1.1.2 *Nanosize and shape*

Another property that is specific to dendrimers is that their size and structures can be controlled which is an advantageous attribute in biomedical application. Controlled size and structures enable dendrimers to cross cell membranes and reduce the probability of premature clearance from the body. They are also of similar sizes to several biological structures. For example, 5.0 G PAMAM dendrimers is approximately the same size and shape as hemoglobin (Hb) (5.5 nm diameter) [18]. There are several classes of dendrimers that have been synthesized with different types of core materials, surface modifications and branching units. The structure of the dendrimer is size dependent, unlike polymeric micelles which are concentration dependent. Low generation numbers tend to have a more open and amorphous structures while higher generations exhibit spherical conformations that are capable of encapsulating drugs. X-ray analysis shows that the structure becomes more globular from lower to higher generations. The structure is such as to minimize the repulsion between the segments by spreading it out [19].

7.3.1.1.3 *Biocompatibility*

Dendrimers are also classified as a type of "smart" carrier due to their ability to overcome biological barriers, enabling them to circulate in the body for a sufficient amount of time, and also due to their targeting ability. The toxicity of the dendrimer is dependent on its end groups. Amine terminated PAMAM and PPI dendrimers usually exhibit concentration-dependent toxicity and hemolysis [11, 20, 21] while neutral or anionic groups terminated dendrimers have proven to be less toxic comparatively [20]. Toxicity caused by the cationic groups can be reduced by introducing negatively charged or neutral groups to its periphery [22].

7.3.1.1.4 *Encapsulation properties*

The open structure of the dendrimer allows the encapsulation of labile and poorly soluble drugs within its branches which improves the stability of the drug in the blood serum. Encapsulating the drugs using dendrimers would also result in controlled drug release in the body [23]. Many studies have been conducted to explore the different structures and types of dendrimers that can be potentially used in the encapsulation of drugs. These dendrimers were also designed to have "smart" properties that respond according to pH changes or other trigger factors. Most of the studies conducted assume that the dendrimers have a hollow core and dense shell structure which enables the encapsulation of the drugs in the hollow core. However, there are also studies which show an alternative structure of the dendrimer which is dense in the core, the density decreasing as it moves outwards [24, 25]. Generally, higher generation dendrimers usually adopt the hollow core structure which is capable of encapsulating drugs whereas the lower generations have a dense core structure. It is also important to note that the dendritic structure is not dependent on the concentration, unlike conventional polymeric micelles. It was shown in a study that the dendritic structures exist as single molecules that have the capability to encapsulate and thus are categorized as unimolecular micelles [26].

7.3.1.2 Dendrimer–drug interactions

7.3.1.2.1 *PEGylated dendrimers*

Poly(ethylene glycol) (PEG) is commonly grafted or conjugated onto the dendrimers to introduce a hydrophilic shell that surrounds the hydrophobic dendritic core, forming a unimolecular micelle [27, 28]. PEG is typically added to the dendrimers and is widely used in pharmaceutical applications as it is highly soluble in water, is biocompatible, and has the ability to modify the biodistribution of carriers [5]. Additionally, PEG coated carriers as compared to non-PEG coated carriers are known to have a lower uptake rate by the liver and a longer half-life in the bloodstream [29]. In a study conducted by Kojima *et al.* [30], it was shown that the drug loading and EE of Adriamycin (Fig. 7.2) increases with the increasing length of PEG grafts. The study compared the EE with another drug, methotrexate and the EE of methotrexate was observed to be significantly higher. This was due to the acidic nature of the methotrexate drug which interacted with the basic interior of the dendrimer. Following this study, a study by Pan *et al.* showed that there the degree of substitution of PEG had little effect on the EE of the dendrimer and similar EE results were obtained as the one reported by Kojima. A dendrimer-based block copolymer, citric acid-PEG-acid, was able to form inclusion complexes with a range of guest molecules such as 5-aminosalicylic acid, mefenamic acid, and diclofenac [31]. The EE for the 5-aminosalycylic was the highest and this is attributed to its small size and high polarity. We can thus conclude that the size and polarity of the drug would influence the EE of the drug delivery system. In particular, smaller drugs would lead to higher EE and a more polar drug would interact well with a polar core of the structure. While most of the studies mentioned above focus on the entrapment of drugs in the core of the dendritic structure, there are studies which showed increased loading capacity of drugs by loading them with the surface molecules/groups of the dendrimer, thus increasing the EE of the drug carrier system. The surface groups available on the surface of the dendrimers doubles with every generation [32]. However, it is important to note that not all of these surface groups can be conjugated to the drugs due to steric hindrance or back folding of chains into the dendrimer.

7.3.1.2.2 *Electrostatic interaction of dendrimer–drug system*

The electrostatic interaction between dendrimers and drugs occurs when there is a large number of ionizable groups present on the surface of the polymers. These groups serve as sites for the electrostatic binding of ionizable drugs. These ionizable groups can also be found in the core of the dendrimers, allowing the same interactions to occur within that core. For example, primary amine groups are found on the surface whereas tertiary amine groups are found at the branching points in the core of full generation PAMAM dendrimers. Ibuprofen, an anti-inflammatory drug, is one of the most commonly used for the complexation with full generation PAMAM. The drug is complexed to the dendrimer via the electrostatic interactions of the weakly carboxyl groups on the drug and the amine groups of the dendrimer. The study [33] estimated that about 40 molecules of ibuprofen can bind with a G4 dendrimer at pH 10.5 which results in a significant enhancement of solubility. However, it is important to note that this enhancement is not solely due to the ionization of the surface amines as it is insufficient to account for the enhanced solubility. Thus, other factors like dendrimer aggregation were considered to explain the significant enhancement in solubility. A study by Dubin and his team showed that binding occurs more easily for larger dendrimers for small carboxyl-terminated dendrimers which do not have any amine groups in the core [34].

7.3.1.2.3 *Conjugation of drugs to polymers*

The conjugation of drugs to polymers occurs via covalent bonds between the drugs and the surface groups of dendrimers. The drugs and dendrimers are connected via hydrolysable bonds or biodegradable linkages and this allows for a more controlled drug release rate.

7.3.2 Solid Lipid Nanoparticles

SLNs, as the name suggests, are made from solid lipids. They usually have a diameter of approximately 50 to 1000 nm and this measurement can be obtained via the dynamic light scattering (DLS) technique [35]. They are prepared by the emulsion of solid lipids. An advantage of using SLNs instead of polymeric nanocarriers is that it can be produced using the high-pressure homogenization identical to parenteral O/W emulsions which is widely used in the pharmaceutical industry. Thus, it would be easy to produce SLNs on a large scale for pharmaceutical applications as the existing lines can be utilized. They also exhibit better potential for a more controlled release of drugs as compared to nanoemulsions of liquid lipids due to their solid matrix [36].

7.3.2.1 Factors affecting loading capacity (EE) of lipids

There are four main factors which affect the loading capacity in a drug: (1) the solubility of the drug in the melted lipid, (2) the miscibility of drug melt and lipid melt, (3) chemical and physical structure of the solid lipid matrix, and (4) the polymorphic state of lipid material. In order to obtain a sufficiently high loading capacity, the solubility of the drug in the lipid melt must exceed the minimum solubility required to solubilize. This is because the solubility will decrease as it cools and this might be even lower in solid lipid. The solubility in the lipid melt can be enhanced by adding solubilizers. It can also be enhanced by using mono- and diglycerides as matrix materials. The chemical properties of the lipid such as its crystallinity is an important factor as a highly crystalline structure would lead to drug expulsion [37]. Complex lipids that contain differing chain lengths of fatty acids would result in the formation of less perfect crystals with ample space to accommodate drugs [35]. The crystal structure thus influences the retention time of the drug in the crystal structure as it determines how firmly the drugs are encapsulated in it. Lastly, the polymorphic state of the lipid material also influences the incorporation of drugs in the crystal structure as the recrystallization of the nanoparticle lipid is different from the crystallization of bulk lipid. The space available for the incorporation of drugs differs as well. In general, bulk transformation would normally promote drug expulsion.

7.3.2.2 Specific SLN interaction with DOX

DOX is one of the most utilized anticancer drug for many types of cancers [38]. It is known to have low penetration and exhibits poor distribution in solid tumors. Being weakly basic, it is protonated in the vicinity of a tumor which is slightly acidic. This causes the cellular uptake to decrease [39]. Thus, due to these properties, the application of DOX is limited clinically. To tackle these limitations, DOX is often used with a drug carrier. However, due to its hydrophilic nature, its EE is usually fairly low [40] when incorporated with drug carriers such as SLNs and polymeric micelles. Thus, specific chemical moieties and interactions have been investigated to increase its EE and to widen its clinical applications. One of the suggested method is by lipophilic contra-ion pairing.

7.3.2.3 Doxorubicin–docosahexaenoic acid (DHA) interactions

Docosahexaenoic acid is a polyunsaturated fatty acid that is known to improve the sensitivity of the tumor cells to chemotherapy when used together with DOX specifically [41]. This is because DHA has been proven to enhance oxidative stress and the subsequent lipid peroxidation of tumor cells [42]. In a recent study [43] DHA was used as it was hypothesized to increase the chemotherapy sensitivity as well as the EE (Fig. 7.5) by the contra-ion pairing theory. The study showed an improvement in the EE to almost 100% from about 36% (without DHA). It also showed improved cytotoxicity of DOX in SLN. The particle size formed was 100 nm which is in the optimum range. There was also an improvement in the antitumor activity which is in concordance with other studies.

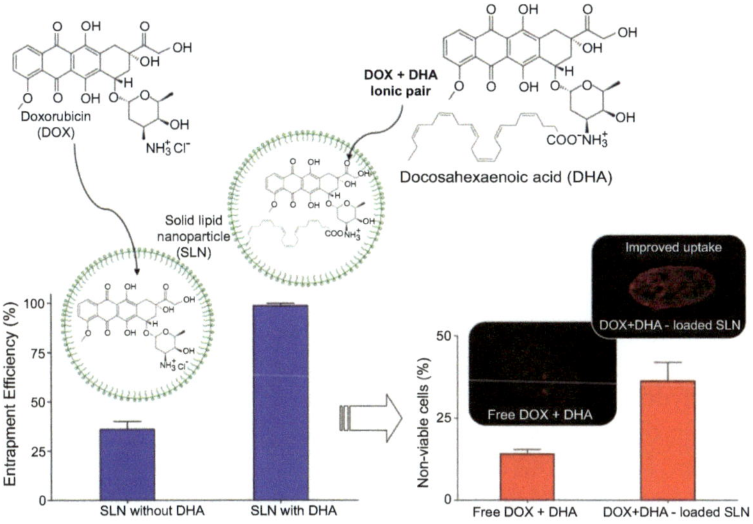

Figure 7.5 The effect of adding DHA on the EE and cell toxicity. Reprinted from Ref. [43], Copyright 2013, with permission from Elsevier.

7.3.2.4 Doxorubicin–alpha-tocopherol succinate (TS)

Similar to the use of DHA, TS is used to enhance the EE of DOX in SLN. Additionally, studies have shown the potential of TS as a material with antitumor activity and is the most effective form of vitamin E that is able to inhibit the proliferation and apoptosis of cancer cells. TS is also expected to enhance the efficiency of DOX [44]. Thus TS can be seen as an alternative for the DHA to improve the EE in SLN. A study by Oliveira and team [40] showed the improved EE of DOX in SLN when TS is added. This can be attributed to the increased lipophilicity of DOX as it forms ion pairing with TS which was similar to the findings with the DOX/DHA combination mentioned earlier.

From the two studies mentioned, we can safely conclude that the contra-ion pairing approach to improve EE of DOX in SLN is very effective. Thus, it is possible to apply these findings to enhance the EE of drugs with similar chemical properties in SLN which are also weakly basic and hydrophilic in nature.

7.3.3 Polymeric Micelles

Polymeric micelles are known for their unique core shell structure which is good for the encapsulation of hydrophobic drugs. Micelles can be formed when a polymer contains both hydrophobic and hydrophilic components. The hydrophilic part of the micelles generally has a brush-like structure and this prevents biological invasion and minimizes protein adsorption. The core–shell structure also allows for a max loading capacity and the ability to carry these drugs with prolonged circulation times. However, improvements in drug loading efficiency, stability in blood and facile transport into the cell membrane is required before it can be used clinically.

The structure and nature of polymeric micelles is similar to that of biological transport vehicles such as viruses and lipoproteins which have a core–shell structure and are generally nanosized [45]. The hydrophilic shell of the micelle would lead to long circulation time in the body as it reduces the uptake by the reticuloendothelial system (RES). Poly(ethylene glycol)–poly(lactic acid) (PEG-PLA) is one of the FDA approved polymeric micelles that is used to encapsulate an anticancer drug, Genexol PM for clinical trials. As such, similar studies on PEG-polyester–based micelles have been conducted as they are seen to possess the potential of being a good candidate for a drug carrier. A series of studies [46] on poly(ethylene glycol)–poly(ε-caprolactone) (PEG-PCL) micelles presented a better loading of the different anticancer drugs when compared to the PEG-PLA micelles.

7.3.3.1 Formation of micelle and encapsulation interactions

Polymeric micelles are generally formed due to the aggregation of the copolymers that are made up of hydrophilic and hydrophobic blocks. These hydrophobic blocks will form the core of the micelle, away from the water, whereas the hydrophilic component will form the shell structure, interacting with the water molecules in the surroundings. When this occurs in the presence of hydrophobic drugs, the hydrophobic core will trap the hydrophobic drugs, reducing its interaction with the hydrophilic surroundings. Thus, these drugs are held in the core via weak hydrophobic interactions [47, 48] and the EE of the drugs are mainly dependent on the encapsulation method. There have been multiple efforts to improve the EE of polymeric micelles by employing various strategies like balancing the hydrophobic–hydrophilic components of the copolymer, varying the hydrophobic architecture of the micelle or by crosslinking and conjugation. However, these strategies did not produce the desired improvement in EE. Consequently, other strategies which focused on the interactions within these polymeric micelles were developed and employed. An increase in interactions between anticancer drugs and the cores of the polymeric micelles would lead to greater EE as it provides a greater thermodynamic driving force for drug loading [49]. Examples of strategies to improve these interactions are stereocomplexation [50], host–guest interaction [51], hydrogen bonding [52], electrostatic interaction, and π–π stacking interaction. π–π stacking interaction is the interaction of molecules with π–π conjugated architectures. It is a form of physical interaction and it is one of the main driving forces for the self-assembly of supramoleculars. This π–π interaction is especially useful in the loading of most anticancer drugs as they are usually hydrophobic and made up of a complex aromatic π–π conjugated structure. Thus π–π stacking interaction can be observed between these drugs and the polymeric micelles with π–π conjugated architectures.

7.3.3.2 Enhancing EE via π–π stacking interactions

A study by Liang and team [53] investigated the effect of modifying the terminal group of methoxy poly(ethylene glycol)–poly(ε-caprolactone) (mPEG-PCL) micelles by adding small molecules of cinnamic acid (CIN), 7-carboxymethoxy coumarin (COU), and chrysin (CHR) with different π–π conjugated architectures. The study showed a significant improvement in the loading capacity of the modified mPEG-PCL micelles and it was proven that this was due to π–π stacking interactions.

7.3.3.3 Hydrogen bonding interactions and crystallinity of PCL-DOX drug delivery systems

Other factors that affect the EE of DOX in PCL copolymer is the hydrogen bonding between DOX and PCL and the crystallinity of PCL. Higher PCL crystallinity leads to lower EE/drug loading as only the amorphous phase is more likely to encapsulate the drug. In the case of PCL, the higher molecular weight would lead to higher crystallinity and hence lower drug loading capacity. However, longer PCL chains might lead to bigger sizes and thus increase the loading capacity. So, it is important to find a balance between increasing the chain segment while making sure that the molecular weight is not too high, such that it compromises its ability to encapsulate more drugs. On another note, the presence of hydroxyl and amino groups in DOX enables the H-bonding interactions with the carbonyl carbons in PCL and this would enhance the DOX-loading in the core of the micelles [54].

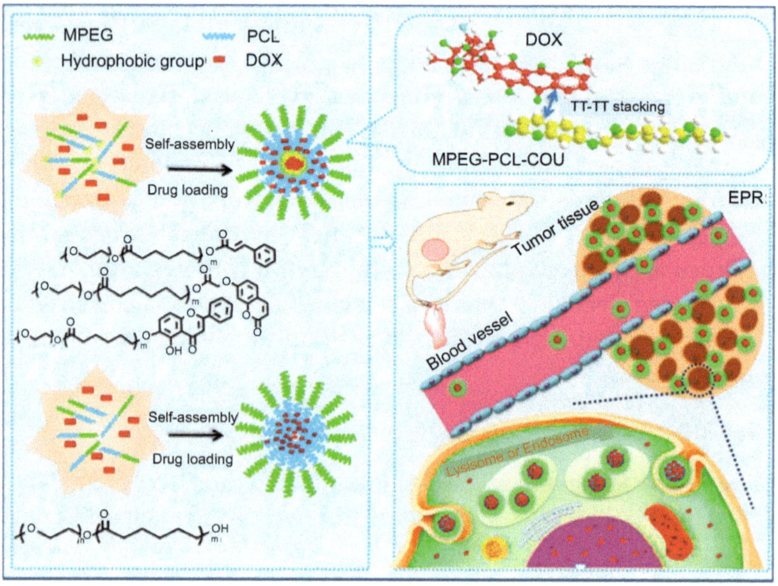

Figure 7.6 Illustration of polymeric micelles with π-conjugated moieties modification for DOX delivery. Reprinted from Ref. [53], Copyright 2015, with permission from Elsevier.

7.3.4 Liposomes

Liposomes are made up of a lipid bilayer and an aqueous core. Lipids are amphiphilic in nature which means that they have both hydrophobic and hydrophilic components in the molecule. When placed in water, the lipid molecule tends to self-assemble due to the presences of hydrophobic and hydrophilic components, resulting in the formation of an aqueous core. This aqueous core is surrounded by a membrane-like lipid bilayer. Liposomes are commonly used as a drug carrier as it has the ability to prolong the half-life of the drug and it reduces toxicity. The biodistribution of the drugs when encapsulated in the liposome is no longer dependent only on the physiochemical characteristics of the drug, but also controlled by the properties of liposomes. Liposomes can be synthetic or natural and are biodegradable and biocompatible, which minimize any toxic side effect they might induce. This characteristic makes them a very important material in the biomedical industry. Additionally, liposomes are able to solubilize both hydrophobic and hydrophilic materials and drugs, making it a valuable carrier in the drug delivery system. Hydrophobic drugs can be encapsulated inside the bilayer of the liposomes while hydrophilic drugs are encapsulated within the aqueous core or at the bilayer interface. This unique capability of liposome, coupled with the other advantages mentioned earlier, make it the more preferred choice as a drug carrier compared to other drug delivery vehicles such as nanoparticles and micro-emulsions. However, liposomes are known to have high colloidal and biological instability and their EE is relatively low when compared to other forms of drug delivery vehicles.

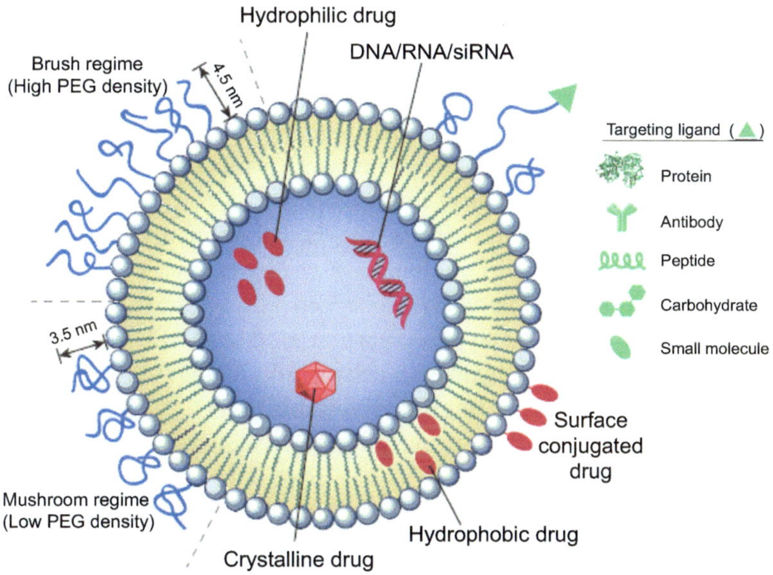

Figure 7.7 Illustration of a liposome.

7.3.4.1 Effect of composition on EE of hydrophilic drugs

The membrane composition of the lipid bilayer affects EE and the partitioning of lipophilic drugs as these drugs are encapsulated within the lipid membrane and thus dependent on the drug's solubility in the bilayer. The presence of additives in the lipid bilayer is suggested to affect its properties such as fluidity, thickness, and polarity. The encapsulation of these drugs might be affected as well. The permeability of the liposome is also a contributing factor to the EE, especially of hydrophilic drugs, as the lipid bilayer in the liquid crystalline phase is more permeable to the encapsulated material than the gel state. Cholesterol (Chol) incorporated into the liposomal bilayers can influence the membrane organization, dynamics, and function. Chol also helps to increase the EE as it reduces the rotational freedom of the phospholipid hydrocarbon chain which helps to reduce the loss of hydrophilic materials. It also helps to stabilize the lipid bilayer. The volume of the aqueous phase encapsulated in the core of the liposome also influences the EE of the liposome and the incorporation of Chol may modify the diameter of the vesicle and thus the EE of the drug. In a study by Taylor and team [55], they suggested that the incorporation of 50% of Chol to dimyristoyl-sn-glycero-3-phosphorylcholine (DMPC) liposomes increased the size of the liposomes by 37% and the EE of sodium cromoglycate increased by 112.7%. Even though the incorporation of Chol might lead to higher EE in some studies, it is not actually suitable for all types of drug compounds.

Another factor that affects the hydrophilic drug loading in liposomes is the carbon tail length of the hydrophobic component. A higher EE of carboplatin was observed when a shorter chain lipid was used as this leads to a decrease in membrane thickness which favors the entry of the hydrophilic drug [56]. A longer lipid chain has been suggested to lead to greater van der Waals interaction which promotes cohesion and prevents drug leakage for alendronate-loaded liposomes.

7.3.4.2 Effect of charge

The interaction between the drug and the lipid bilayer membrane has a significant effect on the EE of the drug [57]. The EE of the drug can be enhanced by modifying the electrostatic interaction between the drug and the lipid bilayer. It can be enhanced by using negatively or positively charged lipids to interact with the oppositely charged drugs. For example, the EE of a positively charged drug, cidofovir, was enhanced when it was encapsulated with positively charged liposomes, composed of *N*-[1(2,3-dioleoyloxy)propyl]-*N*,*N*,*N*-trimethylammonium chloride (DOTAP) and cholesteryl 3(beta)-*N*-dimethyl aminoethyl (DC) with Chol, as compared to neutral liposomes prepared with PC and Chol [58].

7.4 Conclusion

In this paper, the factors affecting the EE of various types of drug delivery systems were reviewed. Four types of drug delivery vehicles and the factors affecting their EE were discussed, proving that the strategies employed to enhance the EE of a drug for each system is different. However, if we look closely at these strategies, the factors that affect the EE of each system can be similar to some extent. For example, the EE of some systems can be enhanced by electrostatic interactions between the drug and the carrier. In the case of dendrimers, this can be done by providing a large amount of binding sites on the surface and in the core of the dendrimers whereas in the case of liposomes, oppositely charged lipid would be selected. However, there are some interactions that are specific to the type of carrier, like the π–π stacking interactions that was discussed for polymeric micelles.

References

1. Craik D. J., Fairlie D. P., Liras S., Price D., *Chemical Biology & Drug Design*, 81 (2013) 136–147.
2. Vrignaud S., Benoit J.-P., Saulnier P., *Biomaterials*, 32 (2011) 8593–8604.
3. Janes K. A., Fresneau M. P., Marazuela A., Fabra A., Alonso M. J., *Journal of Controlled Release*, 73 (2001) 255–267.
4. Ahmad Z., Shah A., Siddiq M., Kraatz H.-B., *RSC Advances*, 4 (2014) 17028–17038.
5. Greenwald R. B., Choe Y. H., McGuire J., Conover C. D., *Advanced Drug Delivery Reviews*, 55 (2003) 217–250.
6. Gowda R., Jones N. R., Banerjee S., Robertson G. P., *Journal of Nanomedicine & Nanotechnology*, 4 (2013) doi:10.4172/2157-7439.1000184.
7. Hurley L. H., *Nature Reviews Cancer*, 2 (2002) 188–200.

8. Çağdaş M., Sezer A. D., Bucak S., Liposomes as potential drug carrier systems for drug delivery, In *Application of Nanotechnology in Drug Delivery*, Ali Demir Sezer, Ed., IntechOpen (2014). Available from: https://www.intechopen.com/books/application-of-nanotechnology-in-drug-delivery/liposomes-as-potential-drug-carrier-systems-for-drug-delivery.
9. De Jong W. H., Borm P. J. A., *International Journal of Nanomedicine*, 3 (2008) 133–149.
10. Nii T., Ishii F., *International Journal of Pharmaceutics*, 298 (2005) 198–205.
11. Jevprasesphant R., Penny J., Jalal R., Attwood D., McKeown N. B., D'Emanuele A., *International Journal of Pharmaceutics*, 252 (2003) 263–266.
12. Kesharwani P., Jain K., Jain N. K., *Progress in Polymer Science*, 39 (2014) 268–307.
13. Hawker C. J., Wooley K. L., Frechet J. M. J., *Journal of the Chemical Society, Perkin Transactions*, 1 (1993) 1287–1297.
14. Jansen J. F. G. A., de Brabander-van den Berg E. M. M., Meijer E. W., *Science*, 266 (1994) 1226–1229.
15. Hawker C. J., Frechet J. M. J., *Journal of the American Chemical Society*, 112 (1990) 7638–7647.
16. Miller T. M., Neenan T. X., *Chemistry of Materials*, 2 (1990) 346–349.
17. Mane P. P., Jadhav V. S., Humbe P. V., Hakke S. C., Kale B. B., *IAJPR*, 4 (2014) 3851–3862.
18. Esfand R., Tomalia D. A., *Drug Discovery Today*, 6 (2001) 427–436.
19. Percec V., Cho W. D., Mosier P. E., Ungar G., Yeardley D. J. P., *Journal of the American Chemical Society*, 120 (1998) 11061–11070.
20. Malik N., Wiwattanapatapee R., Klopsch R., Lorenz K., Frey H., Weener J. W., Meijer E. W., Paulus W., Duncan R., *Journal of Controlled Release*, 65 (2000) 133–148.
21. Roberts J. C., Bhalgat M. K., Zera R. T., *Journal of Biomedical Materials Research*, 30 (1996) 53–65.
22. Chen H.-T., Neerman M. F., Parrish A. R., Simanek E. E., *Journal of the American Chemical Society*, 126 (2004) 10044–10048.
23. D'Emanuele A., Attwood D., *Advanced Drug Delivery Reviews*, 57 (2005) 2147–2162.
24. Ballauff M., Likos C. N., *Angewandte Chemie International Edition*, 43 (2004) 2998–3020.

25. Baars M. W. P. L., Meijer E. W., *Dendrimers II: Architecture, Nanostructure and Supramolecular Chemistry*, (2000), F. Vögtle (ed.), Springer Berlin Heidelberg: Berlin, Heidelberg, pp. 131–182.

26. Newkome G. R., Moorefield C. N., Baker G. R., Saunders M. J., Grossman S. H., *Angewandte Chemie International Edition in English*, 30 (1991) 1178–1180.

27. Ooya T., Lee J., Park K., *Journal of Controlled Release*, 93 (2003) 121–127.

28. Liu M., Kono K., Fréchet J. M. J., *Journal of Controlled Release*, 65 (2000) 121–131.

29. Bazile D., Prud'Homme C., Bassoullet M. T., Marlard M., Spenlehauer G., Veillard M., *Journal of Pharmaceutical Sciences*, 84 (1995) 493–498.

30. Kojima C., Kono K., Maruyama K., Takagishi T., *Bioconjugate Chemistry*, 11 (2000) 910–917.

31. Namazi H., Adeli M., *Biomaterials*, 26 (2005) 1175–1183.

32. Karabulut B., Kerimoðlu O., Uðurlu T., *MÜSBED*, 5 (2015) 31–40.

33. Milhem O. M., Myles C., McKeown N. B., Attwood D., D'Emanuele A., *International Journal of Pharmaceutics*, 197 (2000) 239–241.

34. Zhang H., Dubin P. L., Ray J., Manning G. S., Moorefield C. N., Newkome G. R., *The Journal of Physical Chemistry B*, 103 (1999) 2347–2354.

35. Müller R. H., Mäder K., Gohla S., *European Journal of Pharmaceutics and Biopharmaceutics*, 50 (2000) 161–177.

36. Mehnert W., Mäder K., *Advanced Drug Delivery Reviews*, 47 (2001) 165–196.

37. Westesen K., Bunjes H., Koch M. H. J., *Journal of Controlled Release*, 48 (1997) 223–236.

38. Octavia Y., Tocchetti C. G., Gabrielson K. L., Janssens S., Crijns H. J., Moens A. L., *Journal of Molecular and Cellular Cardiology*, 52 (2012) 1213–1225.

39. Trédan O., Galmarini C. M., Patel K., Tannock I. F., *Journal of the National Cancer Institute*, 99 (2007) 1441–1454.

40. Oliveira M. S., Mussi S. V., Gomes D. A., Yoshida M. I., Frezard F., Carregal V. M., Ferreira L. A. M., *Colloids and Surfaces B: Biointerfaces*, 140 (2016) 246–253.

41. Colas S., Mahéo K., Denis F., Goupille C., Hoinard C., Champeroux P., Tranquart F., Bougnoux P., *Clinical Cancer Research*, 12 (2006) 5879–5886.

42. Mahéo K., Vibet S., Steghens J. P., Dartigeas C., Lehman M., Bougnoux P., Goré J., *Free Radical Biology and Medicine*, 39 (2005) 742–751.
43. Mussi S. V., Silva R. C., d. Oliveira M. C., Lucci C. M., d. Azevedo R. B., Ferreira L. A. M., *European Journal of Pharmaceutical Sciences*, 48 (2013) 282–290.
44. Zhang X., Peng X., Yu W., Hou S., Zhao Y., Zhang Z., Huang X., Wu K., *Cancer Letters*, 307 (2011) 174–181.
45. Kazunori K., Glenn S., Masayuki Y., Teruo O., Yasuhisa S., *Journal of Controlled Release*, 24 (1993) 119–132.
46. Sinha V. R., Bansal K., Kaushik R., Kumria R., Trehan A., *International Journal of Pharmaceutics*, 278 (2004) 1–23.
47. Sun H., Guo B., Cheng R., Meng F., Liu H., Zhong Z., *Biomaterials*, 30 (2009) 6358–6366.
48. Kim J. Y., Kim S., Pinal R., Park K., *Journal of Controlled Release*, 152 (2011) 13–20.
49. Bickerton S., Jiwpanich S., Thayumanavan S., *Molecular Pharmaceutics*, 9 (2012) 3569–3578.
50. Kataoka K., Harada A., Wakebayashi D., Nagasaki Y., *Macromolecules*, 32 (1999) 6892–6894.
51. Kataoka K., Matsumoto T., Yokoyama M., Okano T., Sakurai Y., Fukushima S., Okamoto K., Kwon G. S., *Journal of Controlled Release*, 64 (2000) 143–153.
52. Masayuki Y., Mizue M., Noriko Y., Teruo O., Yasuhisa S., Kazunori K., Shohei I., *Journal of Controlled Release*, 11 (1990) 269–278.
53. Liang Y., Deng X., Zhang L., Peng X., Gao W., Cao J., Gu Z., He B., *Biomaterials*, 71 (2015) 1–10.
54. Shuai X., Ai H., Nasongkla N., Kim S., Gao J., *Journal of Controlled Release*, 98 (2004) 415–426.
55. Taylor K. M. G., Taylor G., Kellaway I. W., Stevens J., *International Journal of Pharmaceutics*, 58 (1990) 49–55.
56. Chaudhury A., Das S., Lee R. F. S., Tan K.-B., Ng W.-K., Tan R. B. H., Chiu G. N. C., *International Journal of Pharmaceutics*, 430 (2012) 167–175.
57. Villasmil-Sánchez S., Rabasco A. M., González-Rodríguez M. L., *Colloids and Surfaces B: Biointerfaces*, 105 (2013) 14–23.
58. Ruozi B., Riva G., Belletti D., Tosi G., Forni F., Mucci A., Barozzi P., Luppi M., Vandelli M. A., *European Journal of Pharmaceutical Sciences*, 41 (2010) 254–264.

Chapter 8

Biodegradable Multilayer Capsules for Functional Foods Applications

Hooi Hong Lau,[a] Ece Kilic,[a,b] and Maxim V. Kiryukhin[a]
[a]*Institute of Materials Research and Engineering, Agency for Science, Technology and Research (A*STAR), 2 Fusionopolis Way, Innovis, #08-03, Singapore 138634*
[b]*Food Science and Technology Programme, Department of Chemistry, National University of Singapore, 3 Science Drive 3, Singapore 117543*
kiryukhin-m@imre.a-star.edu.sg

This chapter is intended for readers interested in the areas of encapsulation, targeted delivery, and controlled release and particularly focuses on biodegradable multilayer films aiming applications in functional foods. The unique advantages of multilayer films made of polysaccharides, proteins, peptides, as well as composite materials with incorporated inorganic minerals are described here. The chapter provides insights into the mechanisms behind multilayer biodegradability and clues on the targeted delivery of bioactive components of functional foods.

Polymer Capsules
Edited by Ye Liu and Xian Jun Loh

ISBN 978-981-4774-60-4 (Hardcover), 978-0-429-42873-9 (eBook)
www.jennystanford.com

8.1 Introduction

Over the last century, food scientists have developed numerous nourishing foods with healthy ingredients in addition to safer, longer-term quality, and enriched taste [1]. By the early 1980s, the focus had shifted increasingly on processing foods in such a way that the bioactive components desired for certain biological activities were preserved during their consumption and assimilation [2]. With the growing number of health-conscious people, consumers are more likely now to prefer food products that have added health benefits. With this in mind, a concept of functional foods has been introduced in Japan [3].

Functional foods are usually defined as those that can be consumed as a part of the regular diet but contain biologically active compounds that provide health benefits beyond basic nutritional functions. Examples of these bioactive compounds are probiotics and prebiotics, functional proteins and peptides, and vitamins and polyphenols. Recent research in this area has concentrated on the relationship between bioactive components and their health impact that is often negated by the effects of stomach digestion and poor targeting to the lower gastrointestinal (GI) tract (small and large intestine) [4]. There are several approaches to address this issue.

One is chemical modification of the bioactive components. For example, 1–2 and 1–3 glucosidic linkages are introduced to maltodextrin (prebiotic). These linkages are not cleaved by amylases, hence making maltodextrin less susceptible to salivary and intestinal digestions [5]. Another example of the chemical process approach is conjugation of a functional whey protein, lactoferrin, with branched 20 kDa poly(ethylene glycol) (PEG) resulting in its significantly higher resistance to pepsin proteolysis and hence improvement in the oral delivery of the bioactive component [6].

Another approach is encapsulation of bioactive components in different formulations where they may be covered by shell material or be dispersed in a continuous matrix [7]. The variety in the morphology of these capsules is demonstrated in Fig. 8.1 [8].

Many encapsulation methods have been developed so far. The most commonly applied ones in the food industry are spray drying, fluidized bed coating, liposome entrapment, complex coacervation etc. [9]. All these methods have their own advantages and disadvantages. Spray drying can be used to produce small micro-sized spheres, while fluidized bed coating provides an even coating [10]. Liposome entrapment is favorable as a carrier system when both hydrophilic and lipophilic ingredients are to be carried [11]. Complex coacervation is used to produce small capsules with liquid functional agents. Knowledge of each method and nature of the component to be encapsulated are essential in order to generate an appropriate encapsulation system [10]. Cost, regulation, ease of use, biologically availability, and compatibility are some other important factors that need to be considered carefully in developing a suitable encapsulation/delivery system for certain bioactive agents [7].

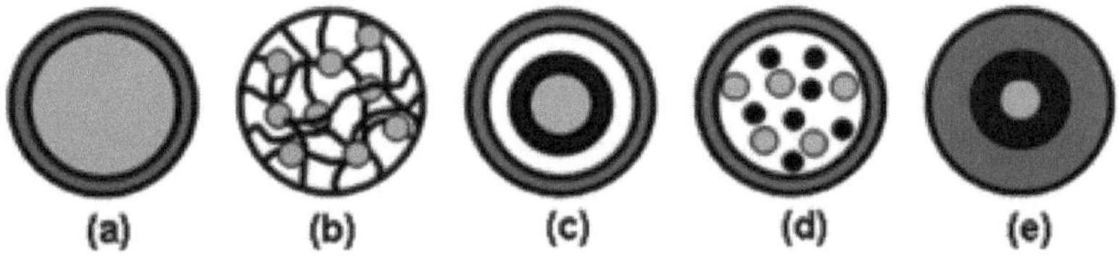

Figure 8.1 Morphologic difference between microcapsules: (a) single-core capsule, (b) dispersed core in continuous matrix, (c) multi-layered capsule, (d) binary core capsule, and (e) single-core–multi-layered capsule. Reproduced from Ref. [8], Copyright 2009, with permission from The Royal Society of Chemistry.

One of very recent advances in this area is encapsulation in microgel particles. Encapsulation of bioactive components (lactoferrin [12], probiotics [13]) in calcium-alginate microgels provides protection from a gastric digestive enzyme, pepsin, and allows the bioactive components' delivery and release in the intestine where the alginate micro-gels degrade in the presence of salts (phosphates) and higher pH (7.5) [12]. Another example is sodium caseinate with added low melting point fats (shortening and margarine widely used in the production of cookies and cakes). It is an effective film-forming material and emulsifier to encapsulate probiotics which could protect it by minimizing heat shock and improve on the survival of the probiotics when spray-dried [14, 15].

To summarize, an ideal delivery system for bioactive compounds should perform several functions, that is, encapsulate an appreciable amount of the compound, protect it from degradation and release it at a specific site of the gastrointestinal tract (GIT) thus maximizing their bioaccessibility [4].

Within the last two decades, an encapsulation technology based on polyelectrolyte multilayer (PEM) films has sprung up [16]. It was first established while studying the electrostatic interaction of counter-charged macromolecules to build multilayered films on a flat surface [17] and then applied around micro-sized particles with subsequent colloidal cores extraction (by dissolution or decomposition) [18, 19]. Nowadays, a number of other intermolecular forces between the components are widely used to form multilayer films, for example, hydrogen bonding or van der Waals forces.

The most important features resulting in PEM as a unique material for capsules shells are its abilities to tune the shell thickness precisely in the range from a few nanometres to tens of micrometres and to tailor multiple functionalities in one shell. For example, providing protection of a delicate cargo from oxidative degradation [20, 21], releasing cargo in response to a number of chemicals (pH, ionic strength, enzymes, etc.), or physical (magnetic field, light, temperature, ultrasound, etc.) triggers [22, 23]. Because of these unique features, researchers have attempted to apply the LbL technique for encapsulation of bioactive compounds [24, 25], fragrances [26], probiotics [27, 28], and drugs [29–31].

Microencapsulation in functional food usually requires biodegradable capsules where the sources are from natural and food grade materials such as polysaccharides (*e.g.*, chitosan (CHI), alginate (ALG), hyaluronic acid (HA), and pectin), proteins (*e.g.*, collagen (COL) and gelatin), and poly(amino acids) such as poly-L-lysine (PLL). When applied for the LbL assembly of multilayer films, the films' thickness may grow either linearly or exponentially, depending on the type of materials and growth conditions. For instance, exponentially growing films were multilayer ones made of PLL-ALG and PLL-HA while linearly growing films were HA-COL [32]. The types of materials chosen to fabricate multilayers, for example, in the form of capsules, are crucial for the desired shells' properties (stability, thickness, and permeability). This review will focus on the types and properties of biodegradable materials for multilayer capsules. Although the LbL technique for encapsulation remains confined to the lab scale, these multilayer capsules are believed to have great potential for bioapplications.

8.2 Polysaccharides-Based Polyelectrolyte Multilayers

Polyelectrolyte multilayers (PEM) fabricated for food bioactive components encapsulation and oral drug delivery usually need to be stable in gastric conditions protecting the encapsulated components, and then degrade in intestinal conditions releasing the encapsulated components. Stability of PEM in a certain condition depends on the types of materials used to fabricate the PEM. The pH, enzymes, or bacterial activity could contribute to PEM degradation. Polysaccharides are one of the most extensively studied natural materials because the side groups (*e.g.*, amino groups from CHI and carboxylate groups from ALG) of the polysaccharides can directly affect their charge density, hydration property and chemical reactivity providing versatility in the construction of biodegradable PEMs. For instance, CHI, ALG, pectin, and HA are among the most reported polysaccharide materials due to their abundant availability, biodegradability, biocompatibility, and versatile properties. Carboxymethyl cellulose (CMC) and dextran sulfate (DEXS) are the other types of polysaccharides typically used as polyanions in PEM fabrication.

CHI is obtained via alkaline deacetylation of chitin which is found in the exoskeleton of crustaceans, in fungal cell walls and other biological materials [33, 34]. ALG is a natural anionic polysaccharide extracted from brown seaweed [35]. Pectin is a complex group of structural polysaccharides found in the cell walls of all land plants [36]. HA was extracted from rooster combs traditionally, and now it is mainly produced via microbial fermentation [37]. CMC is insoluble in acidic stomach fluids but soluble in alkaline intestinal fluids, so it is a good enteric The PEM constituent [38]. DEXS is a derivative of dextran and is synthesized from a number of bacteria [39].

The positive charge of CHI can pair easily with the negative charge obtained from dextran sulfate (DEXS), carboxymethyl cellulose (CMC), ALG, or pectin through the LbL method to form multilayer capsules with a range of thicknesses. These polysaccharides PEM capsules are stable in stomach conditions but degrade in intestinal conditions. The capsule's shell thickness is an important factor to determine the release rate of the encapsulated bioactive components and drugs [40, 41]. The positively charged polysaccharide, chitosan, and three negatively charged polysaccharides (*i.e.*, DEXS, CMC, and ALG) have been reported to be PEM constituents for encapsulation of a model drug (ibuprofen) where the thicker capsules degrade slower and provide less permeability for the drug to be released, that is, the release rate of ibuprofen from the capsules decreases as the capsule's shell thickness increases [41]. Another type of PEM capsule, CHI-pectin multilayer is also able to play a role in protecting aceclofenac (nonsteroidal anti-inflammatory and analgesic colonic drug) from digestion in the gastrointestinal tract [42], delaying the drug's release until it reaches the colon. This prevents side effects of the drug like dyspepsia, abdominal pain, gastritis, ulcerative stomatitis, and nausea that usually occur when the drug is absorbed into the systemic circulation in a short time. The single bilayer CHI-pectin capsules are able to protect the encapsulated drug and be degraded by highly active colonic bacteria [43, 44] in the colon thus releasing aceclofenac at the target site of delivery [42]. Moreover, CHI-pectin multilayer capsules interact strongly with the mucin present in epithelial tissues and hence could further prolong the drug release profile [45].

Polysaccharide solutions at different pH could affect the wetting properties of a surface, which characterize its intermolecular interactions with a liquid. As amino groups from CHI are weak bases and carboxylate groups from ALG are weak acids, the charge density of the side groups could be modulated by adjusting the pH of CHI and ALG solutions. By doing so, a PEM surface with the desired wetting property can be achieved, that is, PEM with chitosan (pH 7 in CHI solution) as a terminal layer was more hydrophobic whereas PEM with ALG (pH 3 in ALG solution) as a terminal layer presented strong hydrophilic behavior [46]. Thus, specific wetting properties of the PEM capsules may be adjusted. This is important in achieving good miscibility with different types of food products.

The charge and charge density of the CHI-ALG multilayer is also an important factor in determining its interaction with human serum albumin (HSA). Crosslinking of CHI-ALG with glutaraldehyde enhances the stability of CHI-ALG films and, simultaneously, enabled the adsorption of HSA. In cellular proliferation, the negatively charged human osteoblast-like cell line (SaOs-2) proliferated significantly in large numbers on PEM with a positively charged CHI as a terminal layer. Both CHI and HA possess good cell adhesion properties. The CHI-HA multilayer with incorporated CHI imidazole-siRNA nanoplexes (NP) functionalize the implant surface and when this specific PEM system degrades, NPs are released over several days and directly taken up by cells localized on top of the implant (Fig. 8.2a) [47]. The biodegradation profiles, sustained and localized drug delivery under physiological conditions, that is, in phosphate buffered saline (PBS), pH 7.4 at 37 °C for 14 days are shown in Fig. 8.2b [47]. Due to the low pKa of CHI, the layer lost most of its positive charges and degraded. However, the CHI degraded slower when the NP was incorporated in the CHI-HA multilayers because the NP strongly interacted with the CHI-HA multilayers.

In encapsulations of food bioactive components, the polysaccharides-based multilayer without crosslinker is sufficient for stomach enzymatic degradation protection. But in cells and proteins adhesion on polysaccharides-based multilayer surfaces, a specific stiffness of surface is required for cell adhesion and proliferation. In this case, the crosslinker is needed.

8.3 Proteins or Poly(Amino Acid)s-Based Multilayers

Proteins and poly(amino acid)s are as widely used materials for biodegradable polymer multilayers as polysaccharides. Proteins and poly(amino acid)s are also abundant and can be obtained from plants and animals. Protein-based multilayers are made of natural macromolecules which can be eventually broken down and resorbed by the body [48]. The most studied protein materials for PEMs are bovine serum albumin (BSA) and poly-L-lysine (PLL). Other types of poly(amino acid)s studied are poly-L-glutamic acid (PLG), collagen (COL), and gelatine (GEL).

Serum albumin is the most abundant protein in the circulatory system and contributes 80% to colloid osmotic blood pressure [49]. It can remain in the body after capsule degradation and the release of encapsulated molecules. BSA, one of the most widely studied albumin, is a globular protein, possessing about 607 amino acids

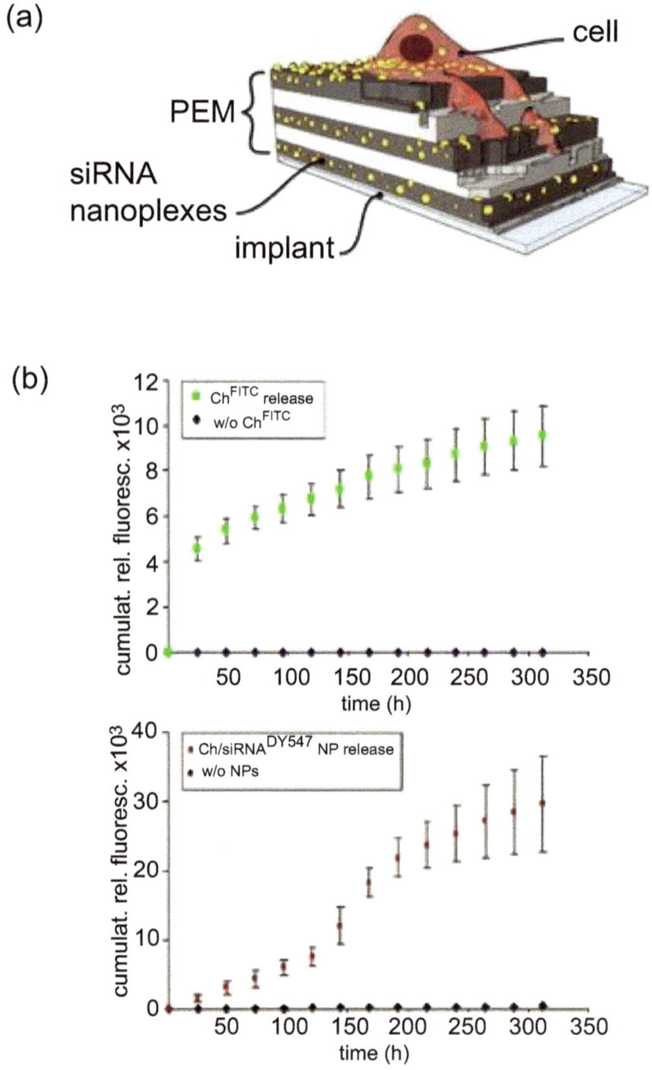

Figure 8.2 (a) An implant surface coated with PEMs incorporating NPs while maintaining stable siRNA complexes. Biodegradation of this specific PEM system allows NPs released over several days and directly taken up by cells localized on top of the implant. (b) Biodegradation of PEM and NPs release. Fluorescein isothiocyanate (FITC) labeled chitosan (ChFITC). DY547 labeled siRNA complexes (siRNADY547). The fluorescence intensity of the supernatant was measured daily and plotted as cumulative values. Reprinted from Ref. [47], Copyright 2013, with permission from Elsevier.

with 17 disulfide bonds and one free cysteine group in its native state. It is highly soluble in water (up to 40 mg/ml) as it contains about 185 ionized groups per molecule at pH 7.0 [50]. Hence, BSA is suitable to the forming of multilayers with different types of compounds like tannic acid (TA) and PLL for various purposes.

For instance, the phenolic groups in tannic acid are excellent hydrogen donors forming hydrogen bonds with BSA carboxyl groups, allowing assembly of enzymatically degradable BSA-TA multilayer capsules [51]. The capsules were not susceptible to trypsin that cleaves peptide chains mainly at the carboxyl side of the amino acids lysine or arginine as their total amount in BSA is 18.72%. However, the capsules degrade under enzymatic activity of α-chymotrypsin which preferentially cleaves peptide amide bonds in phenylalanine tyrosine and tryptophan.

Negatively charged BSA is often paired with positively charged PLL to make multilayers. In one example, they were alternately adsorbed on a polycarbonate porous membrane via the LbL technique [52]. After fabrication of the multilayers, the template was immersed in a mixed solution containing doxorubicin and gelatin. Dissolving the membrane produced multilayer microtubes with gelatin hydrogel remaining in the microtubes' cores. The presence of gelatin greatly improves doxorubicin encapsulation efficiency. These biodegradable BSA-PLL microtubes are responsive towards thermal and enzymatic stimuli. The gelatin hydrogel in the microtubes melts, triggered by near-infrared light releasing its content while the remaining protein-based microtubes are degraded by α-chymotrypsin enzyme completing the doxorubicin release.

Positively charged PLL may bind with negatively charged deoxyribonucleic acid (DNA) to form PEMs. This multilayer is also susceptible toward enzymatic degradation by α-chymotrypsin enzyme solution, serving the purpose of controlled release of therapeutic genes *in vivo* [53].

Multilayers made purely of polysaccharides, proteins, or poly(amino acid)s cannot fulfill all the requirements for many applications. Therefore, they are often combined with synthetic polymers or inorganic materials.

8.4 Composite Multilayers

Some natural macromolecules are inexpensive and abundant, and thus an advantage if cost and time are the limitations [54]. However, it is noted that polysaccharide- and protein-based multilayers are rather soft and weak, and therefore the synthetic polymers are normally incorporated to natural polymers to improve the mechanical properties of polyelectrolyte multilayers for some applications. Good mechanical properties of PEMs are important to ensure that (i) they will be stable indifferent suspending media and (ii) can sustain different stresses (like shear stress).

A combination of natural and synthetic multilayer films like PLL-HA and PSS-PAH is a good construction for long-term and time-bound drug release [55, 56] when the PSS-PAH barrier film is locally degraded by the cells.

With the same materials but different stimuli, Barthes *et al.* had designed complex multilayer films that can respond to stretching [57]. The stretching degree is defined by $\alpha = (l - l_0)/l_0$ where l_0 and l correspond, respectively to the initial and stretched lengths of the silicone sheet. PLL-HA acts as a biodegradable reservoir capped by a synthetic polymer film (PSS-PAH) as a barrier. The glassy PSS-PAH responds to mechanical stretching with α = 30%. The stretching triggers the diffusion of trypsin (TRY) via the openings of the barrier to reach PLL-HA and enzymatic degradation of PLL-HA occurs. With the anticancer drug, paclitaxel as a model remedy, a stretch-responsive drug releasing film is shown in Fig. 8.3.

PLL-HA material is not cytotoxic and is widely used in bio-applications. However, this soft multilayer usually repels cells. As illustrated in Fig. 8.4, the gold nanoparticles coating stiffens the surface of the film significantly and improves the cellular adhesion. Besides, the gold nanoparticles form a semipermeable layer on the film surface, which allows small molecules such as carboxyfluorescein to diffuse rapidly into the film, while diffusion of macromolecules such as enzymes is hindered, preventing the enzymatic degradation of PLL-HA film. At the same time, it allows exchange of metabolites and necessary molecules to take place, either to or from the cell [58].

Another inexpensive strategy to improve the mechanical properties of PEMs would be the incorporation of mineral materials like clays (montmorillonite and halloysite) in the multilayers.

Abundant, inexpensive, and environmentally friendly clays would increase the stiffness of PEM and make them a good surface for cell adhesion and proliferation. Below are some examples of composite PEM materials, mainly for controlled drug release applications.

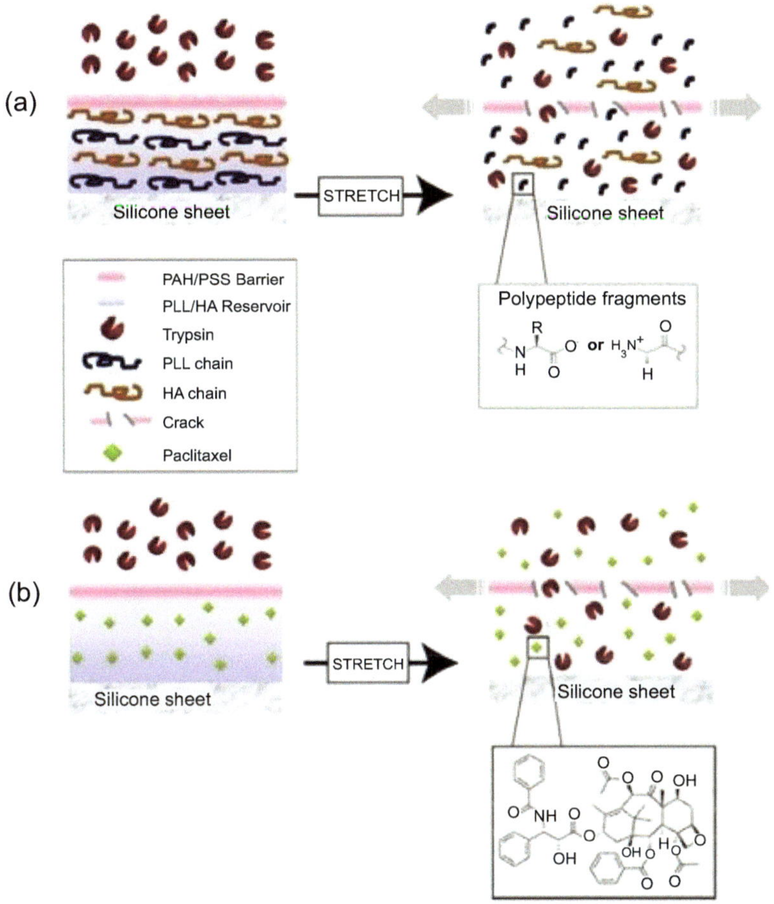

Figure 8.3 (a) Stretch-induced biodegradation of LbL reservoir films. The system is composed of a mechano-sensitive barrier (PAH-PSS) capped onto a PLL-HA reservoir and is put in contact with trypsin (TRY), an enzyme cleaving PLL peptidic chains. Application of a stretch to the system triggers the TRY diffusion via barrier opening and subsequent degradation of the PLL-HA reservoir. (b) Stretch-induced biodegradation of LbL reservoir films applied for the release of embedded drugs. The PLL-HA reservoir is initially loaded with paclitaxel (Taxol) and the drug release in solution is triggered upon stretch. Reprinted with permission from Ref. [57], Copyright 2012, American Chemical Society.

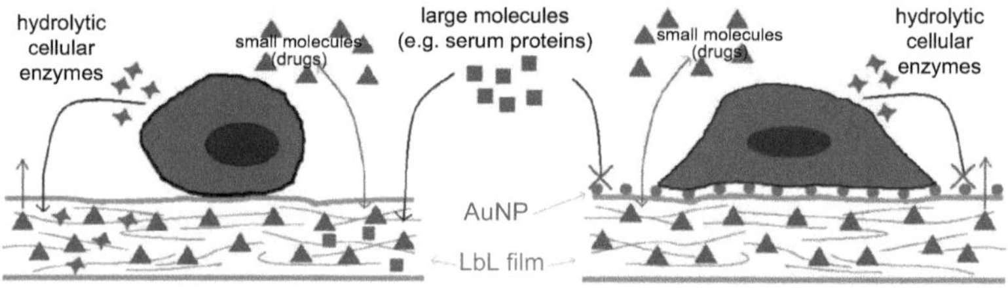

Figure 8.4 A scheme representing cellular adhesion to bare (left) and gold nanoparticles coated (right) PEM films, as well as accessibility of small molecules (drugs) and large (serum proteins, enzymes) to the film. Reprinted with permission from Ref. [58], Copyright 2014, John Wiley and Sons.

Polymer/clay LbL films, poly(*N*-isopropylacrylamide) (PNIPAM) and montmorillonite clay nanosheets respond to multiple stimuli, for example, temperature and salt concentration. An additional pH response was achieved by depositing and crosslinking hybrid, dual-network PNIPAM/clay/PNIPAM/poly(methacrylic acid) (PMAA) multilayers. Both types of films remain stable in a wide pH range and are highly swollen that makes them promising candidates for applications in sensing, actuation, and controlled delivery [59].

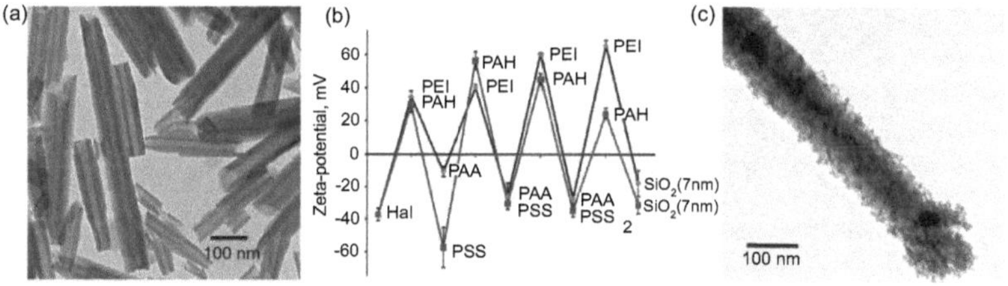

Figure 8.5 (a) TEM images of pristine halloysite from Dragon Mine, Utah, Applied Minerals Inc. Scale bar is 1 mm. (b) Zeta-potential alternation in the processes of LbL assembly of multilayer shells on halloysite clay nanotube (polyacrylic acid (PAA), polystyrene sulfonate (PSS), polyethyleneimine (PEI), polyallylamine hydrochloride (PAH); (c) TEM image of (PAH-PSS-PAH-silica)$_2$ tube coating. Reprinted from Ref. [60], Copyright 2014, with permission from Elsevier.

Besides montmorillonite, halloysite is also a novel type of core nanocontainers with high loading capacity [60]. It consists of 10–15 aluminosilicate layers rolled into a cylinder of a 40–60 nm diameter with an inner lumen diameter of 10–15 nm (Fig. 8.5a). The outer surface of the halloysite tubes is silica, providing a strong negative zeta-potential to the nanocarrier in aqueous dispersions (Fig. 8.5b).

This allows LbL assembly of the PAH-PSS multilayer on the tubes with enhanced coating at the tube ends (Fig. 8.5c). Such capsules provide extended sustained release of drugs over a period from tens of hours to months. This may be useful for oral and transdermal administration of drug formulations.

8.5 Conclusion

Encapsulated food bioactives are beneficial to healthy consumers and even more important to silver age group consumers and young children with poorer digestive systems, as well as patients. LbL deposition is one of the most versatile techniques in encapsulating food bioactive components and drugs. It offers many advantages, such as protecting the encapsulated content during its journey from degradation, release at a specific site of the gastrointestinal tract and thus maximizing bioaccessibility in the body and sustainability of the activity of bioactive components.

A wide range of materials from natural polymers are used in tuning the desired properties of biodegradable multilayer capsules during fabrication to achieve the mentioned purposes of LbL deposition techniques. Polysaccharides, proteins, and poly(amino acid)s are common constituents of multilayers for capsules with lower mechanical strength. To obtain a higher mechanical strength of multilayers, gold nanoparticles and clays are used as a multilayer constituent to increase their stiffness. This research prospect is promising because of its versatility and economical technique for the specific formulation of multilayer capsules.

References

1. Floros J. D., Feeding the world, *Food Technology*, 62 (2008) 11.
2. Floros J. D., Newsome R., Fisher W., Barbosa-Cánovas G. V., Chen H., Dunne C. P., *et al.*, Feeding the world today and tomorrow: The importance of food science and technology, *Comprehensive Reviews in Food Science and Food Safety*, 9 (2010) 572–599.
3. Functional foods. The European food information council. http://www.eufic.org/article/en/expid/basics-functional-foods/accessed on 05.12.2016 (2006).

4. Onwulata C. I., Microencapsulation and functional bioactive foods, *Journal of Food Processing and Preservation*, 37 (2013) 510–532.
5. Kishimoto Y., Oga H., Tagami H., Okuma K., Gordon D. T., Suppressive effect of resistant maltodextrin on postprandial blood triacylglycerol elevation, *European Journal of Nutrition*, 46 (2007) 133–138.
6. Nojima Y., Suzuki Y., Iguchi K., Shiga T., Iwata A., Fujimoto T., *et al.*, Development of poly(ethylene glycol) conjugated lactoferrin for oral administration, *Bioconjugate Chemistry*, 19 (2008) 2253–2259.
7. McClements D. J., Decker E. A., Park Y., Weiss J., Structural design principles for delivery of bioactive components in nutraceuticals and functional foods, *Critical Reviews in Food Science and Nutrition*, 49 (2009) 577–606.
8. Augustin M. A., Hemar Y., Nano- and micro-structured assemblies for encapsulation of food ingredients, *Chemical Society Reviews*, 38 (2009) 902–912.
9. Wang X., Jiang Y., Huang Q., Encapsulation technologies for preserving and controlling the release of enzymes and phytochemicals, Chapter 6, In *Encapsulation and Controlled Release: Technologies in Food Systems*, Lakkis J. M. (ed.), Blackwell Publishing, USA (2007) pp. 135–147.
10. Oxley J., Overview of microencapsulation process technologies, Chapter 4, In *Microencapsulation in the Food Industry,* Gaonkar A. G., Vasisht N., Khare A. R., Sobel R. (eds.), Elsevier (2014) pp. 35–46.
11. Chun J.-Y., Choi M.-J., Min S.-G., Weiss J., Formation and stability of multiple-layered liposomes by layer-by-layer electrostatic deposition of biopolymers, *Food Hydrocolloids*, 30 (2013) 249–257.
12. Bokkhim H., Bansal N., Grøndahl L., Bhandari B., In-vitro digestion of different forms of bovine lactoferrin encapsulated in alginate micro-gel particles, *Food Hydrocolloids*, 52 (2016) 231–242.
13. Sohail A., Turner M. S., Coombes A., Bostrom T., Bhandari B., Survivability of probiotics encapsulated in alginate gel microbeads using a novel impinging aerosols method, *International Journal of Food Microbiology*, 145 (2011) 162–168.
14. Liu H., Gong J., Chabot D., Miller S. S., Cui S. W., Ma J., *et al.*, Protection of heat-sensitive probiotic bacteria during spray-drying by sodium caseinate stabilized fat particles, *Food Hydrocolloids*, 51 (2015) 459–467.
15. Würth R., Hörmannsperger G., Wilke J., Foerst P., Haller D., Kulozik U., Protective effect of milk protein based microencapsulation on bacterial survival in simulated gastric juice versus the murine

gastrointestinal system, *Journal of Functional Foods*, 15 (2015) 116–125.

16. Decher G., Schlenoff J. B. (eds.), *Multilayer Thin Films: Sequential Assembly of Nanocomposite Materials*, 2nd ed., Wiley-VCH Verlag GmbH & Co. KGaA, (2012) pp. 1–21.
17. Decher G., Hong J. D., Buildup of ultrathin multilayer films by a self-assembly process. 1. Consecutive adsorption of anionic and cationic bipolar amphiphiles on charged surfaces, *Macromol. Symp.*, 46 (1991) 321–327.
18. De Temmerman M. L., Demeester J., De Vos F., De Smedt S. C., Encapsulation performance of layer-by-layer microcapsules for proteins, *Biomacromolecules*, 12 (2011) 1283–1289.
19. Sukhorukov G. B., Volodkin D. V., Gunther A. M., Petrov A. I., Shenoy D. B., Mohwald H., Porous calcium carbonate microparticles as templates for encapsulation of bioactive compounds, *Journal of Materials Chemistry*, 14 (2004) 2073–2081.
20. Shchukin D. G., Shutava T., Shchukina E., Sukhorukov G. B., Lvov Y. M., Modified polyelectrolyte microcapsules as smart defense systems, *Chemistry of Materials*, 16 (2004) 3446–3451.
21. Lomova M. V., Sukhorukov G. B., Antipina M. N., Antioxidant coating of micronsize droplets for prevention of lipid peroxidation in oil-in-water emulsion, *ACS Applied Materials & Interfaces*, 2 (2010) 3669–3676.
22. Antipina M. N., Kiryukhin M. V., Skirtach A. G., Sukhorukov G. B., Micropackaging via layer-by-layer assembly: Microcapsules and microchamber arrays, *International Materials Reviews*, 59 (2014) 224–244.
23. De Cock L. J., De Koker S., De Geest B. G., Grooten J., Vervaet C., Remon J. P., *et al.*, Polymeric multilayer capsules in drug delivery, *Angewandte Chemie International Edition Engl.*, 49 (2010) 6954–6973.
24. Dai Z., Heilig A., Zastrow H., Donath E., Mohwald H., Novel formulations of vitamins and insulin by nanoengineering of polyelectrolyte multilayers around microcrystals, *Chemistry: A European Journal*, 10 (2004) 6369–6374.
25. Rivera M. C., Pinheiro A. C., Bourbon A. I., Cerqueira M. A., Vicente A. A., Hollow chitosan/alginate nanocapsules for bioactive compound delivery, *International Journal of Biological Macromolecules*, 79 (2015) 95–102.
26. Sadovoy A. V., Lomova M. V., Antipina M. N., Braun N. A., Sukhorukov G. B., Kiryukhin M. V., Layer-by-layer assembled multilayer shells

for encapsulation and release of fragrance, *ACS Applied Materials & Interfaces*, 5 (2013) 8948–8954.

27. Priya A. J., Vijayalakshmi S. P., Raichur A. M., Enhanced survival of probiotic lactobacillus acidophilus by encapsulation with nanostructured polyelectrolyte layers through layer-by-layer approach, *Journal of Agricultural and Food Chemistry*, 59 (2011) 11838–11845.
28. Priya A. J., Vijayalakshmi S. P., Raichur A. M., Enhanced viability of probiotic saccharomyces boulardii encapsulated by layer-by-layer approach in ph responsive chitosan–dextran sulfate polyelectrolytes, *Journal of Food Engineering*, 136 (2014) 1–8.
29. Pastorino L., Erokhina S., Soumetz F. C., Bianchini P., Konovalov O., Diaspro A., *et al.*, Collagen containing microcapsules: Smart containers for disease controlled therapy, *Journal of Colloid and Interface Science*, 357 (2011) 56–62.
30. Radhakrishnan K., Thomas M. B., Pulakkat S., Gnanadhas D. P., Chakravortty D., Raichur A. M., Stimuli-responsive protamine-based biodegradable nanocapsules for enhanced bioavailability and intracellular delivery of anticancer agents, *Journal of Nanoparticle Research*, 17 (2015) 1–12.
31. Liu F., Kozlovskaya V., Zavgorodnya O., Martinez-Lopez C., Catledge S., Kharlampieva E., Encapsulation of anticancer drug by hydrogen-bonded multilayers of tannic acid, *Soft Matter*, 10 (2014) 9237–9247.
32. Zhang J., Senger B., Vautier D., Picart C., Schaaf P., Voegel J. C., *et al.*, Natural polyelectrolyte films based on layer-by-layer deposition of collagen and hyaluronic acid, *Biomaterials*, 26 (2005) 3353–3361.
33. Di Martino A., Sittinger M., Risbud M. V., Chitosan: A versatile biopolymer for orthopaedic tissue engineering, *Biomaterials*, 26 (2005) 5983–5990.
34. Rinaudo M., Chitin and chitosan: Properties and applications, *Progress in Polymer Science*, 31 (2006) 603–632.
35. Lee K. Y., Mooney D. J., Alginate: Properties and biomedical applications, *Progress in Polymer Science*, 37 (2012) 106–126.
36. Willats W. G. T., Knox J. P., Mikkelsen J. D., Pectin: New insights into an old polymer are starting to gel, *Trends in Food Science and Technology*, 17 (2006) 97–104.
37. Liu L., Liu Y., Li J., Du G., Chen J., Microbial production of hyaluronic acid: Current state, challenges and perspectives, *Microbial Cell Factories*, 10, 99. (2011)

38. Feddersen R. L., Thorp S. N., Sodium carboxymethylcellulose, *Industrial Gums.Polysaccharides and Their Derivatives*, 3rd ed, Chapter 20, Academic Press Inc., (1993) pp. 537–578.

39. Bhavani A. L., Nisha J., Dextran - the polysaccharide with versatile uses, *International Journal of Pharma and Bio Sciences*, 1 (2010) 569–573.

40. Antipov A. A., Sukhorukov G. B., Donath E., Mohwald H., Sustained release properties of polyelectrolyte multilayer capsules, *Journal of Physical Chemistry B*, 105 (2001) 2281–2284.

41. Qiu X., Leporatti S., Donath E., Mohwald H., Studies on the drug release properties of polysaccharide multilayers encapsulated ibuprofen microparticles, *Langmuir*, 17 (2001) 5375–5380.

42. Dharmendra K., Manisha P., Kymonil K., Awasthi S., Aceclofenac delivery by microencapsulation using LbL self-assembly for delayed release, *Pakistan Journal of Pharmaceutical Sciences*, 24 (2011) 495–502.

43. Tozaki H., Komoike J., Tada C., Maruyama T., Terabe A., Suzuki T., *et al.*, Chitosan capsules for colon-specific drug delivery: Improvement of insulin absorption from the rat colon, *Journal of Pharmaceutical Sciences*, 86 (1997) 1016–1021.

44. Hejazi R., Amiji M., Chitosan-based gastrointestinal delivery systems, *Journal of Controlled Release*, 89 (2003) 151–165.

45. Pastorino L., Erokhina S., Ruggiero C., Erokhin V., Petrini P., Fabrication and characterization of chitosan and pectin nanostructured multilayers, *Macromolecular Chemistry and Physics*, 216 (2015) 1067–1075.

46. Acevedo-Fani A., Salvia-Trujillo L., Soliva-Fortuny R., Martin-Belloso O., Modulating biopolymer electrical charge to optimize the assembly of edible multilayer nanofilms by the layer-by-layer technique, *Biomacromolecules*, 16 (2015) 2895–2903.

47. Hartmann H., Hossfeld S., Schlosshauer B., Mittnacht U., Pego A. P., Dauner M., *et al.*, Hyaluronic acid/chitosan multilayer coatings on neuronal implants for localized delivery of siRNA nanoplexes, *Journal of Controlled Release*, 168 (2013) 289–297.

48. Hsu B. B., Myoung-Hwan P., Hagerman S. R., Hammond P. T., Multimonth controlled small molecule release from biodegradable thin films, *Proceedings of the National Academy of Sciences of the United States of America*, 111 (2014) 12175–12180.

49. Carter D. C., Ho J. X., Structure of serum albumin, *Advances in Protein Chemistry*, 45 (1994) 153–203.

50. Giancola C., Sena C. D., Fessas D., Graziano G., Barone G., DSC studies on bovine serum albumin denaturation: Effects of ionic strength and SDS concentration, *International Journal of Biological Macromolecules*, 20 (1997) 193–204.

51. Lomova M. V., Brichkina A. I., Kiryukhin M. V., Vasina E. N., Pavlov A. M., Gorin D. A., *et al.*, Multilayer capsules of bovine serum albumin and tannic acid for controlled release by enzymatic degradation, *ACS Applied Materials & Interfaces*, 7 (2015) 11732–11740.

52. Wu Z., Lin X., Zou X., Sun J., He Q., Biodegradable protein-based rockets for drug transportation and light-triggered release, *ACS Applied Materials & Interfaces*, 7 (2015) 250–255.

53. Ren K., Ji J., Shen J., Construction and enzymatic degradation of multilayered poly-L-lysine/DNA films, *Biomaterials*, 27 (2006) 1152–1159.

54. Pillai O., Panchagnula R., Polymers in drug delivery, *Current Opinion in Chemical Biology*, 5 (2001) 447–451.

55. Garza J. M., Schaaf P., Muller S., Ball V., Stoltz, J.-F., Voegel, J.-C., *et al.*, Multicompartment films made of alternate polyelectrolyte multilayers of exponential and linear growth, *Langmuir, 20 (2004)* 7298–7302.

56. Garza J. M., Jessel N., Ladam G., Dupray V., Muller S., Stoltz, J.-F., *et al.*, Polyelectrolyte multilayers and degradable polymer layers as multicomponent films, *Langmuir*, 21 (2005) 12372–12377.

57. Barthes J., Mertz D., Bach C., Metz-Boutigue M.-H., Senger B., Voegel J.-C., *et al.*, Stretch-induced biodegradation of polyelectrolyte multilayer films for drug release, *Langmuir*, 28 (2012) 13550–13554.

58. Prokopović V. Z., Duschl C., Volodkin D. V., Hyaluronic acid/poly-L-lysine multilayers coated with gold nanoparticles: Cellular response and permeability study, *Polymers for Advanced Technologies*, 25 (2014) 1342–1348.

59. Zhuk A., Robert M., Sukhishvili S., Multiresponsive clay-containing layer-by-layer films, *Acs Nano*, 5 (2011) 8790–8799.

60. Shutava T. G., Fakhrullin R. F., Lvov Y. M., Spherical and tubule nanocarriers for sustained drug release, *Current Opinion in Pharmacology*, 18 (2014) 141–148.

Chapter 9

Essential Oils: From Extraction to Encapsulation

Wen Qi Tan[a] **and Xian Jun Loh**[a,b,c]
[a]*Department of Materials Science and Engineering, National University of Singapore, 9 Engineering Drive 1, Singapore 117576*
[b]*Institute of Materials Research and Engineering (IMRE), 3 Research Link, Singapore 117602*
[c]*Singapore Eye Research Institute, 11 Third Hospital Avenue, Singapore 168751*
lohxj@imre.a-star.edu.sg

There are several beneficial properties of essential oils. They have been used as a disinfectants in the medical industry, as well as preservatives in food. The demand for essential oils is set to keep growing but it is actually a rare commodity, found in minute quantities in plants. It is thus extremely important to ensure that the methods of extraction and encapsulation are highly efficient. The methods used to extract and store essential oils in the industry are discussed in this review paper. Conventional techniques are continuously being improved to keep up with the times, along with the need to minimize the carbon footprint of large-scale production of

Polymer Capsules
Edited by Ye Liu and Xian Jun Loh

ISBN 978-981-4774-60-4 (Hardcover), 978-0-429-42873-9 (eBook)
www.jennystanford.com

essential oils. New and novel methods are being devised in the hope of reducing the extraction time and simultaneously increasing the yield. As essential oils are extremely volatile and fragile, suitable and effective encapsulation methods have been employed to ensure that their quality is not compromised. The advantages and disadvantages to each of these methods are discussed in this chapter.

9.1 Introduction

Essential oils are volatile liquids obtained from plant material and have an aroma distinct to their plant source [1]. As different plants store their oil in different parts of their histological structure, the extraction and synthesis of their respective aromatic liquids vary accordingly. For instance, rose oil is distilled from the petals of the flower; ginger oil is extracted from the roots of the plant [2]. The species and organs of the plant from which the essential oil is extracted from have a direct impact on the resulting yield. The study of the extraction and encapsulation techniques of essential oils is important as these oily substances have a wide range of applications.

9.1.1 Structure of Oil-Secreting Plants

Essential oils are produced by over 17,000 known plant species in the world. They can be found in various plant organs, and are accumulated and stored in the specialized secretory glandules of plants [3]. These secretory structures have a different morphology and functions. There are two main types of such tissues, categorized according to the position of secretion.

The external secretion tissue is, as the name suggests, located outside of the plant, that is on the plant surface. This tissue exhibits exogenous secretion of substances directly outside of the plant. Internal secretion tissue, on the other hand, is located within the plant and undergoes endogenous secretion [3]. The secretion of essential oils (or other materials such as latex, gum, and resins) is a secondary process for plants as it is not essential for their survival, unlike the primary metabolism processes such as photosynthesis and respiration.

Essential oils are highly fragile and volatile compounds, hence the different tissues that specialize in the storage of these oily liquids have been adapted to offer maximum protection against factors external to them. External elements such as light, heat, and moisture will cause essential oils to degrade [2].

External secretion tissue can be found in the petals, stem, and leaves of a plant. An example of such secretion, apart from the aforementioned rose oil, is the essential oil extracted from petals of the lily-of-the-valley (*Convallaria majalis*). The secretion takes place in conical epidermal cells called the epidermal papillae, which are found in such flower petals. Spearmint (*Mentha spicata*) and sage (*Salvia fruticosa*) secrete essential oils from the glandular trichomes of their leaves. These secretory bristles are developed from epidermal cells, and are generally encountered as the accumulation sites of essential oil on the surfaces of stems and leaves [2].

Citrus oil is secreted and stored in internal secretion tissues found on the epicarp (outermost layer) of citric fruits such as orange, grapefruit, and lemon. A spherical secretory pocket holds the essential oils synthesized by the cells surrounding it. Plants such as cinnamon contain cells with intracellular secretion. These are isolated cells that perform exclusive functions, where they accumulate the secreted oily substance inside their vacuoles (a small cavity) and die upon attaining a certain concentration [2].

9.1.2 Chemical Composition and Structure of Essential Oils

Essential oils are complex mixtures comprising volatile compounds of terpenoid and hydrocarbon terpenes (isoprenes) [4]. These compounds are synthesized through different biosynthetic pathways, and with different unique primary metabolic precursors.

9.1.2.1 Terpenes

Terpenes are formed from the condensation of a pentacarbonate unit with two unsaturated bonds of isoprene (2-methyl-1,3-butadiene). Its structure is shown below.

Terpenes are classified into different structural and functional classes according to the number of isoprene units in their structure. Different types of terpenes include hemiterpenes (1 unit), monoterpenes (2 units), and sesquiterpenes (3 units), and so on [4]. Monoterpenes ($C_{10}H_{16}$), in particular, make up more than 80% of the essential oil composition. The second most often found terpenes in essential oil is sesquiterpenes ($C_{15}H_{24}$).

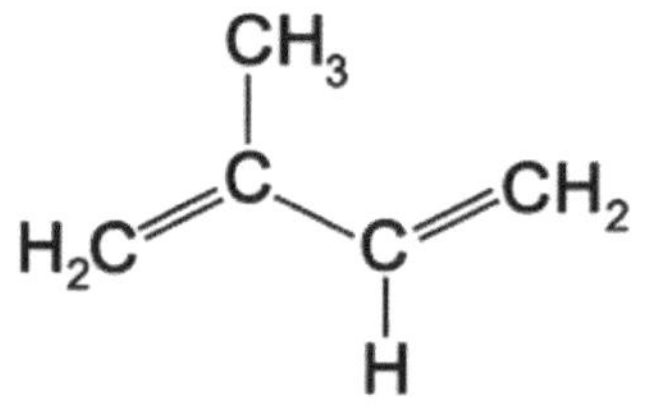

Figure 9.1 Chemical structure of (2-methyl-1,3-butadiene).

The biosynthesis of terpenes involves two universal precursors, namely isopentenyl pyrophosphate (IPP) and dimethylallyl diphosphate (DMAPP) [4].

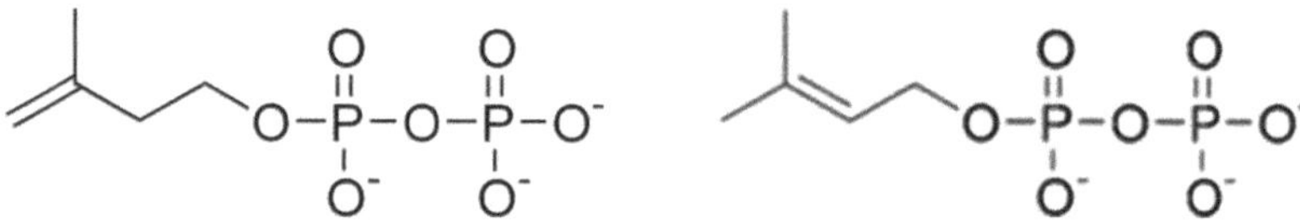

Figure 9.2 Chemical structures of IPP (right) and DMAPP (left).

IPP is synthesized via two routes: the mevalonate pathway (MVA) and the non-mevalonate (MVA-independent) or deoxyxylulose pathway [4]. For the former, mevalonic acid is first formed from the condensation of 3 acetylcoenzyme-A moieties, which then in turn forms IPP. This process usually takes place in the cytoplasm, leading to the formation of most sesquiterpenes.

For the latter synthesis route, 2-*C*-methyl-D-erythritol-4-phosphate (MEP) and 1-deoxy-D-xylulose-5-phosphate (DOXP) resulting from the condensation of glyceraldehyde phosphate and pyruvate are involved. Unlike the MVA pathway, the MVA-independent process occurs mainly in chloroplasts, producing monoterpenes and diterpenes (4 units of isoprene).

The combination of IPP and DMAPP results in the formation of greenly diphosphate (GPP), which is the immediate precursor of monoterpenes. The further condensation of GPP with IPP leads to farnesyl diphosphate (FPP), the immediate precursor of sesquiterpenes. Condensation of FPP with IPP forms geranyl geranyl diphosphate. This is the precursor for other compounds such as phytol, diterpenes, and carotenoids. Heavier terpenes such as diterpenes, may be present in essential oil, but in a much smaller percentage as compared to monoterpenes and sesquiterpenes.

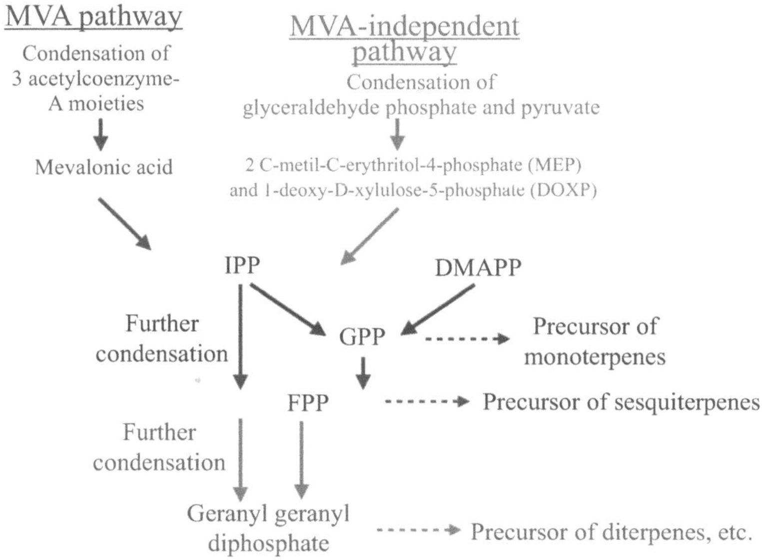

Figure 9.3 Schematic representation of the biosynthesis routes.

9.1.2.2 Terpenoids

Terpenoids, also called isoprenoids, are the oxygenated derivatives of hydrocarbon terpenes [2]. They include alcohols, acids, ketones, phenols, esters, and ethers. It is very difficult to distinguish between terpenes and terpenoids in the chemical composition of essential oils as the two have very similar molecular weights [4].

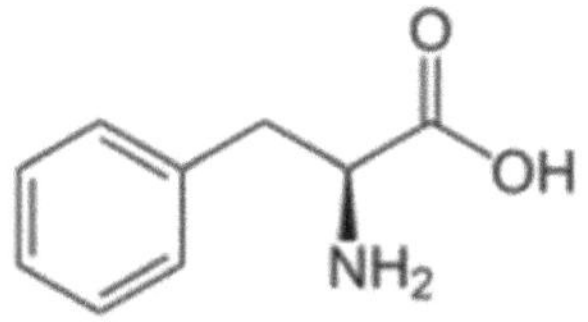

Figure 9.4 Chemical structure of phenylpropanoids.

Vanillin, extracted from vanilla beans, contain a different class of terpenoids called phenylpropanoid and its derivatives (typically phenols or phenol ethers). Phenylpropanoids are made up of one or more benzene rings. Other examples apart from vanillin include essence extracted from cinnamon bark and cloves [2].

9.1.3 Properties and Applications of Essential Oils

The most commonly-known use of essential oils is in perfumes and cosmetics as they provide a pleasant aroma. The psychic comfort and tranquilizing effect from the inhalation of the pleasant aroma is studied in the field of aromatherapy. Aromatherapy is the therapeutic use of these fragrances (essential oils or their individual terpenes) to mitigate or even cure health problems [5]. Cis-jasmonate, which characterizes the aroma of jasmine essence, is suggested to have a calming effect on the brain upon inhalation [5]. This helps to extend both the time and quality of sleep in people who suffer from mild insomnia. Lavender essential oil or linalool (its main component), has similar sedative effects.

Essential oils find many other important applications due to their highly-valued physicochemical characteristics, beyond their aroma. They are known to have anti-microbial effects [6] against a wide spectrum of pathogenic bacteria. As a result, they are widely used in medical industry as disinfectants. Due to their biocidal activities, essential oils are used as cleaning liquid for medical equipment, and aerosol in operating rooms to minimize contaminations. Moving to the pharmaceutical industry, essential oils are included in the composition of various medicinal products in different forms (capsules, emulsions, sprays). Peppermint and eucalyptus oil are examples of essential oils that are used as insect repellents and pest control due to their insecticidal activity. *In vivo* testing on rats have shown that oral administration of a combination of essential oils, including cumin, cinnamon, oregano, fennel, and myrtle was able to enhance insulin sensitivity in type II diabetes [5].

In recent developments, the individual aroma components of essential oils could potentially be used as an anticancer drug, as they show the tendency to induce apoptosis in tumors [5]. Sesquiterpene hydrocarbon, which is a key component of essential oils, has been used in the treatment in one of the most malignant human tumors, glioma. The results show a stimulation of the tumor cell to apoptosis, hence inhibiting their proliferation. Successful development and testing will allow essential oils to complement the chemotherapy treatment used on cancer patients. These are just some examples of the use of such oils in health-enhancement products.

There is also a growing demand for essential oils in the food industry, given their applications in food preservation and packaging. There is a constant need to reduce and eliminate food-borne pathogens, which affects up to 30% of the people in developing countries each year [7]. Coupled with the growing awareness of the negative health impacts from synthetic food preservatives and additives, essential oils as a natural antibacterial additive are being increasingly preferred. In appropriate doses, essential oils demonstrate high efficiency in fighting against bacterial pathogens commonly encountered in meat products [8]. They also appear to be slightly more active against gram-positive than gram-negative bacteria [1]. This is to be expected as the latter has an outer layer of membrane surrounding the cell wall, making them less susceptible to antibacterial activity.

9.2 Extraction Methods

Out of the 17,000 known plant species, only 10% secrete essential oils [2]. It remains a very rare and precious substance; hence extraction methods play an important role in maximizing yield. The conventional method of essential oil extraction is based on water distillation. With technological advancements, new and innovative techniques of extraction have been devised. The latter improves the efficiency of extraction in terms of time required, the amount of yield, as well as product quality. The different methods of extraction will be discussed in this section, followed by an analysis of their respective advantages and disadvantages.

9.2.1 Hydrodistillation

This is the simplest method of extracting essential oil from plants. As its name suggests, this technique involves the distillation of water that is in direct contact with plant materials [5].

The principle of this extraction is based on azeotropic distillation, which is the generation of an immiscible liquid phase with a lower boiling point than the components involved. The plant material is fully macerated in water; the water and the molecules of essential oils form a heterogeneous mixture [2]. The mixture is then treated as a single compound and distilled simultaneously.

Figure 9.5 shows a variation of hydrodistillation, using steam to drive the heating process. It is based on the same principle as hydrodistillation, but without the direct contact between the plant and water. Steam generation occurs outside the distillation vessel, and is introduced to the plant material. Similar equipment is required for both processes, which comprise the condenser and decanter to collect the condensate and to separate the essential oils from water. As the hot steam enters the vessel holding the plant material, it causes the small sacs containing the essential oil to burst. The oil is then carried by the steam towards a condenser kept chilled with cold water running through its external structure. As the steam condenses into a liquid state, the oil and water are separated.

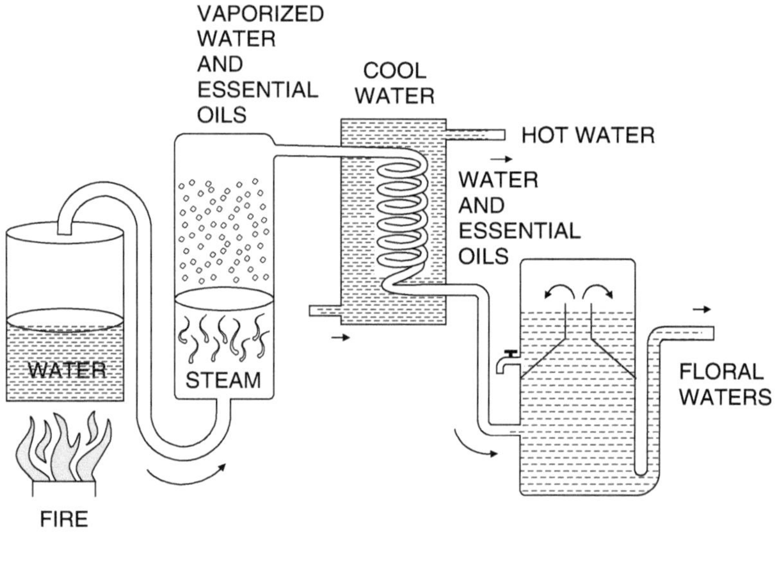

Figure 9.5 Process of steam-distillation to obtain essential oils.

There are several factors to consider when employing this method to extract essential oil from plants. To achieve the best quality, the experimental set-up and procedures are critical. The temperature and pressure of the extraction chamber, for instance, have to be regulated lest they destroy some components of the essential oil. The temperature has to be high enough to vaporize the oil within, but not limitlessly high that it burns the materials. This is especially important for delicate plants such as jasmine and rose.

The advantage of hydrodistillation is straightforward as its main ingredient is water. Apart from being readily available, water is immiscible with the majority of the terpenic molecules of essential oils [2]. This means that the essential oils can be easily separated from condensed water through simple decantation. The plants used in this method do not suffer from any physical damage as they are not clumped and squeezed together during extraction. Hence, this method is suitable for plants that store their essential oil in the petals and flowers, ensuring that their storage unit is not compromised.

However, there are several drawbacks to hydrodistillation. Firstly, it has a long extraction time compared to newer techniques of extraction [2]. It takes on average 3 to 6 hours, and even up to 24 hours for extraction from rose petals. Due to this long extraction duration and prolonged contact, the boiling water may alter and degrade the terpenic molecules in the plant. When steam is utilized instead, there may be less significant chemical alterations observed.

9.2.1.1 Turbo-distillation

The presence of turbines in the turbo-distillation technique improves the process of conventional hydrodistillation [2]. The distillation time is reduced as the turbines allow for more efficient fragmentation and agitation. During turbo-distillation, the plants are soaked in water and steam is circulated through the plant and water mixture. The same aromatic water is continually recycled throughout the entire process. This increases the yield of essential oil as it enables greater recovery of the oil present in the vapor, which would otherwise be discarded. While hydrodistillation is ideal for extraction from delicate petals and flowers, turbo-distillation is suitable for coarse plant material such as barks and roots [9].

9.2.2 Organic Solvent Extraction

Instead of using water as in hydrodistillation, the plant material is immersed in an organic solvent such as methanol, ethanol, or hexane. The solvent first enters the plant to dissolve the oil, and is subsequently removed by distillation under reduced pressure [5]. Hence, this method is employed for plants which are too fragile to be distilled. Tuberose, carnation, and gardenia are some examples. Using organic solvent instead of water minimizes loss of water-soluble fragrance constituents. The solvent used is dependent on the part of the plant being used for extraction [5]. For instance, essential oils within leaves, roots, and fruits are commonly extracted with benzene (with or without a mixture of acetone or petroleum ether). The result is a semisolid concentrate, which is further extracted with absolute ethanol. The second extract is cooled to precipitate the waxes and subsequently filtered.

However, the use of chemical solvent instead of pure water means that the plant species is subjected to chemical alterations. The obtained essential oils may differ significantly from the original essence, especially if the boiling duration and pH level are not well-controlled [2]. The extracted essential oils may also contain residues of the solvent, which runs the risk of polluting the foods and fragrances to which they are added to. There are strict regulations controlling the use of essential oil as flavorings or preservatives in food, to ensure that they post no risk to the health of the consumer [1]. Limonene, menthol, and thymol (to list a few) were previously used as food flavorings, but they have since been removed from the list of permitted flavorings compiled by the European Commission in 2011. They were removed due to their genotoxicity, which refers to the potential of chemical agents damaging the genetic information within a cell and hence leading to genetic mutation.

The disadvantages of this extraction technique can be mitigated by utilizing a combination of an organic solvent with a low boiling point and the steam distillation process [2]. Through this combined method, the risk of contaminating the essential oil can be reduced alongside with the retention of the maximum amount of water-soluble extracts.

9.2.3 Cold Pressing

Cold pressing was devised to extract essential oils from citrus fruit zest [2] such as lemon, lime, and grapefruit. During extraction, the oil sacs of citrus fruit are ruptured and the volatile oils contained within will be released. The oil is then removed mechanically by cold pressing, thereby obtaining a liquid emulsion. Using centrifugation, the oil is subsequently recovered from the emulsion. This technique of extraction is used almost solely for citrus fruits that store their essential oils in tiny pouches. The essence of citrus zest finds a wide range of applications, from home care products (such as dish washing detergents), cosmetics, to the food industry (as flavoring ingredients and additives).

9.2.4 Innovations in Essential Oils Extraction

The conventional techniques of extraction have one major disadvantage related with the thermolability of essential oil components. Due to the high temperatures used in traditional methods of distillation, it is inevitable that the essential oils will undergo chemical alterations. These alterations affect the quality of the oils, and even compromise their safety. Hence, it is important to come up with new ways of extraction to ensure that the chemical composition and natural proportion of essential oils can be maintained to a greater extent.

9.2.4.1 Supercritical fluid extraction (SCFE)

A supercritical state refers to one where fluids are being subjected to well-defined conditions with critical pressure and temperatures [2]. At this state, fluids start exhibiting vastly different properties such as low viscosity, high diffusivity, and density close to that of liquids. In SCFE, carbon dioxide is usually used as the solvent as it is unaggressive on the thermolabile molecules of the plant essence. Its critical point is 72.9 atm (critical pressure) and 31.2 °C (critical temperature) [10], which is easy to attain for the extraction process. Properties such as chemically inert, non-toxic, and non-flammable also make carbon dioxide a more than suitable solvent for this extraction procedure. The purity of carbon dioxide makes it relatively easy to remove its traces from the extracted essential oil through simple depression [2].

The concept of SCFE is based on the recycling of fluid in repeated steps of compression and depression [2]. Carbon dioxide is compressed and heated for it to reach its supercritical state. It is then passed through the raw plant material containing the essential oil to be extracted. Following this, the extract is directed into one or more separators. During this step, the carbon dioxide is gradually depressed, thereby losing its solvent power. The obtained extract is separated from the fluid during depression. The latter is then recycled and used through the repeating cycles.

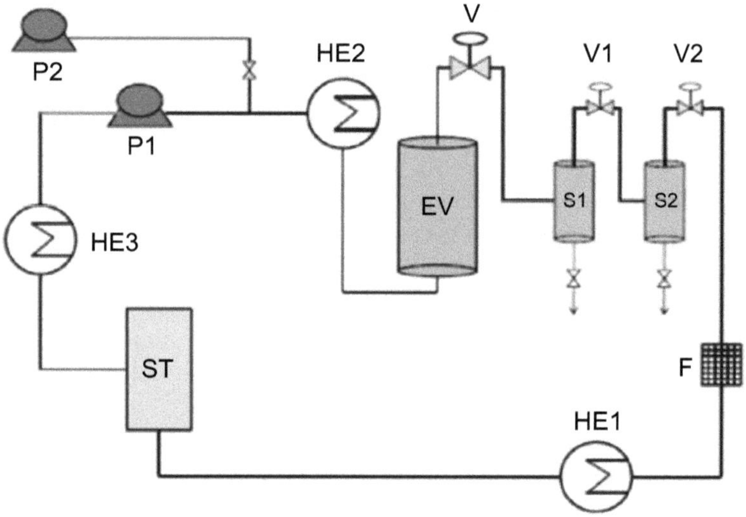

Figure 9.6 A typical scheme for supercritical fluid extraction (SCFE) of plant matrix. P1: CO_2 pump; P2: co-solvent pump; HE1, HE2, HE3: heat exchangers; EV: extraction vessel; S1, S2: separator cells; V, V1, V2: back pressure regulator valves; ST: storage tank of CO_2; F: filter. Reprinted from Ref. [2], Copyright 2015, with permission from Elsevier.

The use of this technique of extraction has grown in the past decades due to the advantages it yields. As mentioned, the use of pure carbon dioxide as a solvent will ensure that the extracted essential oil will not undergo undesirable chemical alterations and hence be of superior quality. It has been demonstrated that carrot essential oil extracted using the SCFE method is more effective against *Bacillus cereus* than that obtained from the conventional method of hydrodistillation [11].

However, this is a costly method of extraction. The equipment used, their installation and maintenance are expensive.

9.2.4.2 Subcritical extraction liquids

The subcritical state is unique and quite distinct from the supercritical state discussed in the previous section. This state is achieved when the pressure is higher than the critical pressure, but the temperature is lower than the critical temperature [2]. Water and carbon dioxide are the most widely used fluids for essential oil extraction. Subcritical extraction using water enables a quick extraction with low operation temperatures. This will prevent any degradation and loss of volatile compounds due to them being exposed to intense heat for a prolonged period of time. The extracted essential oils will be of higher quality as compared to those extracted via conventional techniques. When water is used as a solvent, it generates little to no residue throughout the process. Like the SCFE method, this is an expensive process as it requires very specific equipment and advanced maintenance.

9.2.4.3 Extraction with subcritical carbon dioxide

In order to bring carbon dioxide to its subcritical state, the temperature is ranged between 31 °C and 55 °C, and pressure between 0.5 MPa and 7.4 MPa. Under this controlled condition, carbon dioxide behaves as a non-polar solvent. Using non-polar carbon dioxide prevents the degradation of extracted essential oil as it avoids high temperatures during extraction. Comparing the use of carbon dioxide as a solvent against water, the former appears to ensure the quality of extracts to a greater extent.

9.2.4.4 Ultrasound-assisted extraction (UAE)

Ultrasound allows for the selective extraction of essential oils by accelerating their release from the plant material [2]. This is achieved through a combination of techniques, such as hydrodistillation and solvent extraction, with the assistance of ultrasound. Ultrasound-assisted extraction is especially useful in the extraction of essential oils from seeds. The ultrasonic waves used in this process have a frequency ranging from 20 kHz to 1 MH. Mechanical vibrations of the walls and the membranes of the plant extract will be induced by these waves. Due to these vibrations, there is a rapid release of the oil droplets. They diffuse through the cell walls and are washed out of the cell once the walls are eventually broken by the vibrations. The external walls are usually very thin and are easily ruptured by methods such as sonication. For internal structures, one needs to consider the milling degree of the plant material, which has a direct impact on the yield of essential oil. The size of the plant material has an inverse relationship with the number of cells being exposed to vibrations induced by ultrasonic waves. That is, a reduction in the size of the plant material will increase the number of cells experiencing the vibration motion.

Through the use of ultrasound, the extraction efficiency and duration is improved. The extraction temperature is also reduced, again leading to essential oils of higher quality being obtained. By replacing heat with ultrasound, the essential oil contained within plant materials that are extremely heat-sensitive can be efficiently extracted. Furthermore, the equipment required to carry out UAE is much simpler and hence of lower cost compared to the expensive SCFE.

9.2.4.5 Microwave-assisted extraction (MAE)

Microwaves target the microscopic traces of moisture in plant material and heats them up, causing them to evaporate and expand [12]. This generates intense pressure on the cell well and on the swelling of the plant cell. The pressure pushes the cell wall from inside, and eventually breaks it. The active constituents of the essential oil then seep out to the surrounding solvent. Microwave-assisted extraction operates on methods rather similar to those used in ultrasound.

The choice of solvent in MAE is dependent on the solubility of the target plant material, as well as the microwave-absorbing properties of the solvent [12]. The solvent used is important as it has a direct impact on the eventual yield of the essential oil. The use of an inappropriate solvent will result in an unsatisfactory yield. For instance, the use of hexane in the MAE of essential oil from ginger actually produced less than when using the conventional Soxhlet extraction method [12]. When ethanol is used instead, the former gave a significantly larger yield than the latter. This could be due to the difference in the dielectric properties of the solvent; hexane is unreactive to microwaves and hence does not heat up effectively to rupture the cell walls and release the oil contained within. Ethanol, on the other hand, absorbs microwaves and enhances the entire extraction process.

As with other extraction methods, the extraction time is an important parameter to consider. Intuitively, by increasing the extraction time, one obtains a larger quantity of extract. However, the likelihood of degradation will also increase. Upon longer exposure to microwaves, solvents such as water, ethanol, and methanol may also heat up beyond the optimum temperature, exacerbating the risk of damage to the thermolabile constituents of essential oils [12]. The extraction time varies from plant to plant, and consequently there is no one universally-used timeframe.

The duration of extraction is used to balance out the power of microwaves. While longer extraction time runs the risk of degradation, a combination of low microwave power with longer exposure is considered a wiser approach. In the case of high power coupled with short duration, the rapid rupture of the cell walls occurs. Impurities, along with essential oil, leach out into the solvent under this condition, as it does not allow for selection. When low power is used with a longer extraction time, cell wall rupture takes place at a gradual pace, allowing for selective extraction. The result is an extract of higher purity and quality in the latter process.

There are multiple advantages to essential oil extraction with the assistance of the microwave [13]. Compared to conventional hydrodistillation, there is a significant reduction in the extraction time. Hydrodistillation takes an average of 3–6 hours, as mentioned, while the duration of MAE ranges from a few seconds to a few minutes (up to 20 minutes). Also, much less solvent is required in MAE. The extraction yield is improved due to reduced loss of constituents soluble in the solvent, as well as its increased efficiency in breaking cell walls to obtain the oils within.

9.2.4.6 Solvent-free microwave extraction (SFME)

As the name suggests, there is no solvent used in this technique. It utilizes the microwave heating energy to carry out a dry distillation of a fresh plant at atmospheric pressure [14]. The selective heating of the water content within the plant material will lead to the familiar process of gland swelling and bursting, thereby releasing the essential oil stored within. Similarly, this process requires a much shorter extraction time (approximately 30 minutes [2]) compared to conventional hydrodistillation. As it requires little to no solvent, SFME is considered a green technique in essential oil extraction.

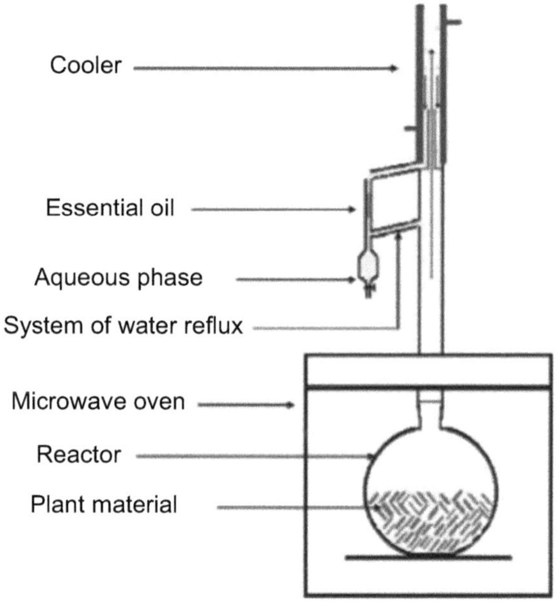

Figure 9.7 Schematic representation of SFME extraction setup. Reprinted from Ref. [14], Copyright 2004, with permission from Elsevier.

9.2.4.7 Microwave hydrodiffusion and gravity (MHG)

In this technique, the plant material is placed in a reversed microwave reactor without any solvent being added [2]. This method was developed by Vian *et al.* in 2008 [15] in which MHG is used to extract essential oil from citrus peels. As discussed earlier, citrus essential oil is usually obtained using the cold pressing of the citrus peels followed by centrifugation. Driven by the need to reduce carbon dioxide emission in conventional methods of extraction, new methods such as MHG are being explored.

The water content within the plant material is heated up via the microwave, which then leads to the rupture of the oil glands and cell receptacles [15]. The oil content is released along with this *in situ* water. As the extract is being diffused out of the plant material via hydrodiffusion, gravity works to draw the extract out of the microwave reactor.

As no solvent will be added, the extraction depends solely on the internal water content of the plant material of interest. Thus, this method is more suitable for fresh plant materials with a minimum of 60% of initial moisture [2].

Similar to all the other novel methods of extraction, MHG extraction requires a shorter extraction (approximately 20 minutes) relative to conventional hydrodistillation. By not utilizing a solvent, this method has the added advantage of reducing waste generated during the process and doing away with post-treatment of waste water [15]. The MHG process emits much less carbon dioxide as it requires only a fraction of the energy consumed by conventional extraction methods. This makes MHG a green and desirable method of extraction. Furthermore, essential oil extracted via this technique has high purity [15].

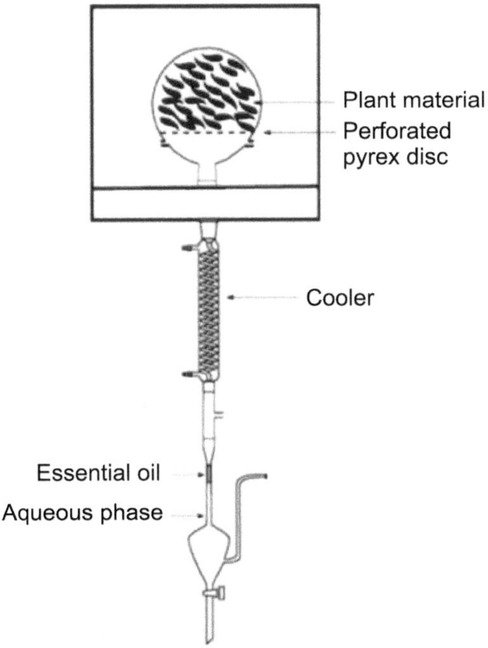

Figure 9.8 Schematic representation of MHG extraction process setup. Reprinted from Ref. [15], Copyright 2009, with permission from Elsevier.

9.2.4.8 Microwave steam distillation (MSD) and microwave steam diffusion (MSDf)

Microwave steam distillation is essentially an improved version of the conventional steam distillation. When compared, MSD performs better in terms of shorter extraction time (approximately 6 minutes) and cleaner features [2]. The MSD process is investigated using lavender flowers [16] irradiated by microwaves. The essential oil within lavender flowers has a higher dielectric loss than the surrounding steam; hence it directly absorbs the microwaves. Microwaves continuously heat the lavender flower while the steam surrounding it remains at a fairly constant temperature. Heating focused on the plant material results in an increase in the extraction rates and efficiency [16]. MSD will be suitable for the routine quality control analysis of essential oils from aromatic plant materials.

Microwave steam diffusion is based on the same principle as MSD, but the vapors flow through the plant material instead. MSDf appears to be more efficient in terms of extraction kinetics [2] where 3 mn is achieved as compared to 6 mn for MSD and 20 mn for conventional steam diffusion. Apart for that, it is also a cleaner and environmentally friendly extraction procedure.

9.3 Methods of Encapsulation

As essential oil is highly susceptible to the damage caused by its external environment, encapsulation is an important procedure to protect it from degradation. Factors such as high temperatures, exposure to UV light, and oxidation all have the potential of compromising the quality of fragile essential oils through the volatilization of the active ingredients within. Hence the different methods of encapsulation will be explored and analyzed in this section.

9.3.1 Encapsulation in Polymeric Particles

9.3.1.1 Nanoprecipitation

Also called solvent displacement, mono-disperse nanoparticles are obtained using the technique of nanoprecipitation. In nanoprecipitation, there are two miscible phases: an organic phase and an aqueous phase [2]. The former is made up of a polymer solution in an organic solvent and the essential oil; the latter contains a non-solvent or a mixture of non-solvents. Non-solvents could be supplemented with one or more naturally occurring or synthetic surfactants. Similarly, the polymer in the organic phase could be either natural or synthetic. Some examples are poly-ε-caprolactone (PCL), poly(lactide) (PLA), and poly(lactide-*co*-glycolide) (PLGA). The first step of nanoprecipitation involves the mixing of the target actives and a polymer in an organic solution to obtain the organic phase [17]. Next, this mixture is added to the aqueous phase containing the surfactant. The resulting dispersion of nanoparticles is vacuum evaporated in order to eliminate the organic solvent, and centrifuged or filtered to obtain the particles. This method of solvent displacement is appropriate for storing hydrophobic materials such as essential oils. The nanoprecipitation technique has been used for the entrapment of curcumin, and a high encapsulation efficiency of 97.5% is reported. Hydrophobicity has been shown to reduce the size of nanoparticles, and in turn increase the efficiency of active molecule entrapment [2].

Nano-encapsulation involves the formation of active-loaded particles, with diameters no larger than 1000 nm [17]. In the case of curcumin-loaded nanoparticles, the particle diameter is about 80.9 nm [17]. This enhances the cellular uptake of the nanoparticles, as well as increases its *in vitro* bioactivity. As compared to free curcumin, the encapsulated churchmen have superior *in vivo* bioavailability.

9.3.1.2 Coacervation

The technique of coacervation is based on the concept of phase separation of one or many hydrocolloids [17]. The complexity of this technique varies vastly when one or multiple polymers are used. The coacervate phase is rich in polymer while the other phase, the equilibrium phase, does not contain polymer [2]. In the case of a simple coacervation, the coacervate phase consists of only one polymer. When more polymers are involved, the process becomes complicated due to the interaction between the two oppositely charged colloids.

When only one polymer is involved, a poor solvent is added to the hydrophilic colloidal solution and the two phases are formed. Complex coacervation, on the other hand, is a spontaneous phenomenon which occurs between two oppositely-charged polymers [2]. A separation between the polymer-rich phase and the aqueous phase will be induced by these charges. Microencapsulation of hydrophobic material is a technique that makes use of the complex coacervation process. Microencapsulation begins with the emulsification of the hydrophobic material in an aqueous solution containing two different polymers. This step is usually conducted at a temperature and pH higher than the gelling and isoelectric point, respectively, of protein [18]. The resulting emulsion is then separated into two liquid phases using electrostatic interaction between oppositely charged polymers. Through controlled cooling of the system below the gelling temperature, the polymer-rich phase will be deposited around the droplets of the hydrophobic material, forming a wall. This wall will harden when a crosslinking agent is added to it, and hard micro-capsules will be obtained. Commonly-used wall materials in complex coacervation are gum arabic and gelatin [2].

9.3.1.3 Spray drying

Since the 1950s, spray-drying encapsulation has been utilized in the food industry [17]. Today, spray drying is arguably the most popular method used to form microparticles as it is easy to conduct at an industrial level. It is an economical process, and is capable of producing particles of high quality. The method consists of liquid atomization into small droplets, which are subsequently dried using hot air and collected as solid particles [2]. These particles are typically spherical, with an average size ranging from 10 μm to 100 μm [17]. Gum arabic, modified starch, and maltodextrin are some of the commonly used wall materials in microencapsulation by spray drying. These materials are selected based on their high solubility, low viscosity, and good emulsifying properties [2]. Often, a combination of these ingredients is used in order to optimize efficiency. For instance, maltodextrin is a popular choice of wall material due to its cheap cost. However, as it has a slightly lower emulsifying capacity, gum arabic or modified starch is used together with it to obtain an effective encapsulation [17].

9.3.1.4 Rapid expansion of supercritical solutions (RESS)

Similar to the conventional methods of extraction, the traditional methods of encapsulation suffer from the disadvantages of requiring a large amount of solvents and solvent residues [2]. To overcome these disadvantages, more environmental-friendly processes based on supercritical fluids are being devised. Supercritical carbon dioxide is used in these processes to design particles suited to the required applications. The reason behind the widespread popularity of supercritical carbon dioxide is its low cost and the little harm it poses to the environment. It also has a low critical temperature, which is useful for the precipitation of thermally-sensitive materials. In the process of RESS, the solutes are dissolved in the supercritical carbon dioxide at high pressures and temperatures [2]. Due to the high temperatures, the solutions expand. The pressure is gradually reduced, causing a fall in the solubility of the solutes to the point where they precipitate. In this process, it is important to ensure that the solutes and the active molecules are soluble in the supercritical carbon dioxide used.

9.3.2 Encapsulation in Liposomes

Liposomes are colloidal particles made up of a membranous system. This system is formed by lipid bilayers, which encapsulates the aqueous cavities within [17]. As it possesses both lipid and aqueous phases, liposomes can be used in the entrapment, delivery, and release of water-soluble, lipid-soluble, and amphiphilic materials. Liposomes are formed by the hydrophobic–hydrophilic interaction between the phospholipids and water molecules. They are categorized according to their size and lamellarity. Liposomes are widely-used as carriers of both hydrophilic molecules in aqueous compartments and lipophilic ones in the bilayers [2]. The use of liposomes for encapsulation of essential oils ensures that the oils remain physicochemical stable. For instance, bioactive agents encapsulated into liposomes can be protected from the digestive juices in the stomach, enhancing their absorption levels in the gastrointestinal tract [17]. Several methods for fabricating liposomes are discussed here.

9.3.2.1 Thin film hydration

The thin film hydration method is also known as the Bingham classical method, developed by Bingham in 1967. This technique is used to form multilamellar vesicles (MLV) with the size ranging up to a few micrometers.

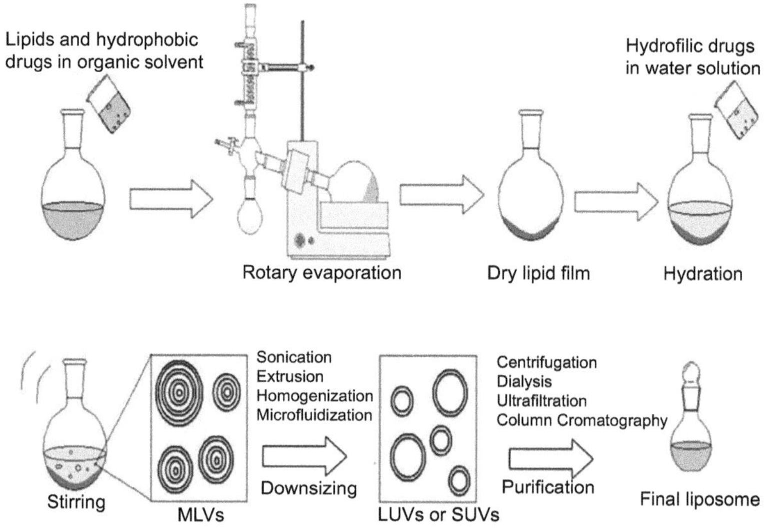

Figure 9.9 Thin film hydration method and size reduction. Reprinted from Ref. [2], Copyright 2015, with permission from Elsevier.

In the formation of MLV using the thin film hydration technique, phospholipids and essential oils are dissolved in an organic phase [17]. A thin phospholipid film is obtained at the bottom of the flask, as depicted in Fig. 9.9, after rotary evaporation of the organic solvent under pressure. The dry lipid film is then hydrated with an aqueous phase. Under agitation via stirring, MLVs will be formed. As stirring is incapable of effective size control, this method will produce vesicles of large size distribution and lamellarity. To obtain liposomes suspension of homogeneous and smaller size, MLVs are converted in small unilamellar vesicles (SUVs) or large unilamellar vesicles (LUVs). This is achieved by subjecting the system to sonication and extrusion, with the former being more widely used.

This is a method with a simple underlying mechanism. However, there are several disadvantages to this method of encapsulation.

Phospholipids and other materials used run the risk of degradation [2]. The liposomes formed using the thin film hydration technique also exhibit low encapsulation efficiency.

9.3.2.1.1 *Extrusion*

As aforementioned, extrusion is used to homogenize the size and lamellarity of MLVs produced by thin film hydration. Extrusion is carried out by running the mixture of MLVs multiple times through a track-etched polycarbonate membrane with pores of varying diameters [2]. The key parameter in this process is the membrane diameter as it has a direct impact on the final liposome size and size distribution. Beside the diameter, the pressure of the extrusion process has to be taken into account as well [19]. The various parameters have to be optimized in order to maximize the encapsulation efficiency. The presence of essential oil in the formation has also been reported to lead to polydispersity [19].

9.3.2.1.2 *Freeze thaw*

The freeze-thaw technique is also a way to homogenize and reduce the liposome size formed by thin film hydration. Through the freeze-thaw process, encapsulation efficiency grows due to the increase in the interactions between the lipid film and the essential oil during the freezing and thawing cycle [2]. In order to form small liposomes, the freezing time should be short with good homogenization [20].

9.3.2.2 Reverse phase evaporation

Reverse phase evaporation is used to obtain larger unilamellar vesicles [17]. This method consists of the preparation of an oil-in-water emulsion obtained by the mixing of the phospholipids organic phase in an aqueous phase [2]. When the organic phase is evaporated, LUVs are produced. In the case where MLVs of heterogeneous sizes and lamellarity are produced, the mixture will be subjected to 3–5 freeze-thaw cycles to transform them into unilamellar vesicles. The size of liposomes obtained from the reverse phase evaporation ranges from 8 μm to 10 μm. These larger vesicles provide high encapsulation efficiency for certain hydrophobic substances such as *E. globulus* (69.2%) and *M. alternifolia* (47.1%).

9.3.2.3 Supercritical fluid technology

9.3.2.3.1 *Modified rapid expansion of supercritical solution technique (RESS)*

In the conventional RESS process as discussed earlier, solutes have to be dissolved in the supercritical solvent. The solution is then rapidly expanded into the atmosphere for the solute to precipitate as microparticles [2]. However, phospholipids have low solubility in supercritical carbon dioxide. This greatly limits the use of conventional RESS in the formation of liposomes. Modifications would have to be done in order to adapt the process for liposome formation. Since phospholipids are insoluble in supercritical carbon dioxide, the phospholipids and essential oil will be pre-dissolved in ethanol. For self-assembly of phospholipids in liposomes, an aqueous phase will be needed. The advantages of using supercritical fluids apply in the modified process as well.

9.3.2.3.2 *Particles from gas saturated solution (PGSS)-drying*

The PGSS-drying process is another type of precipitation method utilizing supercritical fluids. The effectiveness of this method depends on a few parameters. One of them is the gas-to-product ratio (GPR); an increase in GPR will reduce the size of the liposomes formed [2]. The pre-expansion temperature and pressure have a direct relationship with the liposome size, where the decrease in either of the former will result in the corresponding decrease of the latter. The reason behind this could that there is a better atomization when there is an increased carbon dioxide concentration in the emulsion. Conversely, particle size is increased when the concentration of phospholipids increases as that will make the emulsion more viscous, generating an opposition to atomization. When GPR is increased, the entrapment efficiency is decreased as the essential oil undergoes evaporation [2].

9.4 Encapsulation in Solid Lipid Nanoparticles

Solid lipid nanoparticles (SLNs) are nanocarriers which contain lipids which are in the solid phase at room temperature [2]. Triacylglycerols and waxes are examples of the lipid component. SLNs have many advantages: physical stability, good protection of encapsulated material from degradation, and controlled release. Furthermore, the toxicity of the lipid matrix is low as it is made from physiological lipids. These properties make encapsulation of essential oil in SLNs a suitable method for medical and pharmaceutical applications, in cases where the encapsulated essential oil is to be ingested into the body.

9.5 Conclusion

Essential oils have many different types of beneficial properties, from the soothing effect it provides from simply smelling its pleasant aroma to the antibacterial activity it exhibits. There are many applications of essential oils in various industries today. For instance, it is used as a disinfectant in the medical industry as well as preservatives in the food we consume. The demand for essential oils is set to keep growing, as the world looks to harness their natural benefits in order to dispense with harmful chemicals. However, as essential oil is actually a rare commodity, usually found in very small quantities in plant material, it is extremely important to ensure that the methods of extraction and encapsulation are highly efficient. Many methods used to extract and store essential oils in the industry are discussed in this review paper. Conventional techniques are continuously being improved to keep up with the times, along with the need to reduce the carbon footprint of large-scale production of essential oils. New and novel methods are being devised in the hope of reducing the extraction time and increasing the yield at the same time.

As essential oils are extremely volatile and fragile, suitable and effective encapsulation methods will have to be employed to ensure that their quality is not compromised. There are advantages and disadvantages to each of these methods. Ultimately, which method or combination of methods is used is dependent on the type of essential oil one is looking to produce. This is because there is no one-size-fits-all method that can be employed for all types of essential oils.

References

1. Burt S., Essential oils: their antibacterial properties and potential applications in foods—a review, *International Journal of Food Microbiology*, 94 (2004) 223–253.
2. Asbahani A. E., Miladi K., Badri W., Sala M., Addi E. H. A., Casabianca H., *et al.*, Essential oils: From extraction to encapsulation, *International Journal of Pharmaceutics*, 483 (2015) 220–243.
3. Svoboda K. P., Greenaway R. I., Investigation of volatile oil glands of Satureja hortensis L. (summer savory) and phytochemical comparison of different varieties, *International Journal of Aromatherapy*, 13 (2003) 196–202.
4. Zuzarte M., Salgueiro L., Essential oils chemistry, In: *Bioactive Essential Oils and Cancer* (2015) Springer pp. 19–61.
5. Hamid A. A., Aiyelaagbe O., Usman L. A., Essential Oils: Its Medicinal and Pharmacological Uses, *International Journal of Current Research*, 3 (2011) 86–98.
6. Inouye S., Takizawa T., Yamaguchi, H. Antibacterial activity of essential oils and their major constituents against respiratory tract pathogens by gaseous contact, *Journal of Antimicrobial Chemotherapy*, 47 (2001) 565–573.
7. WHO. Food Safety and Foodborn Illness. *World Health Organization Fact Sheet*, 237 (2002).
8. Oussalah M., Caillet S., Saucier L., Lacroix, M. Inhibitory effects of selected plant essential oils on the growth of four pathogenic bacteria: E. coli O157:H7, Salmonella Typhimurium, Staphylococcus aureus and Listeria monocytogenes, *Food Control*, 18 (2007) 414–420.
9. Kumar K. S., Extraction of essential oil using steam distillation, *National Institute of Technology Rourkela* (2010)
10. Reverchon E., Supercritical fluid extraction and fractionation of essential oils and related products, *The Journal of Supercritical Fluids*, 10 (1997) 1–37.
11. Xu L., Recent advances on supercritical fluid extraction of essential oils, *African Journal of Pharmacy and Pharmacology*, 5 (2011) 1196–1211.
12. Mandal V., Mohan Y., Hemalatha S., Microwave assisted extraction—an innovative and promising extraction tool for medicinal plant research, *Pharmacognosy Reviews*, 1 (2007) 11.

13. Golmakani M.-T., Rezaei K., Comparison of microwave-assisted hydrodistillation with the traditional hydrodistillation method in the extraction of essential oils from Thymus vulgaris L, *Food Chemistry*, 109 (2008) 925–930.

14. Lucchesi M. E., Chemat F., Smadja J., Solvent-free microwave extraction of essential oil from aromatic herbs: comparison with conventional hydro-distillation, *Journal of Chromatography A*, 1043 (2004) 323–327.

15. Bousbia N., Vian M. A., Ferhat M. A., Meklati B. Y., Chemat F., A new process for extraction of essential oil from Citrus peels: Microwave hydrodiffusion and gravity, *Journal of Food Engineering*, 90 (2009) 409–413.

16. Sahraoui N., Vian M. A., Bornard I., Boutekedjiret C., Chemat F., Improved microwave steam distillation apparatus for isolation of essential oils: Comparison with conventional steam distillation. *Journal of Chromatography A*, 1210 (2008) 229–233.

17. Fang Z., Bhandari B., Encapsulation of polyphenols – a review, *Trends in Food Science & Technology*, 21 (2010) 510–523.

18. Piacentini E., Giorno L., Dragosavac M. M., Vladisavljević G. T., Holdich R. G., Microencapsulation of oil droplets using cold water fish gelatine/gum arabic complex coacervation by membrane emulsification, *Food Research International*, 53 (2013) 362–372.

19. Celia C., Trapasso E., Locatelli M., Navarra M., Ventura C. A., Wolfram J., *et al.*, Anticancer activity of liposomal bergamot essential oil (BEO) on human neuroblastoma cells, *Colloids and Surfaces B: Biointerfaces*, 112 (2013) 548–553.

20. Ong S. G. M., Ming L. C., Lee K. S., Yuen K. H., Influence of the encapsulation efficiency and size of liposome on the oral bioavailability of griseofulvin-loaded liposomes, *Pharmaceutics*, 8 (2016) 25. doi:10.3390/pharmaceutics8030025

Chapter 10

Semipermeable Polymeric Envelopes for Living Cells: Biomedical Applications

Marina V. Novoselova and Maria N. Antipina
*Institute of Materials Research and Engineering, Agency for Science, Technology and Research (A*STAR), 2 Fusionopolis Way, Innovis, #08-03, Singapore 138634*
antipinam@imre.a-star.edu.sg

Entrapment of living cells in semipermeable membranes or envelopes has already made a significant impact on medicine and various biotechnological applications. The encapsulation of cells isolates and protects them from an outside environment and allows maintaining a normal cellular mode of action due to selective permeability that grants access to vital nutrients and excretes metabolites. The controlled extended delivery of curative aids to the patient by encapsulated cells has been proven as a powerful tool in the treatment of a large number of diseases. Treatment protocols that involve implantations especially benefit from the immunoisolation of implants provided by capsules. Encapsulation also appears to be an effective and low-cost method to deliver human symbiotic

Polymer Capsules
Edited by Ye Liu and Xian Jun Loh

ISBN 978-981-4774-60-4 (Hardcover), 978-0-429-42873-9 (eBook)
www.jennystanford.com

microorganisms curing illnesses or in supporting wellbeing. This chapter summarizes the main requirements in semipermeable membranes suitable for biological agents and highlights the features of living cells that they must possess for successful enveloping via the existing encapsulation technologies. It also comprehensively describes encapsulation strategies for different cell types and illustrates some successful applications of encapsulated cells in biomedicine and advanced foodstuff products.

10.1 Introduction

Inclusion of biologics in physical membranes, which initially was designed and developed as a purely research-oriented tool over six decades ago, has now found its therapeutic usefulness in regenerative medicine [1, 2]. Transplantation of allogenic or xenogeneic cells in the encapsulated form is often a last hope for the successful treatment of disorders when other available therapies fail. This approach results in medicines being transported to the desired physiological sites of action, renewing functionalities that were lost or hampered due to disease or degeneration. The key obstacle to cellular transplantation is a negative host reaction, resulting in the rejection of the transplant. Encapsulation envelops the transplanted cells from the initial hostile environment, providing protection which is critical in order to achieve therapeutic functions following implantation. This so-called immunoisolation of cells is obtained by specially engineered semipermeable coating membranes, which allow nutrients, oxygen, waste, and biotherapeutic products to freely diffuse across boundaries. At the same time, they reject various high molecular weight components of the immune system, that is, antibodies and immunocytes from penetration [3].

As for bacterial cells, probiotics and bacterium-based artificial bioreactors secreting therapeutic products are also gaining benefits from encapsulation in semipermeable biodegradable membranes [4]. Such capsules improve the viability of microorganisms upon storage and on the way to their target organ, the intestine, protecting them from harsh conditions in the stomach and minimizing the undesired immune response of the host [5, 6].

10.2 Properties of Semipermeable Envelopes

The main desired feature of semipermeable membranes for cell encapsulation is high permeation selectivity. In addition, they have to maintain a stable structure over the residence time of the delivered curing aid, be biologically inert and degrade to non-toxic metabolites [3]. Thus, several factors have to be considered when designing such membranes.

10.2.1 Permeation Selectivity

Membrane design has to rely on an important factor, that is, selectivity of membrane permeation. Multiple functions, for example, the efflux of metabolites, wastes, and therapeutic factors, concomitant with the intake of nutrients and oxygen, have to be permitted by the membrane, while detrimental components of the recipient immune system, such as antibodies or cytokines, have to be prevented from entering [3, 7, 8].

The ability of membrane to transport various entities both inside (intake of the substances vital for cell existence) and outside (efflux of metabolites) processes, and so, the rates of these processes define cell fate. Variations in the metabolite efflux rate are quite drastic for different cell types; optimal permeability of membrane has to rely on the selection of cells [1]. Molecular cut-off (MWCO) is the application-dependent maximum molecular weight of a molecule that is allowed to pass through the membrane pores [1, 4], and it confines the desired permeability of the membrane. Figure 10.1 depicts the MWCO for various types of microcapsule membranes. It is easy to see that transplantation favors MWCO at the levels that would allow the sizeable intake of nutrients, but at the same time restricts MWCO values at those levels, which would lead to the rejection of antibodies [4]. The use of implants for restorative therapy does not differentiate the MWCO in relation to either xenogeneic or allogeneic tissue types employed in the engrafting process [1], and the MWCO of the capsules for oral delivery is governed by the efficient membrane passage of metabolites, together with either efflux or extended storage of the altered molecules [1, 9].

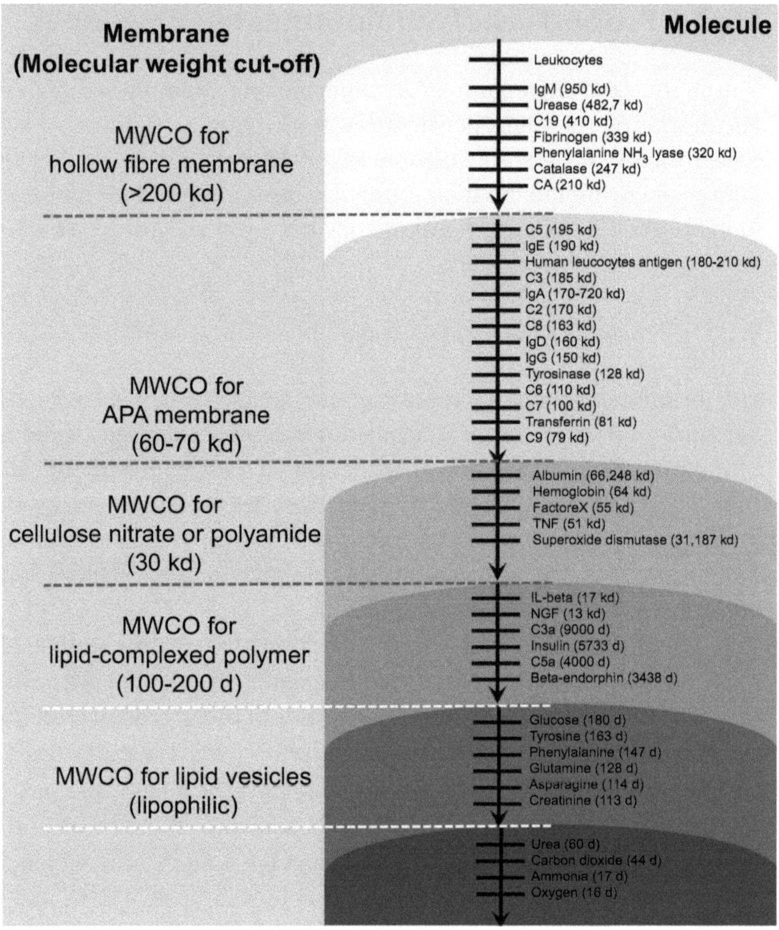

Figure 10.1 MWCO of different types of cell enveloping membranes. Right column contains molecular weights of various relevant biologically active species, such as cells, enzymes, antibodies, complement components, proteins, peptides, and metabolites. Abbreviations: C2–C9 and C19, various components of the complement cascade; Ig, immunoglobulin; IL-1, interleukin 1; NGF, nerve growth factor; APA, calcium alginate beads encapsulated into a bi-layer of poly-L-lysine. Reproduced from Ref. [4] under Creative Commons Attribution License.

The outcome of the cell-derived biomolecules, that is, whether they have to accumulate inside or have to pass through the cell wall, determines the range of the MWCO for bioreactor cultures. Overall, the upper cut-off for solute passage through the membrane is defined by the MWCO, which is closely looked at as a function of membrane permeability for the cell encapsulation devices; values of MWCO ranged from 50 to 150 kDa [7, 9–11]. Macromolecules that could permeate the membrane are typically characterized by the range of molecular weights, and therefore these values for macromolecules could deviate in wide margins. At the same time, species that have exactly the same molecular weight could have quite different properties (size, shape, polarity, etc.), and this can grossly affect their permeation profiles [12]. The introduction of permeability and diffusion coefficients has improved the accuracy of evaluating the process of cross-membrane transport [13]. The latter is driven by a thermodynamic parameter, that is, equilibrium partition coefficient, and by the kinetic parameter, that is, diffusion coefficient [14–17]. These two parameters are determined by a number of factors, for example, type and size of the solute, membrane thickness, and interaction of the solute and the membrane, which define the permeation properties of the membrane. Additionally, a concerted exchange of the substances across the membrane and efficient diffusion are also essential to maintain a healthy membrane environment [14].

10.2.2 Biocompatibility and Biostability

While selecting material for implant manufacture, biocompatibility is the main consideration because it involves the ability of material to comply with the appropriate host response. Thus, the prerequisite of biocompatible material selection for assembling semipermeable membranes is also important to grant an optimal influx and dissemination of the oxygen and nutrition. If this condition is not fulfilled, the surface of the membrane will be covered with the overgrowing fibrous tissue that may lead to a weakened supply of both nutrients and oxygen, and the graft will be rejected due to necrosis of the transplant. Similarly, the biostability of encapsulating material is an essential factor in restraining biodegradation during the life span of the implant [3, 7, 12].

10.2.3 Mechanical Stability

Resistance to an applied force as a parameter of cell mechanical stability depends on the geometry and the chemical composition of the semipermeable membrane. Since a cell transplantation device operates *in vivo* for a significant time, structural elements of the membrane have to possess significant resilience, that is, stiffness (resistance to deformation) and toughness (resistance to fracture). This is necessary to endure the compression forces and stresses caused by shear which are typical in the implantation site [3, 12, 18, 19]. At the same time, fracture, crack, or winding of the membrane during use could lead to the failure of the whole device. More importantly, the rigidity of the membrane would affect the encapsulated cell response and the ability to regenerate.

The implementation of mechanically stable membrane covers was suggested as one approach to alleviate this issue [20]. The loss of mechanical rigidity and integrity of hydrogel-based devices, which are known for their susceptibility to deformation, substantially affects their life span; however, changes in their architecture, such as denser crosslinking and increase of polymer concentration, directly affect mechanical properties by the increasing stiffness of the membrane [21–23].

The use of multilayers, the grafting of the surface of membranes with strengthening polymers, the reinforcement of the membranes with a polyester mesh network and minimization of the device size are reported among other approaches to improve the mechanical strength of semipermeable membranes [3, 24–26].

10.3 Types of Semipermeable Envelopes

The isolation of living cells has been achieved with semipermeable envelopes of different designs (Fig. 10.2):

(i) *Macro-isolation systems*:
  (a) Intravascular devices
  (b) Extravascular diffusive devices (e.g., hollow fibers, tubular membranes, flat sandwich pouches, islet sheets, and islet patches)

(ii) *Micro-isolation systems*: Microencapsulation, typically involving small spherical vesicles ranging in size from 0.3 to 1.5 mm in diameter containing individual cells or small cell masses [27]

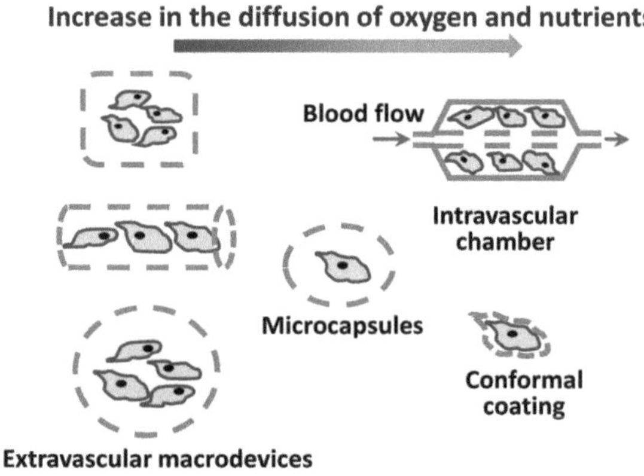

Figure 10.2 Types of capsules designed for immunoisolation of biological cells (semipermeable membranes are schematically shown by dotted lines). Reprinted from Ref. [3], Copyright 2011, with permission from Elsevier.

10.3.1 Macro-isolation Systems

10.3.1.1 Intravascular devices

Intravascular device (Fig. 10.3) includes a shunt that is connected to the bloodstream, which facilitates convenient delivery of nutrients to regulate the physiological cycle. The device contains a piece of microporous tube, which provides unrestricted blood flow through its lumen, and a housing, on which the tissue is fixed. Using vascular anastomoses, the device is implanted into the bloodstream of the recipient. A matrix which facilitates the cell growth into tissue-like substances is provided by the artificial capillary beds, which are perfused with cell culture units. The compact nature and efficiency of such a device are granted by the high surface area to volume ratio, which also allows large numbers of cells to be cultured [27].

The modified diffusion chamber, another popular intravascular device, consists of a core of multiple synthetic capillaries made of polyacrylonitrile and polyvinylchloride (PAN-PVC) copolymer fibers. The fibers are surrounded by islets of Langerhans from normal rats, and the capillaries allow for the transport of low-molecular weight nutrients or blood but not high-molecular weight proteins (e.g., antibodies). When this system was used for short-time glucose maintenance, it was quite efficient, but longer uses have led to the formation of thrombus, even upon co-administration of anticoagulants [28]. Besides, the extended use of diffusion chambers leads to the death of xenogeneic islets due to the severe local accumulation of inflammatory cells and reduced vascularization. These problems have limited the clinical use of such efficient devices [29].

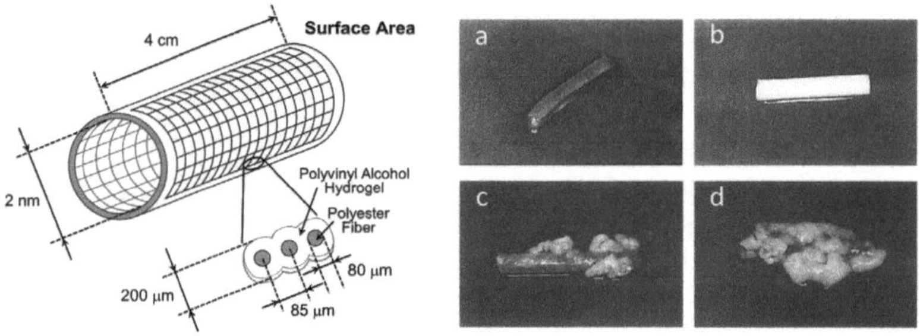

Figure 10.3 Left: Illustration of the schema of MRPT (mesh-reinforced polyvinyl alcohol tube). Right: A Light microphotograph of a PVA (polyvinyl alcohol) hydrogel rod made of 97 % water content intrapentoneally implanted for 18 weeks. Microphotographs of other synthetic polymers: B, polytetrafluoroethylene rod; C, polyethylene rod; D, silicone rod. Reprinted with permission from Ref. [25], Copyright 1992, Lippincott-Raven Publishers.

PAN-PVC–based single coiled membrane with a high flow rate, which prevents adhesion of the cells to the membrane, is a key component of another type of intravascular device that has been used in several applications, such as, tubular membranes. Thrombus formation presented a problem for the early versions of these devices, and attempted alterations in order to reduce this negative effect have led to a restricted diffusion of essential nutrients, thus limiting the use of tubular membranes (Fig. 10.3) [25].

Intravascular devices have been applied in the past to a certain degree, but a set of drawbacks has limited their usage. In short, these devices are thrombogenic, often leading to clot formations in the recipient. The main reason for thrombogenicity and for induced inflammatory responses is believed to be associated with the type of materials used for assembling these devices and those for vascular anastomosis. This affects the viability and long-term stability of the transplanted substances such as islets of Langerhans.

10.3.1.2 Extravascular devices

The advantage of these devices is that they are implanted into the body without the use of anastomosis. Native or recombinant cells are encapsulated within the hollow fibers or, more preferably, within semipermeable membranes that are assembled in the flat, circular or tube-like elements. Minimal invasion is required for implanting these devices, in contrast with intravascular analogs [3]. The implantation site, however, has to be carefully chosen to surround the implanted cells with healthy blood vessels to extend the viability of the implant [1, 3]. Thrombus formation is also eliminated as a disadvantage, as is the need for immunosuppressants [27].

10.3.2 Micro-isolation Systems

Microencapsulation is a preferred method in comparison to the use of macrocapsules/intravascular devices. The morphology of microcapsules attains a spherical nature, which has higher surface-to-volume ratio, enhances selective permeability, guarantees cell viability and assists prompt secretory responses to external stimuli. Microcapsules are mechanically more stable than macrocapsules due to well-defined size parameters. They also have more cost-effective assembly methods than their macro-analogs [30, 31].

Microcapsules can be assembled by a variety of methods, each of them containing steps or options to optimize their fabrication processes [31–43]. Although they prevail over many problems that are typical for their macro-analogs, such systems have their own drawbacks. Micro-immunoisolation systems have been tested in large animal models; however, many of these experiments have used immunosuppressive agents or were performed on spontaneous diabetic subjects [30]. Unfortunately, the lack of experimental control and the skewed consistency of results obtained for these models disallow their proper scientific interpretation and limits their applicability [44].

10.4 Fabrication Methods

Several techniques were developed for the successful encapsulation of cells or cell masses, for example, conformal coating methods, more specifically, complex coacervation, interfacial precipitation, interfacial polymerization, whereby the membrane is positioned in direct contact with the cells [1]. A more stable membrane around the cell mass core can be obtained by thermoreversible gelation, crosslinking, polymerization, etc. Such a core may be designed and assembled to include components, which promote cell survival or specific cell functions, for example, intake of nutrients and trophic factors of the extracellular matrix (ECM) [27, 45].

10.4.1 Conformal Coating

The large size of the cell containing capsules could hamper the diffusion of nutrients and oxygen to the enclosed cells and also emerges as a limiting factor to their potential transplantation sites. For example, a large volume of encapsulated islets prohibits their placement into the portal vein of the liver [3]. Conformal coating of cells with thin surface layers of various compounds have been developed to tackle both problems of larger implants.

Conformal coating can be considered as a special case of microencapsulation when a barrier is formed directly on a small cell mass or a small piece of tissue [1]. This approach has proven to be useful when the amount of implantable cells is maximized while minimizing the device volume [1]. Conformal coating is primarily aimed at improving the transport of nutrients and metabolites between the capsule exterior and enclosed cells, which helps to increase the efficiency of cell packing, thus minimizing the implant size. Conformal coating is useless for single cells, since it ends up with a big membrane material–cell ratio; however, it can be effectively applied for small clusters of cells, where diffusion limitations are significant [1]. The vast majority of conformal coatings reported in the literature so far possess limited stability, tending to dissociate from the cell surface after only a few days [1].

10.4.2 Layer-by-Layer Technique

Multilayered materials assembled via the alternate adsorption of complementary polymeric molecules and/or nanosized species are very promising systems for the development of cell-enclosing envelopes owing to their permeation control and versatility in material selection.

Unlike many other cell encapsulating methods, the layer-by-layer (LbL) shell assembly is performed on a single cell, allowing a smaller total size of the capsule and tailoring the desired functionalities and species to each capsule layer. The LbL coating can be performed on the surface of almost all variety of living cells. The technique has been applied for the encapsulation of microbial cells, including bacteria and yeast cells and mammalian cells, such as blood cells (erythrocytes, echinocyte, and blood platelets), human pancreatic islets, breast cancer cells, and mouse fibroblasts and mesenchymal stem cells [46]. A nanostructured enveloping assembly comprising layers of biological or synthetic polymers and/or inorganic nanoparticles represents an easy way of manipulating the cell surface charge and spatial location. It can also help to construct multicellular tissue-mimicking clusters, protect the cells from aggressive pH and enzymatic conditions, control bacterial metabolism and pathogenicity while providing the encapsulated cells with all necessary nutrients [46].

The polymeric multilayers are generally permeable for ions and molecules with the molecular weight smaller than 1 kDa [47]; thus they fulfill the main requirement of semipermeability to provide the living cells with enough nutrients but rejecting big and potentially harmful molecules, for example, enzymes. Although the diffusion of big molecules trough polymeric multilayer films is still possible, it appears to be a very slow process controlled by the concentration gradient across the shell and can be additionally inhibited via increasing the number of comprised multilayers [48].

Depending on the application, the capsule building blocks can involve synthetic or natural polymers including polysaccharides (alginate, chitosan, dextrans, etc.), proteins, polypeptides, polyphenols, and nucleic acids [49]. Grafted polymers, such as poly-L-lysine-graft-polyethylene glycol and chitosan-graft-phosphorylcholine have also been reported for encapsulation of living cells [50, 51]. Additionally, the layered polymeric assemblies can be decorated and functionalized with various inorganic nanomaterials, for example, magnet nanoparticles [52].

A vast majority of multilayer assemblies used for cell enveloping are formed via electrostatic interactions of oppositely charged polymers or doping additives. The surface of the living cells possesses a negative charge at physiological conditions, thus, the first capsule layer must be cationic (Step 1 in Fig. 10.4). At the next step (Step 2 in Fig. 10.4) a polyanion is adsorbed and multiple alternate adsorptions of the complementary polymers lead to the formation of a multilayer shell. Centrifugation process can be used to remove the uncoupled polymeric molecules. If the cells are too fragile, microfiltration commonly used for encapsulation of oil droplets can be advised [53, 54].

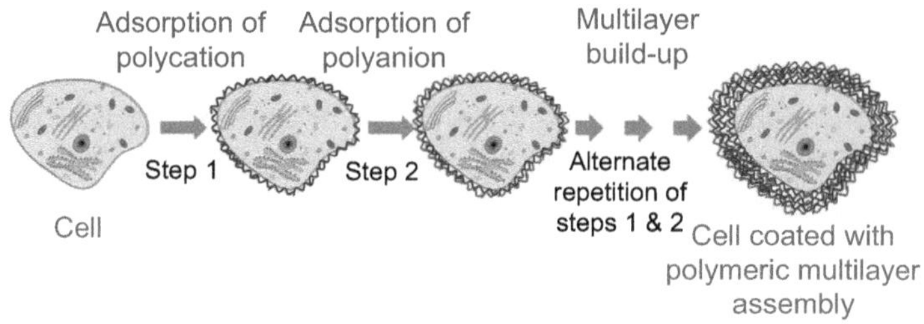

Figure 10.4 General scheme of the polyelectrolyte multilayer assembly on the surface of living cells.

One of the most exploited attractive features of the electrostatically assembled multilayers is their changeable permeability in response to environmental pH [55, 56]. This property is attributed to weak polyelectrolytes whose chargeable groups can be protonated or deprotonated at different values of pH, causing uncompensated positive or negative charges within the assembly [57]. The multilayer films comprising weak polyelectrolytes can change their thickness and significantly increase permeability to macromolecules when the pH shifts from neutral to more acidic or to more alkaline. Extreme pH conditions may even cause such multilayers to disassemble. Applied to the encapsulation of cells, this feature is definitely not desirable if the task is to protect the living species in the stomach and release them in the intestine. Of substantial help here may be the use of strong polyacids and polybases or the exploration of another type of interlayer interactions, such as hydrogen bonding.

A biocompatible polymer, poly(*N*-vinylpyrrolidone) (PVP), was widely studied in pH-responsive hydrogen bonded assemblies as a hydrogen acceptor in combinations with weak polyacids serving as hydrogen donors in their protonated state. The multilayers of PVP and polyacids, such as poly(methacrylic acid) (PMA) [58] and tannic acid (TA) [59], are assembled at pH as low as the pH of the stomach (pH 2) but disintegrate on pH increase up to intestinal (pH 7.4) due to deprotonation of hydroxyl groups responsible for uncompensated negative charges and electrostatic repulsion within the assembly. For example, yeast cells have been encapsulated in PVP/TA multilayered envelopes with an anchoring cationic layer of polyethyleneimine [60].

Another attractive pair for biodegradable multilayer encapsulating assemblies can be selected from the respective arrays of proteins and polyphenols. Some of these capsules are stable to acidic pH and the enzymatic content of the stomach but degrade enzymatically in the intestine. Additionally, the surface of encapsulated cells can be easily decorated with donor serum albumins via the polyphenol linking layer to eliminate the immune response to cell implants [61].

The coating materials and conditions upon the assembly can play a crucial role in preserving cell viability. For instance, the best survival rate (>80%) for the MELIN cell line was achieved using the polydiallyldimethyl ammonium chloride/polystyrene sulfonate polymeric pair. Moreover, the cell viability was affected in buffers containing sodium ions, which were advised to be replaced with potassium ions. The cell viability was also better when cells were allowed to recover for 2 h prior to each bilayer deposition [62].

10.4.3 EMC Formation

Either cores or membranes can fulfill the task of functioning as a synthetic ECM [45]. ECM represents a very complex cellular environment containing a range of solutes, for example, proteins and proteoglycans. The primary role of ECM is supplying structural support and a regulatory environment for mammalian cells, including functions such as cell assembly to form organs and tissues, growth control and regulation, and cell–cell interaction. [63].

Cells can benefit from the addition of ECM components in the organization of the cell mass and the expression of differentiated capsule functions [1]. Most alginate membranes possess a hydrophobic nature, which in general disfavors cell attachment and spreading; therefore, the utilization of synthetic ECM was studied in relation to adherent cells [45].

10.4.4 Thermoreversible Gelation

Cell encapsulation in the hydrogels is a popular and well-studied method. The hydrophilic nature of hydrogels is attractive for biomaterial applications, since it mimics the tissue environment for the cells, promoting transport and enabling cell longevity. In addition, the viscoelasticity of hydrogels favors the use of minimally invasive techniques and small surgical openings, when the injectable matrix is inserted in the human body. Easy incorporation (simple mixing in the matrix before injection) of bioactive molecules or cells is yet another advantage of these systems [64]. A schematic of the typical hydrogel formation process is shown in Fig. 10.5 [65]. Thermosensitive hydrogels are particularly efficient as injectable biomaterial-based matrices because they spontaneously gelate at physiological conditions.

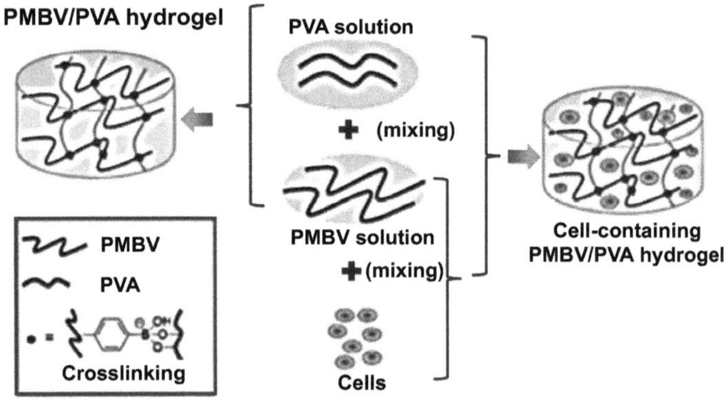

Figure 10.5 Schematic diagram of typical hydrogel formation (poly(2-meth acryloyloxyethylphosphorylcholine-*co*-n-butyl methacrylte-*co*-p-vinylphenyl boronic acid/polyvinyl alcohol) and fabrication of cell-containing hydrogel [65].

In vivo biological signaling to cells, which is supported by natural hydrogels, is vital for cellular proliferation and development, essential features of an efficient implant. Natural hydrogels, however, are prone to certain deficiencies such as variability in physical and mechanical properties as compared to synthetic hydrogels and the presence of impurities that are hard to eliminate; they can substantially compromise the biocompatibility of natural hydrogels.

A valuable alternative to natural hydrogels is their synthetic analogs. Their advantages are the consistency in chemical composition and related physical and mechanical properties that are easily managed. Synthetic hydrogels are missing an important aptitude of natural counterparts, that is, the ability to convey the biological signals required for adequate cell functioning after *in vivo* implantation. Typical polymers applied in the assembling of synthetic hydrogels are poly(vinyl alcohol) (PVA) and polyethylene glycol (PEG). In the field of tissue engineering, an effective solution to overcome the problems of synthetic and natural hydrogels has been the development of a hybrid crossover, biosynthetic hydrogels, which have bioactive moieties inserted into the synthetic framework. Biosynthetic hydrogels have gained some popularity owing to inherited advantages of both types of hydrogels [3].

Nonetheless, when used in cell encapsulation technology, hydrogels exhibit a number of drawbacks. In order to reach a certain mechanical stability, hydrogels are often crosslinked by either high-energy irradiation or application of low-molecular weight crosslinking agents (e.g., aldehydes), such conditions and reagents are incompatible with cell encapsulation. Besides, the inborn bulky chemical structure of hydrogels affects the diffusion processes that are vital for a healthy cell environment. All these limitations affect the more widespread use of hydrogels in cell-based assays. It is also essential to have a certain flexibility and yet full control in guiding the polymerization processes during hydrogel formations, because once the hydrogel framework is synthesized, any further alterations of its structure are barred and therefore the design of hydrogels as biomaterials must be carefully planned in order to attain the desired evolution of the final network structure.

10.4.5 Interfacial Polymerization

In this method, the microcapsules can be prepared using an emulsion of the w/o (water-in-oil) type. A typical procedure includes emulsification of an aqueous suspension of the cells in an organic solvent, where the two biocompatible monomer reagents are dissolved in non-miscible liquid phases [32]. The monomers will react upon combining the liquids at the interface to yield a primary membrane. The reaction rate gradually declines as the diffusion of monomers gets restricted on the assembly of the polymeric shell. The reaction reaches its completion when a sufficient wall thickness is achieved as a result of emulsion polymerization; further monomer diffusion is almost (if not completely) stopped. The process is illustrated schematically in Fig. 10.6. Interfacial polymerization allows the creation of cells enveloped within a thin but fairly robust shell.

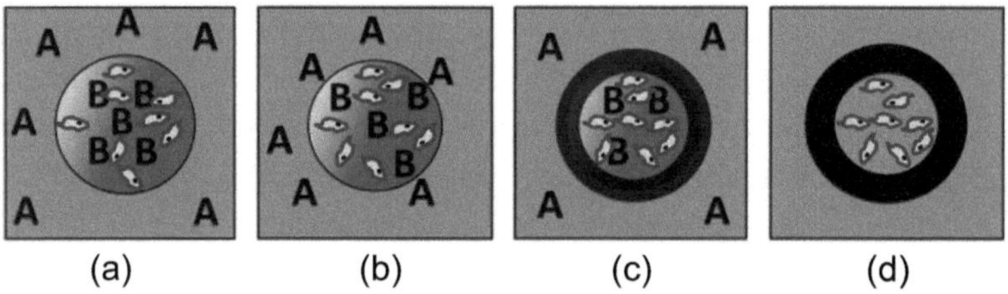

Figure 10.6 Schematic diagram of interfacial polymerization: (a) biocompatible monomers dissolved in organic phase (A) and aqueous phase (B) with resuspended cells, (b) monomers diffused and met at interface, (c) polymerization and formation of primary membrane, and (d) continued polymerization and microcapsule formation).

10.4.6 *In Situ* Polymerization

This method involves employing an emulsion droplet carrying a dissolved monomer. At the outset of the polymerization process, initially formed oligomeric fragments do precipitate in the aqueous medium to create the primary nuclei. With the polymerization reaching the steady growth stage, these primary nuclei continue to develop, leading eventually to the separation of the polymeric phase at the interface, thus creating the shell of the capsule. Entrapment of the core material forms the final microcapsules (Fig 10.7). This strategy has been used, for instance, to encapsulate single mammalian cells, HeLa cells, hMSCs, and BACs that showed high viability [66].

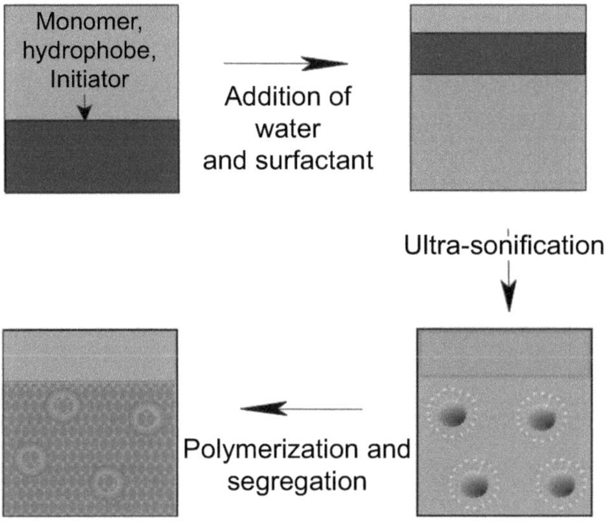

Figure 10.7 *In situ* polymerization method overview.

10.4.7 Interfacial Precipitation

Interfacial precipitation is used to assemble the microcapsules carrying mammalian cells (Fig. 10.8). A suspension of cells and a solution of polymer are extruded through two concentrically configured needles, which leads to the formation of a core that contains a suspension of a cell surrounded by a liquid polymeric shell. Such core–shell droplets are continuously dripping into an aqueous medium, where the polymer solvent is extracted, thus forming a solid shell as a precipitate. Aprotic organic solvents such as dimethyl sulfoxide (DMSO), dimethyl formamide (DMF), and dimethyl acetamide (DMAc), are used to dissolve the organic polymers. Since cell contact with organic solvents is hard to avoid, it can be minimized using various co-extrusion schemes [67].

10.4.8 Coacervation

Polymer coacervation is a well-known and widely used physicochemical method of gelation and microencapsulation of chemicals and biological materials, including cells [68–71]. An illustration of the coacervation process is presented in Fig. 10.9. By definition, coacervation is a partition of colloidal systems between two liquid phases. Polymer coacervation consists of three basic steps. During the first step, the core substance in the oil form is dispersed in the aqueous solution of polymer, for example, a polycationite. When a second polymer, for example, polyanionite, is added in the form of aqueous solution to the dispersion, the polymers form a complex, which is insoluble in water and thus shells out on the oil droplets. Various initiators are used to begin this process, for example, change of pH or temperature, dilution of the solvents, or change of dispersion's ionic strength by adding a salt. The shell thickness can be varied by the amount of the added second polymer. The stabilization of resulting microcapsules is achieved by crosslinking, increased temperature, desolvation, etc.

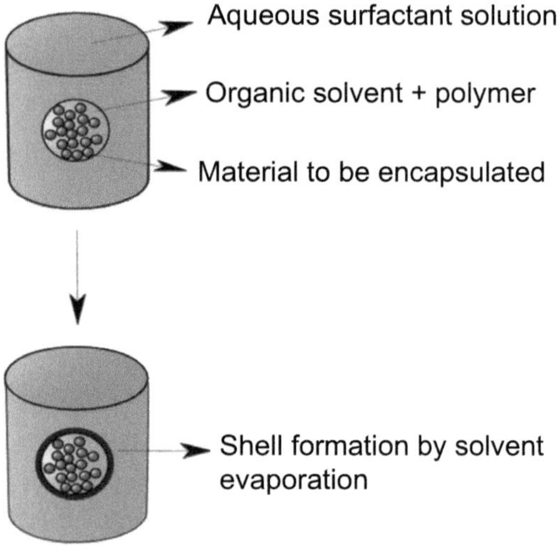

Figure 10.8 Schematic representation of microencapsulation by solvent evaporation technique. Reprinted from Ref. [67], Copyright 2008, with permission from Elsevier.

10.4.9 Suspension Crosslinking

When it comes to the preparation of protein and polysaccharide microcapsules, suspension crosslinking is a reliable and well-developed method [72, 73]. A polymer in aqueous solution, which contains a core material, is dispersed in an organic solvent immiscible with water (suspension/dispersion medium) in the form of small droplets. A suitable compound to maintain the stability of the droplets is a part of the suspension medium. In the next step, the droplets are crosslinked, forming the respective microcapsules. To achieve crosslinking, the mixture can be heated above 500 °C; alternatively, formaldehyde, terephthaloyl chloride, etc. are used as a crosslinker.

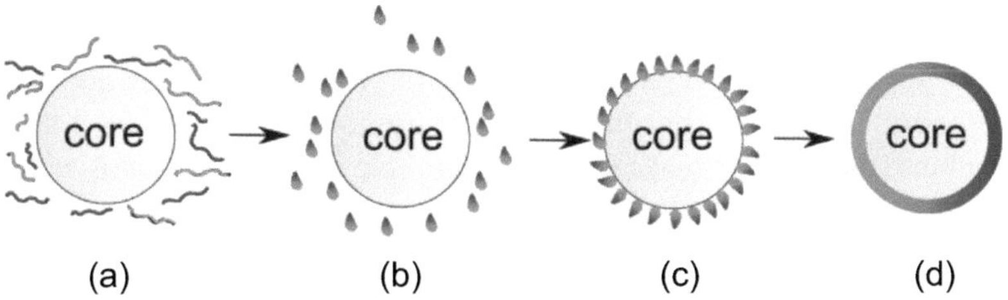

Figure 10.9 Schematic representation of the coacervation process: (a) particles of core material dispersed in solution of shell polymer, (b) separation of coacervate from solution, (c) adsorption of coacervate microdroplets on the surface of core particles, and (d) coalescence of coacervate microdroplets and capsule shell formation.

Suspension crosslinking is a robust technique with a broad range of applications and potential payloads. Beyond the living cells, it can be of use for the enclosure of other solid or water insoluble species as well as molecules possessing good solubility in water. This technique is also beneficial in the fabrication of micro- and nanocapsules; while using this methodology, albumin nanocapsules that contain doxorubicin and magnetite particles have been successfully assembled [44, 74].

10.4.10 Coloidosomes

Colloidosomes are comprised of a tightly packed layer of colloidal particles that are connected, thus creating a particle wall. The empty space between the particles is arranged as an array of uniform pores. The size of these pores is controlled over the nanometer to a micrometer scale, serving as a tool to adjust the shell permeability. These capsules are called "colloidosomes," similar to liposomes, and are assembled from phospholipid bilayers. Colloidosomes can enclose living cells, providing them with a rigid supportive scaffold, while simultaneously protecting them from the immune system and allowing free passage of nutrients and gases. Moreover, colloidosomes can be used as a three-dimensional template to grow tissues in a protected environment, which allows the exchange of nutrients and metabolites [67, 75].

Several ways to control the rupture and permeability are available for the system of colloidosomes. Among those are varying the size of coating particles, sintering of the formed colloidosomes and further filling of the holes with smaller particles or polymers. The release strategies in the case of colloidosomes generally either imply the permeability control to achieve a slow, albeit sustained, release or control of their rupture stress for shear-induced collapse [67].

10.4.11 Incorporation of Porins

Incorporation of porins, bacterial membrane proteins allowing passive diffusion of specific molecules, in the walls of polymeric capsules mimics a natural bacterial way to enhance the efficacy of the passage of nutrients and gases in response to various stress factors. The outer cell wall of bacteria *E. Coli* allows for the permeation of solutes with a molecular weight of about 400 Da or lower. Substances of such size can freely transfer through the bacteria cell wall under a concentration gradient. In the case of lack of nutrition, the pure diffusion process is helpless in overcoming the deficiency due to its slow speed. To complement these, nature has designed porins, a range of specific and highly sophisticated membrane channels, which function (open and close) in response to a specific trigger. One of the most common diffusion porins is OmpF (Outer membrane protein F) [76-78]. The other widely studied type of membrane channels in *E. coli* is the malto-oligosaccharide specific channel LamB, or maltoporin.

Porins are attractive subjects for advanced material science owing to their high stability. For example, OmpF can be found intact in 4 M GuHCl at 70 °C or in 2% solution of sodium dodecyl sulfate. In addition, the advances in recombinant technology have enabled reasonably high-yield production of porins in *E. coli* (tens of mg per liter of culture). The ability of porins to crystallize, forming a

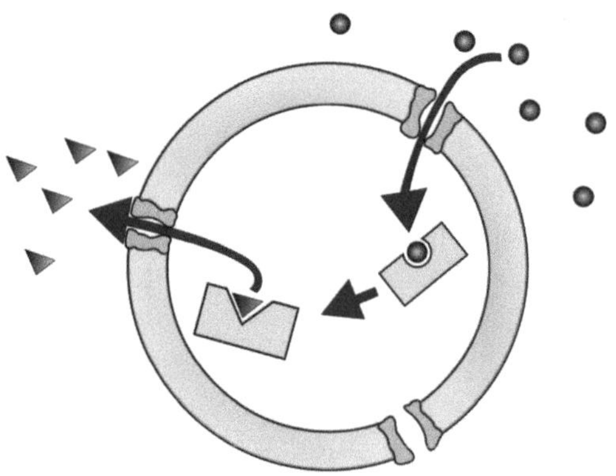

Figure 10.10 Porin-based polymer-stabilized nanoreactor. Reprinted from Ref. [76], Copyright 2001, with permission from Elsevier.

high-resolution 3D crystal structure with detailed depiction of binding sites, adds to the advantage of their use in the design of new materials being significantly impactful on molecular engineering aiming for new proteins [76]. Another attractive feature of porins is an established access to the information on thermodynamics and kinetics of substrate transfer across the porin channels [76].

Incorporation of porins into artificial membranes emerges as an effective tool to control the selectivity and rate of substrate diffusion in and out of the capsule void. Winterhalter *et al.* have shown a potential use of *E. Coli* porins in liposome walls. In addition, encapsulation in the liposomes or polymer nanocapsules shields porins from proteases preventing their denaturation [76]. Importantly, the use of porins as functional elements in block copolymers or lipid-based capsule membranes allows to follow the passage of a single molecule through the porin channel (Fig. 10.10) [76]. One essential trait of such biomimetic systems is that they provide a sustainable release of payload, lasting for a long time.

10.5 Polymers for Cell Encapsulation

A range of requirements needs to be satisfied in order for material to be applied as a biological object scaffold, for example, biocompatibility, biodegradability in the timeframe scheduled for the application, permeability of the engineered architectures to grant the exchange of nutrients and metabolic waste, compliance with and support of cell growth and proliferation [79] and a reasonable and adequate mechanical stability [14].

Macromolecular properties of hydrogels are similar to natural ECM, which allows for their use in cell and tissue engineering [80, 81]. Examples of hydrogel-forming polymers include alginate, collagen, hyaluronic acid, gelatin, chondroitin sulfate, fibrin, chitosan, and agarose [82].

Perhaps one of the most widely used natural polymers is alginate. Alginates are linear block-polymers that are found on the cell walls of brown seaweed. The composition of alginates includes unbranched binary copolymers of 1-4 linked β-D-mannuronic acid (M) and α-L-guluronic acid (G) of substantially versatile arrangement and sequential structure (MMM-blocks, GGG-blocks, and MGM-blocks). Salts of alginates, for example, alkali-, ammonium-, and magnesium-alginates, are soluble in water. Gelation of alginates occurs when the carboxyl groups of the polymers are crosslinked with multivalent cations (e.g., Ca^{2+}, Ba^{2+}, La^{3+}, Fe^{3+}) and polyelectrolytes [83].

Even with a number of advantageous properties, alginates may not be considered as ideal contenders for tissue engineering as they are unstable and decompose under various, often mild, conditions. For instance, acidic media is known to affect the alginates. These materials could also crack and become mechanically unstable in the presence of lactic acid. When alginate is gelated with calcium ions, its integrity deteriorates as it is treated with chelating agents capable of forming complexes with calcium cation (e.g., phosphates, lactates, and citrates). Other disadvantages of alginate hydrogels include the formation of surface defects such as pores and cracks. Surface-based structural abnormalities lead to a relatively fast diffusion of liquids through the hydrogel network, which significantly weakens the protective function of the respective capsules. The important undesired consequence of using alginate-based implants is also inflammation and fibrosis in the implantation site due to the ratio between L-guluronic (G) and D-mannuronic (M) acids in this molecule. A common complication of fibrosis is necrosis of the enclosed cells caused by the lack of oxygen and nutrients. The strategies to compensate these defects are (i) admixing of other polymers to alginate, (ii) additional coating of the alginate hydrogel beads with other polymeric or composite compounds, and (iii) using various additives for structural modification of the alginate [82, 84].

Various polyelectrolytes have been tried in the alginate-based formulation with the purpose of addressing the issues of pure alginate hydrogels. The most promising mixed formulation in terms of the stability and toleration by the host, often mentioned as APA capsules, is calcium alginate beads encapsulated into a bi-layer of poly-L-lysine (PLL) and alginate [4, 14]. In the APA capsule system, the core is formed with the cell loaded alginate hydrogel, while PLL forms a sandwiched layer between the hydrogel core and another layer of alginate on top. However, inflammatory reactions are also not excluded for this formulation [85]. Other studies warn that APA capsules have poor stability in the GI tract [86], while yet more research states that APA composition suffers from permeability issues [4, 87]. Table 10.1 summarizes some modifications of chemical composition beyond the classical APA aiming to improve the durability and mechanical stability of cell encapsulating membranes [3, 14].

Polycations that exhibit compatibility with the immune system represent an important area of research in biopolymers. Nonetheless, the potential of clinical application for polycation-based microcapsules remains to be proven. While poor biocompatibility of PLL is believed to limit its success, some *in vivo* studies have claimed promising results when replacing PLL with poly(methylene-*co*-guanidine) (PMG) hydrochloride or poly-L-ornithine (PLO). Beyond biocompatibility, replacing PLL with PLO helped to reduce the size of the capsules formed from 800 to 400 mm [14]. Complexes of PMG and alginate (week interactions) or cellulose sulfate (strong interactions) have been also proven to provide capsules with substantial mechanical strength and durability. Recent advances include the development of biodritin, a polymer formed by chondroitin sulfate and sodium alginate. It was demonstrated that the mechanical and chemical stabilities of the respected capsules can be controlled by implementing strong or weak interactions of other polymeric components with biodritin [14].

Table 10.1 Encapsulation systems beyond the classical APA

Design	Advantages	Ref.
Alginate–poly-L-ornithine	Increased biocompatibility	[88]
Alginate–cellulose sulfate–PMCG	Independent adjustment of capsule parameters	[89]
HEMA–MMA	Improved mass transfer, stability and durability	[90]
Agarose–polystyrene sulfonate	Blocks the activation of complement via enhancing the activity of C1 inhibitors	[91]
Alginate–agarose	Increased mechanical stability	[92]
Barium–alginate	Increased mechanical and chemical stability	[93]
PDADMAC	Increased mechanical stability	[94]
Alginate–chitosan	Increased biocompatibility	[95]
PVA	Improved nutrient transport	[96]
Photopolymerised PEG–diacrylate	Increased biocompatibility	[97]
Genipin crosslinked AC microcapsules	Covalent link created by naturally occurring genipin. Enhanced membrane strength and durability	[98–100]
Alginate–PMCG–alginate capsules	Improved membrane strength Easy adjusted capsule size and wall thickness	[101, 102]
A-CS-PMCG system	Improved mechanical strength Easy control over membrane thickness	[103–107]
Polyvinylamine hydrochloride–based microcapsules	Mechanically stable Permeability can be controlled over a wide range	[108]
PHEMA-MMA system	Better stability and durability Limited cell survival and mass transfer in some application	[109–111]

Design	Advantages	Ref.
Chitosan core-poly(MAA-HEMA-MMA) shell	Prepared in physiological conditions Controllable mechanical strength and permeability Supported hepatocytes growth Maintained cell functions	[112]
Collagen core complexed with terpolymer HEMA-MMA-MMA-shell	Improved mechanical strength and chemical stabilizing Enhanced cell functions	[113–115]
NaCD-PDMDAAC	MWCO is less than 2 kDa if prepared by adding of a porogen (starch). Increased cutoff to 70 kDa allows secretion and release of therapeutics by the encapsulated cells.	[94, 116–120]

Alginate-based capsules for isolation of cells can be also crosslinked with barium ions instead of calcium. Such hydrogels have been investigated for encapsulation of allograft islets followed by transplantation in non-obese diabetic (NOD) mice [121]. Other reports describe the use of cell loaded alginate particles as a template for coating with a multilayer composed of different polyanions and polycations, optimization of the alginate–PMG–alginate microcapsules, and incorporation of an inorganic silica-based polycation [121].

The surface of alginate microcapsules can be enhanced with polyethylene glycol (PEG) grafted copolymers of PLL and polyethylene oxide for improved shielding of loaded cells from the immune system [122–124].

To overcome the innate problems of alginate, sizeable research focusing on the new encapsulation techniques rather than on improvement of the APA encapsulating system is required.

Natural or derived polymers such as polypeptide homopolymers (poly-L-arginine, poly-L-aspartic acid, poly-L-glutamic acid) as well as polysaccharides (hyaluronic acid, quaternized chitosan, heparin, and dextran sulfate) are the materials of choice for the assembly of tolerable biodegradable capsules to deliver biological cells *in vivo* [125].

In order to mimic the membrane features of natural viruses or somatic cells, polymeric capsules can be coated with lipid layers. In general, such coatings are aimed to enhance capsule targeting and cellular uptake by reducing non-specific interactions between capsules and the surrounding solutes. The formation of lipid layers has been extensively investigated by Moya *et al.* in the case of polymeric multilayer capsules [126]. Studied phospholipids (dipalmitoyldiphosphatidic acid and dipalmitoyldiphosphatidylcholine) and sphingosine were suggested to form intermolecular complexes with the capsule polymers. As a result of this proposed phenomenon, such hybrid lipid/polymer complexes possessed high morphological stability.

It was shown that lipids such as dipalmitoyldiphosphatidic acid, dipalmitoyldiphosphatidylcholine, and sphingosine adsorb on the outer layer or incorporate within the shell of polyelectrolyte capsules, forming either lipid bilayers or multilayers. The formation of asymmetric lipid bilayers on polyelectrolyte multilayer-coated colloids was accomplished. The stability of lipid coatings on polymeric capsules can be improved by the introduction of inorganic-based synthetic lipids. Thus, the *N*-[*N*-(3-triethoxysilyl) propylsuccinamoyl]dihexadecyl-amine (Si-lipid) was reported to enhance the layer properties [127].

FDA approved polyethylene glycol (PEG) is one of the most studied polymers for the purpose of forming chemically crosslinked hydrogels [128–131]. Biodegradable hydrogels can be obtained via copolymerization of PEG with poly(propylene fumarate), poly(lactic acid), or poly(glycolic acid) [132–134]. In addition, many naturally occurring biopolymers such as hyaluronic acid, fibrinogen, chitosan, and heparin are also widely studied in combination with biodegradable PEG hydrogels [135–140]. The latter have been recommended for medical applications which require cellular adhesion and as delivery vehicles [82, 141].

It is important to mention the major role of PEG and poly(2-methyloxazoline), another hydrophilic polymer, in the surface passivation of implants and delivery systems to prevent them from getting opsonized by host antibodies, complement proteins, or leukocytes, eventually preventing the immune response and causing inflammation [127]. In the area of tissue engineering, encapsulation is related to the extensive use of synthetic hydrophilic polymers such as poly(propylene fumarate) and poly(vinyl alcohol) [82].

Thermoplastic polymers possessing long linear hydrophobic chains are another good alternative to hydrogels for the encapsulation of living cells in macrocapsules and vascular devices owing to their superior mechanical and chemical stability. The capsules of this kind of materials are formed by sequential thermal treatment and cooling. Acrylonitrile/vinyl chloride copolymer (PAN-PVC) is one of the most popular thermoplastic polymers used for encapsulation of cells mainly in the form of hollow fibers. The severest drawback of thermoplastics significantly limiting their clinical applications is a poor level of water permeability that hampers penetration of nutrients to the enclosed cells, thus compromising their viability [3].

Polymeric composite films containing silicon or ceramics, possess a controllable and highly uniform porous structure even in the nanometer range. Thus, they are also among the attractive candidates for the fabrication of semipermeable membranes in cell encapsulation. These materials typically enable such important capsule features as good stability and excellent mechanical strength. The following composites have been used to build up capsules for living cells: nanoporous silicon membranes (for immunoisolation of islets) [142], porous TiO_2/Ti composite membranes (for attachment of hypothalamic neurons) [143], and bioceramics (for bone cell transplantation) [144]. In the case of TiO_2/Ti composites, it was possible to control the porosity and associated permeability by changing the sintering temperature. Despite the useful features of noncarbon-based polymeric materials, their engineering remains quite complex, limiting their current applications to macro-level architectures, which in turn does restrict their clinical potential.

10.6 Applications

10.6.1 Mammal Cells

Now being rapidly developed in the field of biomedicine, gene and cell therapies represent two closely interrelated approaches to the treatment of human diseases based on genetic defects or those illnesses causing overexpression of malfunctioning cells of a certain type that have to be killed for wellbeing of the patient. There is a great deal of hope and optimism on cellular transplantation as an efficient tool to cure various human disorders caused by hormone or protein deficiencies such as diabetes, Parkinson's and Huntington's disease, liver ailments and other afflictions. The potential of this approach includes the development of bioreactor-type delivery systems which *in situ* supply recipients with a wide range of therapeutic products aimed at compensating for deficiencies or halting the pathological process. Unfortunately, a number of persistent challenges have hampered the efficiency of cellular transplantation. In particular, protection of the transplanted cells from immune attack by the transplant recipient is of paramount importance. This is often achieved by the administration of potent immunosuppressants. These therapeutic agents possess considerable toxicity, causing a wide variety of serious side effects in patients, the most dangerous of which are the lowered resistance to infection, and higher chances of cancer and uncontrolled bleeding. The immunosuppressive drugs may also affect the transplanted cells, disturbing their metabolism and functioning [30].

Encapsulation represents a much safer alternative to immunosuppression: the shell creates an immunologically inert barrier, so that the transplanted cells are hidden from the immune system of the recipient. Depending on the nature of the disease, the enveloped cells can be introduced into the recipient's body either locally (to treat solid tumors), regionally (to treat brain disorders), orally or intraperitoneally (to achieve a systemic effect).

Encapsulation is also a cost-efficient approach to cell therapy due to the abundance and low cost of conventionally applied polymeric materials [14]. An added advantage is that the immunologically inert enveloping polymers create no need to customize the capsule for specific human leukocyte antigen (HLA) classes of different patients.

A variety of developed methods makes it feasible to encapsulate practically all types of cells; however, the success of therapy with encapsulated cells requires careful consideration of the cell line.

Cell functionality is of key importance. In this regard, the aim is to identify the cell line with the most beneficial functionality to solve the particular task in therapy or biotechnology, that is, the pancreatic islet cells are the obvious candidates for the treatment of diabetes; encapsulated hepatocytes would cure hepatic failure; neural mediators secreting cells of natural or engineered origin would serve in the treatment of neurological disorders; cells secreting enzymes or hormones are important in replacement therapy or in treating cancer (e.g., antiangiogenic factors secreting cells); hybridoma cells can be used for the enrichment of monoclonal antibodies; etc. [87].

The following consideration deals with the cell source. From the practical point of view, the procedure of cell encapsulation and transplantation is no different for autogenic, allogenic, or xenogeneic cells; however, the use of autogenic cells has much less immunological repercussions than allogeneic or xenogeneic cells. In addition, therapies with xenogeneic cells put patients at risk of infection with animal viruses which is also interrelated with ethical concerns [121].

The rate of cell proliferation implies limitations on the capsule size and the life time of transplants. Clearly, the proliferating cells will fill the entire capsular space at some point of time that will eventually compromise their viability. In contrast, non-proliferating cells, for instance, myoblasts, would be a part of more durable implants producing therapeutic products after encapsulation and transplantation [14].

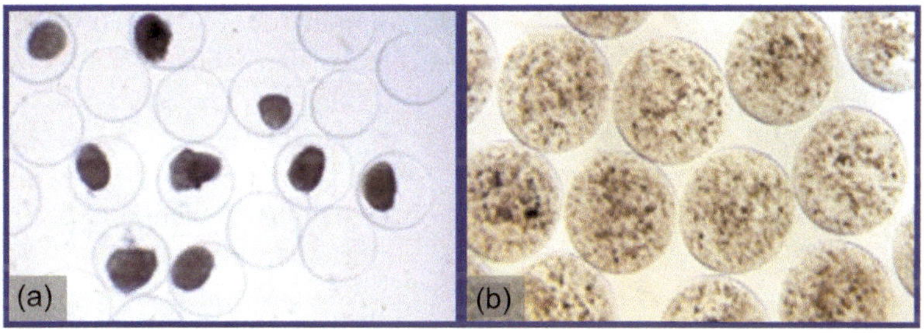

Figure 10.11 Encapsulated cells: the islets of Langerhands (a), stem cells hepatocytes (b). Reprinted from Ref. [14], Copyright 2004, with permission from Elsevier.

Different cell types have been encapsulated with the purpose of biomedical applications, ranging from the islets of Langerhands (Fig. 10.11a) to stem cells hepatocytes (Fig. 10.11b) [14]. Cell types commonly used for encapsulation also include kidney cells, BHK cells, C2C12 cells, human immortalized cells, fibroblasts, myoblasts, pancreatic islets, ovary and parathyroid cells, hepatocytes, chondrocytes, leydig cells, adrenal chromaffin cells, stem cells, PC12 pheochromocytoma and myeloma cells, besides hybridoma and virus producer cells [14].

The most common cell lines and enveloping capsules employed for biomedical purposes are summarized in Table 10.2.

Table 10.2 Cell lines, applications and capsule materials used for biomedical purposes

Cell type	Application	Material
Fibroblasts	Metabolic deficienciencies, neutrophic factors, epilepsy	Alginate, HEMA-MMA
Myoblasts	Metabolic deficienciencies, neutrophic factors, epilepsy	Alginate, HEMA-MMA
Kidney cells	Hemophilia, neurotophic factors, antiangiogenesis	Alginate
Pancreatic islets	Diabetes	Conformal coating: nanothin-PEG, PVA/PEG, polyion complex, chitosan/ alginate/ chondroitin sulfate; Macrocapsules: PEG, agarose, polysulfone, PVA, PVA/PAA, polyurethane, chondrocyte cell sheet; Microcapsules: alginate, agarose-polystyrene/sulfonic acid, AN69, cellulose sulfane, biodritin
Red blood cells	Production of universal red blood cells	Four bilayers of alginate and chitosan-graft-phosphorylcholine surrounded by two bilayers of AL and poly-L-lysine-graft-polyethylene glycol
Ovary cells	Fabry disease, hormone replacement therapy	Alginate, HEMA-MMA

Cell type	Application	Material
Parathyroid cells	Artificial organs	Alginate
Hepatocytes	Liver transplantation	Macrocapsules: PEG, polyacrylonitrile-sodium-methallylsulphonate; Microcapsules: poly-L-lysine, alginate, polyelectrolyte copolymer/ modified collagen, chitosan, HEMA-MMA
Chondrocytes	Bone and cartilage regeneration	Alginate
Leyding cells	Hormone replacement	Alginate
Adrenal chromattin cells	Parkinson's disease, chronic pain	Alginate
Stem cells	Bone regeneration	Alginate
PC12 pheochromocytoma cells	Neurotrophic factors, neurotransmitters,	Macrocapsule: polyurethane; Microcapsules: alginate, HEMA-MMA
Myeloma cells	Hepatic growth factor	Alginate
Hybridoma cells	Antibody production	Alginate, acetate, alginate-agarose
Tumor cells	Cancer vaccine, interleukins	Alginate, chitosan
Virus producer cells	Cancer	Cellulose sulfate
293 cell line	CNS/Glioma	Alginate
	Pancreas cancer	Cellulose sulfate
Baby hamster kidney cells	CNS/Huntington's	Macrocapsule: poly-ether-sulfone
G8P285 hybridoma cells	Retroviral neurodegeneration	Cellulose sulfate
J558/TNF-alpha cells	Cancer	Alginate
HEK 293 cells	Cancer	Alginate
CYP2B1 cells	Cancer	Agarose
Mesenchymal stem cells	Myocardial infarction	Alginate

Source: Refs. [3, 14, 51, 145].

10.6.1.1 Cell therapy

Rapid advances in biomedical research have uncovered a number of potent treatments of severe human disorders through cell therapy. Cell therapy is now applied in regenerative medicine, in the treatment of genetic diseases, autoimmune diseases, cancer, and in cardio-vascular and neurological degenerative disorders. In clinical practice, the protocols of cell therapy have contributed to the cure of some forms of leukemia, extended the lifespan of patients with organ transplants, and improved the quality of life of patients suffering from diabetics and multiple sclerosis. A vast majority of cell therapy protocols relies on an *ex vivo* approach with cells that allow easy isolation, culture, and reimplantation [146–149]. The method involves extraction of cells from the patient (or another host) followed by their genetic modification *in vitro* and introduction or reintroduction into the patient to deliver a therapeutic product.

The most important complication of the *ex vivo* approach is the immune response of the recipient to the reimplanted non-autologous cells. Thus, the advances in cell therapy have created a growing demand for robust technologies for mechanical protection and immunoisolation of the modified cells. Polymer-based biocompatible envelopes capable of hiding the transplant from the recipient's immune system while allowing secretion of the therapeutic product emerge as the most suitable processes to cope with possible complications of cell therapy and to improve its overall outcome.

Examples of successful attempts include the encapsulation of mouse fibroblasts producing human growth hormone (hGH) in an alginate-poly-L-lysine-alginate shell. The cells were genetically modified with the human growth hormone gene (Ltk-GH). After the ability of enzyme-secreting encapsulated cells to survive in a culture over a reliably long period of time was confirmed, they were transplanted into mice. The encapsulated transplants were then proven to be immunologically inert by showing no sign of damage by the recipient's immune system for one year. Importantly, the transplanted cells continued producing hGH over the whole period of observation [150]. The same type of membrane was used for the coating of mouse fibroblasts and myoblasts transfected with an engineered fusion gene to secrete a protein, human adenosine deaminase. As a result, the encapsulated cells remained viable and functional for 5 months [151].

The encapsulated delivery of genetically modified cells secreting therapeutics represents a solid platform to treat neurological disorders, for example, Parkinson's disease. The approach involves fabrication of transplants consisting of genetically modified cells secreting trophic factors. When introduced into a patient, the transplant will maintain the viability of the host neurons and increase the survival rate of co-grafted cells. The main role of the encapsulating system related to this task is to increase the life span of the transplant. In one of the examples, modified hamster kidney (BHK) fibroblasts producing the human nerve growth factor (hNGF) were encapsulated in poly(acrylonitrile vinyl chloride) and transplanted in the striatum of hemiparkinsonian rats with co-grafted adrenal chromaffin cells. Encapsulation was proved to increase the duration of trophic support to co-grafts [152].

Ex vivo cell therapy has a huge potential in the treatment of various forms of solid tumors by enabling a sustainable release of cytotoxic drugs at the tumor site. Löhr *et al.* managed to prolong the median survival for patients with confirmed inoperable adenocarcinoma of the pancreas from 22 weeks to 44 weeks through taking advantage of chemotherapy combined with therapy with encapsulated transfected xenotransplants. Patients in clinical studies were systematically administered with a low dose of the anticancer prodrug, ifosfamide, which was converted to its active forms (phosphoramide, mustard, and acrolein) at the site of tumor by 2B1 isomer of the rat cytochrome P450 gene expressed by genetically modified L293 cells injected into a vessel leading into the tumor. To avoid immune rejection, the injected cells were enclosed in polyelectrolyte bi-layer capsules (d = 0.7 mm) of sodium cellulose sulfate (NaCS) (polyanion) and poly(dialyldimethyl ammonium chloride) (PDADMAC) (polycation). Clinical studies were forerun with animal studies in the nude mouse, that showed a significant antitumor effect of the therapy. The feasibility of intraarterial placement of the capsules into the pancreas was confirmed by angiographic experiments in pigs. The procedure was well tolerated by patients without any complications, allergic reactions, or pancreatitis [153].

Antiangiogenic endostatin-secreting Chinese hamster ovary (CHO) cells were tested for the treatment of melanoma. The recombinant cells were enclosed in semipermeable polymeric microcapsules (~200 μ in diameter) and administrated in melanoma mice. The treatment with encapsulated cells inhibited the melanoma growth and enhanced the survival rate in treated animals without causing immune response [154].

In a less successful example of delivery of the recombinant product, alginate encapsulated transfected rat hybridoma cells secreting cytokine-neutralizing monoclonal antibodies directed against murine IL-4 were implanted intraperitoneally and subcutaneously into mice. Unfortunately, deterioration of the alginate hydrogel didn't allow maintaining high levels of antibodies in the bloodstream. Another important issue was ascite developed by all treated animals in response to the cell released from the capsules into the intraperitoneal cavity [155].

10.6.1.2 Cell transplantation

Transplantation of autologous cells is the favored approach to avoid immune rejection of transplants; however, it is often difficult to obtain somatic cells in the amount sufficient to achieve curing of the disease. Thus, encapsulating systems for transplantation of allogeneic cells in an immunologically safer way has emerged as a highly promising tool.

Transplantation of Langerhans islets usually requires lifelong immunosuppression that chronically exposes patients to the side effects and complications especially dangerous in the case of juvenile recipients with type 1 diabetes. To minimize the immune response (both cellular and humoral immunities) and avoid the immunosuppression, cell clusters are usually encapsulated in semipermeable membranes that also protect the inner islets from mechanical stress. The first study on the encapsulated islet was reported in 1977 by Gates and Lazarus [156], and since then encapsulated islets have been developed for clinical settings.

The most popular material for islets encapsulation is alginate hydrogel obtained from high-viscose ultrapure alginate stabilized with divalent cations such as barium or calcium. Alginate beads have been actively used to house islets since 1980, when Lim and Sun [71] first succeeded in achieving normoglycemia in diabetic rats for a time span of three weeks. Modification of alginate beads with poly(ethylene glycol) allows the reduction of plasma absorption onto the capsule surface and improve the viability and insulin releasing function of the housed islets [157, 158]. Additional LbL coating of alginate beads with poly-L-lysine and alginate has been proven to provide better immunoisolation to the transplanted islets by blocking the diffusion of serum immunoglobulin, albumin and hemoglobin [159]. When the thus obtained capsules were administered into the peritoneal cavities of diabetic rats, they promoted normoglycemia for 9 months with no sign of destruction of the islets and no fibrous adhesions around the capsule [160].

The issues of alginate hydrogels described earlier in this chapter have stimulated the exploration of other methods and materials for this cell type. One of the examples of successful encapsulation of Langerhans islets is their LbL coating with synthetic polyelectrolytes [poly(allylamine hydrochloride) or poly(diallyldimethyl ammonium chloride) used as polycations and poly(styrene sulfonate) used as polyanion]. Insulin release confirmed the function of the islets after encapsulation. The assembled capsules provided good protection to cells from the immune system which was demonstrated by means of an immune assay. Three coating layers were enough to affect the antibody recognition of the islets [161].

Successful encapsulation of Langerhans islets has been also reported with polyvinyl alcohol [162] and low molecular weight dextran sulfate [163].

Semipermeable macrocapsules for Langerhans islets were obtained of hydroxy-methylated polysulfone (PSU) hollow fibers, the material which has been proven to absorb large amounts of insulin. In such devices, islets showed a good insulin releasing function similar to naked islets and their transplantation into diabetic rats allowed normoglycemia for over 1 month [164–166].

Despite the promising results achieved with a few different encapsulating materials, the quest for an enveloping system enabling islets that can be utilized permanently is still on. Thus, there is a strong demand for polymeric semipermeable systems capable of maintaining the functions of the encapsulated islets and also providing them a proper immunoisolation for the long term [167].

Patients with acute liver failure can fully benefit from the transplantation of allogeneic hepatocytes. The proof of the concept was demonstrated, for instance, on rat hepatocytes encapsulated in a two-layer polymeric membrane. The capsules formed via the complex coacervation method contained two polymeric layers. The outer layer with a thickness of 2- to 3-mm was a 1:1:2 mixture of 2-hydroxyethyl methacrylate, methacrylic acid, and methyl methacrylate. Positively charged modified collagen was chosen to form the inner layer since this material is proven to enhance cellular functions. The permeability of the obtained membrane was selective to molecules smaller than albumin, creating promises for its *in vivo* use. Moreover, encapsulation was proven to enhance cellular functions of hepatocytes up to three times compared to uncoated cells grown as a monolayer [168].

Encapsulation of hepatocytes in alginate/poly-L-lysine shells helps to maintain their viability and enhance metabolic activity [169, 170]. Aoki *et al.* observed that metabolic activity of hepatocytes inside the polymeric envelopes grew with time relative to that at the point immediately after their encapsulation. After transplantation into rat spleen, the encapsulated hepatocytes survived well and retained their hepatic function. Importantly, the group of animals with intrasplenic transplantations of hepatocytes in the encapsulated form showed drastic liver regeneration 48 h after the transplantation [169]. However, the fact of liver regeneration on receiving the transplantations of hepatocytes encapsulated in the alginate/poly-L-lysine membrane was not confirmed in mice with acute liver failure [170]. The improved survival rate was solely attributed to the metabolic support provided by the capsules.

10.6.1.3 *In vivo* gene therapy by viral vectors

Despite the extensive development of lipoplexes and polyplexes for gene therapy, the transfection efficacy of those vectors is still low, and viral vectors based on the replication deficient retroviruses are still the most commonly used delivery vehicles *in vivo* [14].

However, a sufficient level of gene transfer often requires multiple applications of the viral vector. Multiple rounds of gene delivery with this vector involves a laborious procedure of gene packaging and a low success rate because of a high likelihood of immune responses directed towards the viral particles.

An efficient way to avoid repeated injections of the therapeutic gene carrying viruses could be implantation of transfected cells in which the viral particles can replicate, being continuously secreted for a sufficiently long period of time. The implantation of cells producing viral particles containing the therapeutic genome in close proximity to the diseased organ or tissue would allow their easy targeting, thereby also improving the overall efficacy of treatment. The described approach has shown some success in brain tissue, where the immune response to the implanted cells is not significant, so that the cells are able to produce the virus long enough to provide the desired therapeutic effect [171, 172]. Successful implantation at the other sites in the body would require the transfected virus producing cells to be hidden from the immune system, which will normally get activated to eliminate transfected cells, especially if those originate from different species as is usually the case in such protocols.

Semipermeable polyelectrolyte bilayer films composed of cellulose sulfate (polyanion) and poly(diallyldimethyl ammonium chloride) (polycation) were used to coat cells producing a retrovirus with the purpose of preventing the development of the immune response after their implantation. It was proven that capsules didn't hamper the cell viability and allowed them to release the virus into the culture medium for at least six weeks. *In vivo* experiments confirmed the *in vitro* data explicitly showing the advantages of the cell encapsulation over multiple deliveries of therapeutic viral particles in terms of safety, simplicity, and efficiency [14].

10.6.1.4 Stem cell therapy

Mature primary cells often possess reduced regenerative capacity, and consequently, stem cell therapy emerges as a promising method to produce aids to treat Alzheimer's, Parkinson's, and other diseases like arthritis, cardiac infarction, and cancer. The potency of stem cells to improve tissue function and regeneration is due to their ability in self-renewal, differentiation into the desired cell types (chondrocytes, cardiac cells, neurons, etc.), and secretion of trophic factors. Beyond prevention of the immune rejection, microencapsulation of stem cells in semipermeable membranes offer such benefits as the creation of a defined microenvironment capable of modulating the phenotype of the mature cells [173], mechanical protection, and prevention of agglomeration during proliferation.

Factors such as cell–cell and cell–substrate interactions play an important role in the proliferation and differentiation of stem cells, so that the production of a specific desired cell type via differentiation of stem cells inside microcapsules often fails. Recently, some of the intrinsic and extrinsic signals regulating the fate of stem cells have been identified [174]. Simultaneous progress in material engineering enables fine-tuning of the above-mentioned factors by regulating the size and stiffness of the microcapsules as well as the density of the encapsulated cells [175, 176].

Alginate and collagen [177, 178] hydrogels are among materials that are most commonly used for encapsulation of stem cells. It was proven, however, that alginate hydrogels need an additional LbL coating with poly-L-lysine and alginate to minimize the immune rejection because of the mechanical and chemical instability of the hydrogel particles and their poor resistance to osmotic swelling often causing cells to escape [179]. Agarose has also been found beneficial in maintaining the viability and potency of stem cells within microcapsules either on its own or in combination with collagen [180], alginate [181], and chitosan [182].

10.6.1.5 Assisted reproduction technologies

Encapsulation of follicles, sperm, and oocytes is a highly promising approach to improving the success rate of assisted reproduction. In this regard, encapsulation of follicles primarily aims to investigate basic biology, mainly serving as a research tool to model their development. The outcome of this fundamental study can be later translated to clinical use. In contrast, encapsulations of sperm and oocytes are directly targeted towards the development of clinically approved protocols for greater *in vitro* fertilization outcomes [183].

For successful encapsulation of follicles, it is of particular importance to simulate their native microenvironment. The encapsulating system should capture and maintain the 3D architecture of ovaries. It must also enable natural cell–cell interactions and the paracrine signaling critical for proper follicle development. Another critically important feature of such capsules is their ability to support follicles as they increase in volume which in normal conditions can undergo greater than the required 10^6-fold increase. A suitable encapsulating system has been composed of fibrin-alginate interpenetrating network-based hydrogels. The environment of the hydrogel provided a rigid matrix for small follicles and its degradable component, fibrin, which stood down to them upon growing in size [183].

Sperm encapsulation primarily aims to protect the cells in refrigerated storage, shield them from female immunoglobulins and microorganisms and allow better retention in the uterus in the intra uterine insemination procedure, simultaneously enabling the passage of glucose, the main component of the refrigerated semen extenders. The other important goal of encapsulation of sperm and oocytes is their protection from damage upon cryopreservation that will be discussed below.

10.6.1.6 Biosensors

Encapsulation of mammalian cells appears as a powerful tool to work them into biosensors as the functional units. One of the most exciting examples of these is the biosensor detecting exogenous estrogens. Whereas xenoestrogens of plant origin (also known as phytoestrogens) could help in preventing various illnesses such as hormone-dependent cancers, osteoporosis, neurological, and cardiovascular disorders, other compounds with a structure similar to endogenous estrogens, for example, parabens are associated with problems in the reproductive system and may be a cause of breast and testicular cancer. MELN cells transfected with a gene encoding the estrogen regulated luciferase could serve as sensing elements for xenoestrogens; however the cells require extra protection with semipermeable membranes to improve the durability of the sensor. Germain and co-authors have found that the polymeric envelope composed via alternation of positively charged layers of poly-diallyl dimethyl ammonium chloride with negatively charged layers of poly-styrene sulfonate enable the cell survival range by more than 80%. The encapsulated cells were found metabolically active with undisturbed essential functions, such as synthesis of RNA and proteins [62].

10.6.1.7 Advanced tissue engineering

Tissue engineering is another important area that takes advantage of exploiting the polymer enveloped cells. The technology is generally aimed at assembling isolated cells in highly organized multilayered structures mimicking human tissues, resulting in single cell encapsulation methods enabling better stability of cells, besides tailoring tools for their spatial manipulations to become more relevant and beneficial to tissue engineering than those enveloping cell clusters and groups of cells.

A viable 25 mm thick tissue comprising four layers of mouse fibroblast cells has been grown on a glass substrate. While the first cell layer was attached directly to the glass surface, the following layers were sandwiched with the EMC-mimicking anchoring multilayers (6-nm thick) assembled of gelatin and fibronectin in the LbL fashion. In this architecture, fibronectin interacts with gelatin through a collagen binding domain and with the $\alpha_5\beta_1$ integrin receptor on the cell surface, "gluing" the biopolymer and cell layers together. The produced four-layered tissue was stable even after it was peeled off from the substrate [184]. Using the same approach of sandwiching the cell layers with seven consecutive layers of fibronectin/gelatin, an architecture mimicking the blood vessels has been achieved. A xenogeneic tissue comprised human smooth muscle and endothelial cells "glued" with this thus fabricated EMC [184].

10.6.1.8 Minimizing of cell injuries upon cryopreservation

The advances in cell encapsulation are envisioned to find extensive applications in stem cell therapy and the reproductive medicine helping to protect the mesenchymal stem cells, sperm, oocytes, and embryos from injuries related to cryopreservation.

Without any special care, cryopreservation can easily cause cell damage due to the formation of intracellular and intracellular ice, solution effects and dehydration. The common practice to minimize cell injuries and damage is to subject the cells to vitrification which is achieved by their exposure to special compounds, the so-called cryoprotectants. To promote vitrification, the cryoprotectants have to penetrate the cells, increasing the viscosity and decreasing the freezing temperature inside them. Unfortunately, commonly used cryoprotectants, for example, glycerol and dimethyl sulfoxide possess high cellular toxicity at concentrations required for vitrification.

Small alginate hydrogel microcapsules (approximately 100 μm in diameter) were found very promising in the protection of living cells in the cryopreservation process. The approach has been demonstrated on mesenchymal stem cells that poorly tolerate high concentrations of cryoprotectants, being highly prone to damage caused by ice formation. The encapsulating alginate hydrogel helped to achieve cell vitrification and enabled their intact morphology after cryopreservation. The strength of the alginate hydrogel matrix was found to be an important factor in the outcome of the cryopreservation process. Thus, it is advisable to add up to 0.15 M of Ca^{2+} cations in the system to support a certain level of hardness of the cell-carrying microcapsule [185].

A recent study by Choi *et al.* has revealed that oocytes can better survive the cryopreservation process due to the presence of the zona pellucida, the hydrogel-like outer shell. In this regard, fabrication of the outer shell by the microencapsulation process is a promising approach to minimize the damage due to cryopreservation of many mammalian cells that don't naturally possess the zona pellucida. Various polymeric hydrogels a appear to be the most suitable materials to serve this purpose [186].

10.6.1.9 Other bioapplications

Besides the above-mentioned facts, a variety of other biomedical applications can benefit from advances in cell encapsulation. Cell encapsulation technologies now assist in the synthesis of cell-derived molecules [187–189], clonogenic cytotoxicity testing [190] and the obtaining of monodispersed cells for improved tissue oxygenation [191]. Encapsulation helps to retain close cell–cell contact, facilitating the *in vitro* culture of contact depending species [192–194]. *In vivo* cell culturing [195] and *in vivo* drug testing have also gained advantages from encapsulation in semipermeable envelopes [196–198].

10.6.2 Bacteria

Microorganisms are used for the improvement of human health in two different ways: (1) as probiotics enhancing intestinal flora or recipient with bacteria promoting a healthy digestive system and (2) as a bioreactor for the delivery of therapeutic products.

10.6.2.1 Probiotics

Gastrointestinal microbiota (or gut microbiota) in humans is a community of microorganisms (ca. 5 trillion in average number), a majority of them being bacteria belonging to 30–40 different species [199]. The role of yeasts that are normally present in the gut in a small amount is not yet understood. In contrast to that, the role of the bacterial part of the gut microbiota in maintaining the host organism in healthy conditions is quite important. The beneficial functions of gut microorganisms include anaerobic digestion of waste [200], development of the immune system, enteric protection by affecting the proliferation of pathogens [201], help in absorbing calcium, magnesium, and iron [202] and synthesis of vitamins [203]. The abundance of each member of the gut flora largely depends on diet. It is also known that certain bacteria can contribute to disorders such as inflammatory bowel disease (IBD), obesity, colon cancer, hypercholesterolemia, and nonalcoholic fatty liver disease (NFLD) [204]. Thus, a healthy diet aimed at maintaining a proper balance between the normal and potentially harmful bacteria in the gut is of a paramount importance for overall health and well-being [199, 205]. Beyond diet, alternations in gut microbiota can be caused by the intake of antibiotics. Antibiotics can change the levels of gut microorganisms, provoking the growth of pathogens and antibiotic-associated diarrhea (AAD) [206].

According to the FAO/WHO definition, probiotics are "live microorganisms that, when administered in adequate amounts, confer a health benefit on the host" [207]. Various probiotics have shown positive effects in the treatment of AAD [208], bacterial vaginosis [209], high blood pressure [210], eczema [211], IBD [212], and some other conditions. Although not fully scientifically proven, foodstuff and food supplements containing probiotics are actively produced and marketed as potentially beneficial to our health [213].

The poor effects of probiotics *in vivo* are often related to extensive loss of microorganisms during the product manufacturing process, during storage, and on its passage into the intestine.

Fermentation appears critical for the survival of some microorganisms because of their poor compatibility with traditional starter cultures. A vast majority of probiotics are also affected by mild acidic pH that they are exposed to in fermented products (e.g., yogurts) for an extended period of time. Moreover, once consumed, passing through the stomach exposes them to even more acidic pH. Permeation of oxygen through the packaging materials can be crucial for anaerobic microorganisms. Some probiotics show instability in dried or frozen products. Among the other factors causing the loss of probiotics are hydrogen peroxide production, storage temperatures, and poor growth in milk due to the lack of proteases breaking down milk proteins.

Microencapsulation in protective semipermeable polymeric biodegradable membranes emerges as a promising solution to the problem of poor survival of probiotic microbial cells [214]. Table 10.3 shows some of the capsules that have been successfully used to increase the viability of certain strains of probiotics.

Table 10.3 Microcapsules for preservation and delivery of bacteria to intestine

Bacteria	Types of microcapsules	References
L. rhamnosus, B. longum, L. salivarius, L. plantarum, L. acidophilus, L. paracasei, L. casei, B. lactis, L. reuteri	Alginate beads	[215–219]
B. lactis, L. acidophilus	Alginate–cellulose acetate phthalate	[220]
B. animalis subsp. lactis, L. bulgaricus, L. rhamnosus GC, L. acidophilus NCFM	Alginate–chitosan	[221, 222]
B. animalis subsp. lactis	Alginate–chitosan–Acryl-EZE	[221]
B. bifidum, L. casei	Alginate–chitosan–alginate	[216]
B. animalis subsp. lactis	Alginate–chitosan–Sureteric	[221]
B. adolescentis, B. pseudolongum	Alginate–coated gelatin	[223]
B. bifidum, L. reuteri, L. casei	Alginate–poly-L-lysine–alginate	[216, 224]
L. acidophilus, B. lactis, B. infantis, L.casei	Alginate–starch	[225–227]
B. infantis, B. longum	Gelatin–gum arabic–soluble starch	[228–230]
L. lactis	Gelatin–toluene-2-4-diisocyanate	[231]

(*Continued*)

Table 10.3 (*Continued*)

Bacteria	Types of microcapsules	References
B. bifidum	Gellan–alginate	[232]
B. adolescentis, B. bifidum, B. breve, B. infantis, B. lactis, B. longum	Gellan–xanthan	[233–235]
L. plantarum	Genipin–crosslinked alginate–chitosan	[236]
B. lactis, L. acidophilus	Pectin–casein	[237]
B. longum	Potato starch granules–amylose	[238]
B. breve R070, *B. longum* R023, *L. rhamnosus GC*	Whey protein	[239–241]
B. longum, S. thermophilius, L. bulgaricus, S. lactis	k-carageenan	[242, 243]
L. acidophilus, B. lactis (Bb-12)	Cellulose acetate phthalate	[222]

Prebiotics, the substances that induce the growth or activity of microorganisms and improve their viability are often added to bacterial products. The examples of prebiotics used with probiotics include resistant starches or cryoprotectants (e.g., glycerol) [228]. Prebiotics were found to promote the intact form of probiotic upon storage in food products; however, they showed no support to microorganisms under simulated gastro intestinal conditions [228]. A big advantage of encapsulation is the possibility of simultaneous hosting of probiotics and prebiotics in one entity, or the so-called co-encapsulation. For instance, symbiosis of probiotics and prebiotics co-encapsulated in calcium alginate hydrogel was shown to enhance the viability of microorganisms in various food products and over their residence time in the gastrointestinal tract [244–247].

10.6.2.2 Bioreactor for delivery of therapeutic products

The curative effect is commonly achieved by delivery of a single drug or biopharmaceutical. However, as far as the gastrointestinal tract is concerned, the breakdown of an unwanted metabolite or the delivery of recombinant products are often better carried out by microorganisms. In such cases, various microorganisms with their complex metabolism are used as a microreactor-performing therapy [4].

Advances in gene engineering have enabled us to obtain microorganisms capable of producing a big variety of products possessing special therapeutic properties including specific enzymes, peptides, and cytokines [4, 248]. Beyond the functional versatility of engineered microorganisms, mass production of bacterial pharmaceuticals appears as a cost-efficient and sustainable method because of the short replication times and hardy growth characteristic of various bacterial strains, making the therapy popular and affordable. The general concept of using bacterial cells for therapy is shown in Fig. 10.12.

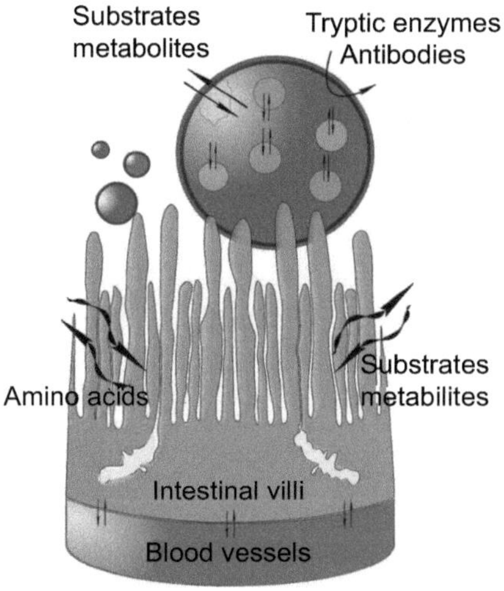

Figure 10.12 Oral delivery of encapsulated genetically engineered microorganisms to gut for disease treatment. The capsule protects bacteria by excluding components of the immune system and digestive enzymes allowing the low–molecular weight substrates, metabolites, and bacteria produced therapeutics to diffuse out for *in situ* action. Reproduced from Ref. [4] under Creative Commons Attribution License.

Despite the promising strategy, oral cell therapy with living, lyophilized, and immobilized entities have been attempted with little success. One of the recognized possible reasons for that was a failure to deliver microorganisms to the correct target tissues due to the severe damage of bacterial cells in the gastrointestinal tract where they are exposed to high acidity in the stomach and to enzymatic attacks. Given orally, many therapeutic cells can also provoke immunogenic responses and be eliminated by the patient immune system [249].

Recent studies on genetically engineered *E. coli* DH 5 cells confirmed that the viability of cells in the gut and their controlled delivery can be improved by microencapsulation in APA envelopes. It was found that the capsules didn't hinder cell growth kinetics. The semipermeable membranes of the microcapsules allow the immobilized microorganisms to metabolize small molecules but serve as a proper protection during their passage through the gut. Stool samples collected from experimental rats ca. 24 h after oral administration revealed the microcapsules were in the intact form [250]. It is important that the encapsulated cells were also excreted with the microcapsules, so that the safety concerns about the clinical use of live bacterial cells can be eliminated.

Beyond cell therapy, encapsulated bacteria are demanded in bioelectronic devices and can be used for the development of novel biosorbents and biocatalysts and for fundamental research in the form of engineered 3D assemblies mimicking the structure of primitive colonial and multicellular organisms [46].

10.7 Conclusion

The chapter summarizes the materials, methods, properties, functions, and main advances in the enveloping of living cells: mammalian somatic and reproductive cells, stem cells, and therapeutic microorganisms. Despite the proven benefits and significant progress foreseen in the treatment of many degenerative and infectious diseases, encapsulation of cells that involves molecular self-assembly on the micro- and nanoscale creates ethical concerns on the transition of the research into clinical practice due to the reproducibility of encapsulation protocols. Thus, a major effort is expected to be made to ensure safety of the polymeric cell envelopes for clinical applications.

References

1. Uludag H., De Vos P., Tresco P. A., Technology of mammalian cell encapsulation, *Advanced Drug Delivery Reviews*, **42** (2000) 29–64.
2. de Vos P., Marchetti P., Encapsulation of pancreatic islets for transplantation in diabetes : the untouchable islets, *Trends in Molecular Medicine*, **8** (2002) 363–366.
3. Nafea E. H., Poole-Warren A. M. L. A., Martens P. J., Immunoisolating semi-permeable membranes for cell encapsulation: focus on hydrogels, *Journal of Controlled Release*, **154** (2011) 110–122.
4. Prakash S., Jones M. L., Artificial cell therapy: new strategies for the therapeutic delivery of live bacteria, *Journal of Biomedicine and Biotechnology*, **2005** (2005) 44–56.
5. Chávarri M., Marañón I., Villarán M. C., Chapter 23: Probiotics, *Encapsulation Technology to Protect Probiotic Bacteria*, Rigobelo E. C. (ed.), (2012), (InTech)doi: 10.5772/50046.
6. Gisela G., Leonardo A. E., Lucia P., Rodrigo V., Eduard G., Angeles C. M., Enhancement of the viability of *Lactobacillus plantarum* during the preservation and storage process based on the response surface methodology, Food and Nutrition Sciences, **5** (2014) 1746–1755.
7. Li R. H., Materials for immunoisolated cell transplantation, *Advanced Drug Delivery Reviews*, **33** (1998) 87–109.
8. De Vos P., Bucko M., Gemeiner P., Navrátil M., Svitel J., Faas M., *et al.*, Multiscale requirements for bioencapsulation in medicine and biotechnology, *Biomaterials*, **30** (2009) 2559–2570.
9. Cohen S., Bafi M. C., Visscher K. B., Chow M., Allcock H. R., Langer R., Ionically cross-linkable polyphosphazene: anovel polymer for microencapsulation, *Journal of the American Chemical Society*, **112** (1990) 7832–7833.
10. Brissová M., Petro M., Lacík. I., Powers A. C., Wang T., Evaluation of microcapsule permeability via inverse size exclusion chromatography, *Analytical Biochemistry*, **111** (1996) 104–111.
11. King G. A., Daugulis A. J., Faulkner P., Goosen M. F. A., Alginate-polylysine microcapsules of controlled membrane molecular weight cutoff for mammalian cell culture engineering, *Biotechnology Progress*, **3** (1987) 231–240.
12. Lacík I., Polymer chemistry in diabetes treatment by encapsulated islets of Langerhans : review to 2006, *Australian Journal of Chemistry*, **59** (2006) 508–524.

13. Emerich D. F., Winn S. R., Immunoisolation cell therapy for CNS diseases, *Critical Reviews in Therapeutic Drug Carrier Systems*, **8** (2001) 265–298.

14. Orive G., Hernández R. M., Rodríguez Gascón A., Calafiore R., Chang T. M., De Vos P., *et al.*, History, challenges and perspectives of cell microencapsulation, *Trends in Biotechnology*, **22** (2004) 87–92.

15. Chipara M. D., Chipara M., Shansky E., Zaleski J. M., Self-healing of high elasticity block copolymers, *Polymers for Advanced Technologies*, **20** (2009) 427–431.

16. Sukhorukov G. B., Fery A., Brumen M., Moehwald H., Physical chemistry of encapsulation and release, *Physical Chemistry Chemical Physics*, **6** (2004) 4078–4089.

17. Antipov A. A., Sukhorukov G. B., Polyelectrolyte multilayer capsules as vehicles with tunable permeability, *Advances in Colloid and Interface Science*, **111** (2004) 49–61.

18. Schmidt J. J., Rowley J., Kong H. J., Hydrogels used for cell-based drug delivery, *Journal of Biomedical Materials Research, Part A*, **87** (2008) 1113–1122.

19. Hou Q. P., Bae Y. H., Biohybrid artificial pancreas based on macrocapsule device, *Advanced Drug Delivery Reviews*, **35** (1999) 271–287.

20. Moussy F., Rouchette J., Reach G., Cannon R., Jaffrin M. Y., *In vitro* evaluation of a bioartificial pancreas under various hemodynamic conditions, *Artificial Organs*, **13** (1989) 109–115.

21. Hanu L. G., Alessi P., Kilzer A., Kareth S., Manufacturing and characterization of water filled micro-composites, *Journal of Supercritical Fluids*, **66** (2012) 274–281.

22. Anseth K. S., Bowman C. N., Brannon-Peppas L., Mechanical properties of hydrogels and their experimental determination, *Biomaterials*, **17** (1996) 1647–1657.

23. Martens P., Blundo J., Nilasaroya A., Odell R. A., Cooper-White J., Poole-Warren L. A., Effect of poly(vinyl alcohol) macromer chemistry and chain interactions on hydrogel mechanical properties, *Chemistry of Materials*, **19** (2007) 2641–2648.

24. Schneider S., Feilen P. J., Slotty V., Kampfner D., Preuss S., Berger S., *et al.*, Multilayer capsules : a promising microencapsulation system for transplantation of pancreatic islets, *Biomaterials*, **22** (2001) 1961–1970.

25. Inoue K., Fujisato T., Gu Y. J., Burczak K., Sumi S., Kogire M., *et al.*, Experimental hybrid islet transplantation : application of polyvinyl alcohol membrane for entrapment of islets, *Pancreas*, **7** (1992) 562–568.

26. Kung I. M., Wang F. F., Chang Y. C., Wang Y. J., Surface modifications of alginate/poly(L-lysine) microcapsular membranes with poly(ethylene glycol) and poly(viny1 alcohol), *Biomaterials*, **16** (1995) 649–655.

27. Stewart D. J., Courtman D., Karoubi G., Encapsulated cell therapy, US20060251630 A1, Nov. 9 (2006).

28. Silva A. I., Mateus M., Development of a polysulfone hollow fiber vascular bio-artificial pancreas device for *in vitro* studies, Journal of Biotechnology, **139** (2009) 236–249.

29. Lelli L., Barbani N., Bonaretti A., Giusti P., Marchetti P., Giannarelli R., Navalesi R., Viacava P., Preparation and characterization of permselective, biocompatible membranes for the macroencapsulation of pancreatic islets, *Journal of Materials Science: Materials in Medicine*, **5** (1994) 887–890.

30. Wang G. T., Immunoisolation patch system for cellular transplantation, US8673294 B2, Mar. 18 (2014).

31. Xavier A. J., Cell encapsulation as a potential non-dietary therapy for maternal phenylketonuria, *MS (Master of Science) thesis, University of Iowa*, (2012).

32. Kailasapathy K., Microencapsulation of probiotic bacteria: technology and potential applications, *Current Issues in Intestinal Microbiology*, **3** (2002) 39–48.

33. Aebischer P., Wahlberg L., Cell culture-containing tubular capsule produced by co-extrusion, US5283187 A, Feb. 1 (1994).

34. Aebischer P., Wahlberg L., Method of encapsulating cells in a tubular extrudate, US5284761 A, Feb. 8 (1994).

35. Goosen M. F. A., O'Shea G. M., Sun A. M. F., Microencapsulation of living tissue and cells, US4689293 A, Aug. 25 (1987).

36. Hommel M., Sun A. M. F., Goosen M. F. A., Microcapsule composition suitable for cardiovascular injection, US4789550 A, Dec. 6 (1988).

37. Goosen M. F. A., O'Shea G. M., Sun A. M. F., Microencapsulation of living tissue and cells, US4673566 A, Jun. 16 (1987).

38. Pick J. M., Creed B. T., Protecting an icemaker against over current damage, US4791523 A., Dec. 13 (1988).

39. Jarvis A. P., Microencapsulation with polymers, US4803168 A, Feb. 7 (1989).

40. Lim F., Encapsulation of biological material, US4352883 A, Oct. 5 (1982).

41. Lim F., Microcapsules containing viable tissue cells, US4391909 A, Jun. 5 (1983).

42. Ronel S. H., D'Andrea M. J., Dobelle W. H., Klomp G. F., Hashiguchi H., Porous hydrophilic materials, chambers therefrom, and devices comprising such chambers and biologically active tissue and methods of preparation, US4298002 A, Nov. 3 (1981).

43. Sefton M. V., Encapsulation of live animal cells, US4353888 A, Oct. 12 (1982).

44. Dubey R., Shami T. C., Bhasker Rao K. U., Microencapsulation technology and applications, Defence Science Journal, **59** (2009) 82–95.

45. Freyman T. M., Levendusky J. A., Microencapsulated compositions for endoluminal tissue engineering, US8518682 B2, Aug. 27 (2013).

46. Fakhrullin R. F., Zamaleeva A. I., Minullina R. T., Konnova S. A., Paunov V. N., Cyborg cells: functionalisation of living cells with polymers and nanomaterials, *Chemical Society Reviews*, **41** (2012) 4189–4206.

47. Sukhorukov G. B., Donat E. H., Moya S., Susha A. S., Voigt A., Hartmann J., Möhwald H., Microencapsulation by means of step-wise adsorption of polyelectrolytes, *Journal of Microencapsulation*, **17** (2000) 177–185.

48. She Z., Antipina M. N., Li J., Sukhorukov G. B., Mechanism of protein release from polyelectrolyte multilayer microcapsules, *Biomacromolecules*, **11** (2010) 1241–1247.

49. Antipina M. N., Kiryukhin M. V., Skirtach A. G., Sukhorukov G. B., Micropackaging via layer-by-layer assembly: microcapsules and microchamber arrays, *International Materials Reviews*, **59** (2014) 224–244.

50. Wilson J. T., Cui W. X., Kozlovskaya V., Kharlampieva E., Pan D., Qu Z., *et al.*, Cell surface engineering with polyelectrolyte multilayer thin films, *Journal of the American Chemical Society*, **133** (2011) 7054–7064.

51. Mansouri S., Merhi Y., Winnik F. M., Tabrizian M., Investigation of layer-by-layer assembly of polyelectrolytes on fully functional human red blood cells in suspension for attenuated immune Response, *Biomacromolecules*, **12** (2011) 585–592.

52. Antipina M. N., Sukhorukov G. B., Remote control over guidance and release properties of composite polyelectrolyte based capsules, *Advanced Drug Delivery Reviews*, **63** (2011) 716–729.

53. Lomova M. V., Sukhorukov G. B., Antipina M. N., Antioxidant coating of micronsize droplets for prevention of lipid peroxidation in oil-in-water emulsion, *ACS Applied Materials & Interfaces*, **2** (2010) 3669–3676.

54. Sadovoy A. V., Kiryukhin M. V., Sukhorukov G. B., Antipina M. N., Kinetic stability of water-dispersed oil droplets encapsulated in a polyelectrolyte multilayer shell, *Physical Chemistry Chemical Physics*, **13** (2011) 4005–4012.

55. Antipov A. A., Sukhorukov G. B., Leporatti S., Radtchenko I. L., Donath E., Möhwald H., Polyelectrolyte multilayer capsule permeability control, Colloids and Surfaces A: Physicochemical and Engineering Aspects, **198–200** (2002) 535–541.

56. Antipina M. N., Kiryukhin M. V., Chong K., Low H. Y., Sukhorukov G. B., Patterned microcontainers as novel functional elements for microTAS and LOC, *Lab on a Chip*, **9** (2009) 1472–1475.

57. Mauser T., Dejugnat C., Sukhorukov G. B., Reversible pH-dependent properties of multilayer microcapsules made of weak polyelectrolytes, *Macromolecular Rapid Communications*, **25** (2004) 1781–1785.

58. Kozlovskaya V., Kharlampieva E., Mansfield M. L., Sukhishvili S. A., Poly(methacrylic acid) hydrogel films and capsules: response to pH and ionic strength, and encapsulation of macromolecules, Chemistry of Materials, **18** (2006) 328–336.

59. Erel-Unal I., Sukhishvili S. A., Hydrogen-bonded multilayers of a neutral polymer and a polyphenol, *Macromolecules*, **41** (2008) 3962–3970.

60. Kozlovskaya V., Harbaugh S., Drachuk I., Shchepelina O., Kelley-Loughnane N., Stone M., *et al.*, Hydrogen-bonded LbL shells for living cell surface engineering, *Soft Matter*, **7** (2011) 2364–2372.

61. Lomova M. V., Brichkina A. I., Kiryukhin M. V., Vasina E. N., Pavlov A. M., Gorin D. A., *et al.*, Multilayer capsules of bovine serum albumin and tannic acid for controlled release by enzymatic degradation, ACS Applied Materials & Interfaces, **7** (2015) 11732–11740.

62. Germain M., Balaguer P., Nicolas J. C., Lopez F., Esteve J. P., Sukhorukov G. B., *et al.*, Protection of mammalian cell used in biosensors by coating with a polyelectrolyte shell, *Biosensors & Bioelectronics*, **21** (2006) 1566–1573.

63. Geckil H., Zhang X., Moon S., Demirci U., Engineering hydrogels as extracellular matrix mimics, *Nanomedicine (Lond.)*,**5** (2010) 469–484.

64. Yu L., Ding J., Injectable hydrogels as unique biomedical materials, *Chemical Society Reviews*, **37** (2008) 1473–1481.

65. Kobayashi M., Hyu H. S., Development and evaluation of polyvinyl alcohol-hydrogels as an artificial atrticular cartilage for orthopedic implants, *Materials*, **3** (2010) 2753–2771.

66. Yang J. M., Li J. C., Wang X. L., Li X. M., Kawazoe N., Chen G. P., Single mammalian cell encapsulation by *in situ* polymerization, *Journal of Materials Chemistry B*, **4** (2016) 7662–7668.

67. Li M., Rouaud O., Poncelet D., Microencapsulation by solvent evaporation: state of the art for process engineering approaches, *International Journal of Pharmaceutics*, **363** (2008) 26–39.

68. Chaikof E. L., Engineering and material considerations in islet cell transplantation, *Annual Review of Biomedical Engineering*, **1** (1999) 103–127.

69. Read T., Sorensen D. R., Mahesparan R., Enger P. O., Timpl R., Olsen B. R., *et al.*, Local endostatin treatment of gliomas administered by microencapsulated producer cells, *Nature Biotechnology*, **19** (2001) 29–34.

70. Joki T., Machluf M., Atala A., Zhu J., Seyfried N. T., Dunn I. F., *et al.*, Continuous release of endostatin from microencapsulated engineered cells for tumor therapy, *Nature Biotechnology*, **19** (2001) 35–39.

71. Lim F., Sun A. M., Microencapsulated islets as bioartificial endocrine pancreas, *Science*, **210** (1980) 908–910.

72. Arshady R., Microspheres and microcapsules: a survey of manufacturing techniques. Part 1: suspension cross-linking, *Polymer Engineering and Science*,**29** (1989) 1746–1758.

73. Katti D., Krishnamurti N., Preparation of albumin microspheres by an im proved process, *Journal of Microencapsulation*,**16** (1999) 231–242.

74. Widder K. J., Senyei A. E., Sears B., Experimental methods in cancer therapeutics, *Journal of Pharmaceutical Sciences*, **71** (1982) 379–387.

75. Park J. K., Chang H. N., Microencapsulation of microbial cells, Biotechnology Advances,**18** (2000) 303–319.

76. Winterhalter M., Hilty C., Bezrukov S. M., Nardin C., Meier W., Fournier D., Controlling membrane permeability with bacterial porins: application to encapsulated enzymes, *Talanta*, **55** (2001) 965–971.

77. Schirmer T., Keller T. A., Wang Y., Rosenbusch J. P., Structural basis for sugar translocation through maltoporin channels at 3.1 A resolution, *Science*, **267** (2016) 512–514.

78. Saint N., Lou K. L., Widmer C., Luckey M., Schirmer T., Rosenbusch J. P., Structural and functional characterization of OmpF porin mutants selected for larger pore size. II. Functional characterization, *Journal of Biological Chemistry*, **271** (1996) 20676–20680.

79. Gunatillake P. A., Adhikari R., Biodegradable synthetic polymers for tissue engineering, *European Cells & Materials*, **5** (2003) 1–16.

80. Tan H., Gong Y., Lao L., Mao Z., Gao C., Gelatin/chitosan/hyaluronan ternary complex scaffold containing basic fibroblast growth factor for cartilage tissue engineering, *Journal of Materials Science: Materials in Medicine*,**18** (2007) 1961–1968.

81. Awad H. A., Wickham M. Q., Leddy H. A., Gimble J. M., Guilak F., Chondrogenic differentiation of adipose-derived adult stem cells in agarose, alginate, and gelatin scaffolds, *Biomaterials*, **25** (2004) 3211–3222.

82. Tan H., Marra K. G., Injectable, biodegradable hydrogels for tissue engineering applications, *Materials*, **3** (2010) 1746–1767.

83. De Vos P., Hamel A. F., Tatarkiewicz K., Considerations for successful transplantation of encapsulated pancreatic islets, *Diabetologia*, **45** (2002) 159–173.

84. Mortazavian A., Razavi S. H., Ehsani M. R., Sohrabvandi S., Principles and methods of microencapsulation of probiotic microorganisms, *Iran Journal of Biotechnology*, **5** (2007) 1–18.

85. Kulseng B., Skjåk-Braek G., Ryan L., Andersson A., King A., Faxvaag A., *et al.*, Transplantation of alginate microcapsules: generation of antibodies against alginates and encapsulated porcine islet-like cell clusters, *Transplantation*, **67** (1999) 978–984.

86. Prakash S., Chang T. M., Preparation and *in vitro* analysis of microencapsulated genetically engineeredE. coli DH5 cells for urea and ammonia removal, *Biotechnology and Bioengineering*, **46** (1995) 621–626.

87. Awrey D. E., Tse M., Hortelano G., Chang P. L., Permeability of alginate microcapsules to secretory recombinant gene products, *Biotechnology and Bioengineering*, **52** (1996) 472–484.

88. Calafiore R., Basta G., Luca G., Boselli C., Bufalari A., Bufalari A., *et al.*, Transplantation of pancreatic islets contained in minimal volume microcapsules in diabetic high mammalians, *Annals of the New York Academy of Sciences*, **875** (1999) 219–232.

89. Wang T., Lacík I., Brissová M., Anilkumar A. V., Prokop A., Hunkeler D., *et al.*, An encapsulation system for the immunoisolation of pancreatic islets, *Nature Biotechnology*, **15** (1997) 358–362.

90. Tse M., Uludag H., Sefton M. V., Chang P. L., Secretion of recombinant proteins from hydroxyethyl methacrylate-methyl methacrylate capsules, *Biotechnology and Bioengineering*, **51** (1996) 271–280.

91. Iwata H., Takagi T., Kobayashi K., Oka T., Tsuji T., Ito F., Strategy for developing microbeads applicable to islet xenotransplantation into a spontaneous diabetic NOD mouse, *Journal of Biomedical Materials Research*, **28** (1994) 1201–1207.

92. Orive G., Hernández R. M., Gascón A. R., Igartua M., Rojas A., Pedraz J. L., Microencapsulation of an anti-VE-cadherin antibody secreting 1B5 hybridoma cells, *Biotechnology and Bioengineering*, **76** (2001) 285–294.

93. Zekorn T., Horcher A., Siebers U., Schnettler R., Klöck G., Hering B., *et al.*, Barium-cross-linked alginate beads: a simple, one-step method for successful immunoisolated transplantation of islets of Langerhans, *Acta Diabetologica*, **29** (1992) 99–106.

94. Dautzenberg H., Schuldt U., Grasnick G., Karle P., Müller P., Löhr M., *et al.*, Development of cellulose sulfate-based polyelectrolyte complex microcapsules for medical applications, *Annals of the New York Academy of Sciences*, **875** (1999) 46–63.

95. Bartkowiak A., Hunkeler D., New microcapsules based on oligoelectrolyte complexation, *Annals of the New York Academy of Sciences*, **875** (1999) 36–45.

96. Iwata H., Amemiya H., Hayashi R., Fujii S., Akutsu T., The use of photocrosslinkable polyvinyl alcohol in the immunoisolation of pancreatic islets, *Transplantation Proceedings*, **22** (1990) 797–799.

97. Cruise G. M., Hegre O. D., Lamberti F. V., Hager S. R., Hill R., Scharp D. S., *et al.*, *In vitro* and *in vivo* performance of porcine islets encapsulated in interfacially photopolymerized poly(ethylene glycol) diacrylate membranes, *Cell Transplant*, **8** (1999) 293–306.

98. Chen H., Ouyang W., Lawuyi B., Lim T., Prakash S., A new method for microcapsule characterization: use of fluorogenic genipin to characterize polymeric microcapsule membranes, *Applied Biochemistry and Biotechnology*, **134** (2006) 207–222.

99. Chen H., Ouyang W., Lawuyi B., Prakash S., Genipin cross-linked alginate-chitosan microcapsules: membrane characterization and optimization of cross-linking reaction, *Biomacromolecules*, **7** (2006) 2091–2098.

100. Chen M. L., Ge Z., Fox J. G., Schauer D. B., Disruption of tight junctions and induction of proinflammatory cytokine responses in colonic epithelial cells by *Campylobacter jejuni*, *Infect. Immun.*, **74** (2006) 6581–6589.

101. Orive G., Hernández R. M., Gascón A. R., Igartua M., Pedraz J. L., Survival of different cell lines in alginate-agarose microcapsules, *European Journal of Pharmaceutical Sciences*, **18** (2003) 23–30.

102. Orive G., Hernández R. M., Gascón A. R., Igartua M., Pedraz J. L., Development and optimisation of alginate-PMCG-alginate microcapsules for cell immo- bilisation, *International Journal of Pharmaceutics*, **259** (2003) 57–68.

103. Bucko M., Vikartovsk A., Lac I., Koll G., Brygin M., Immobilization of a whole-cell epoxide-hydrolyzing biocatalyst in sodium alginate–cellulose sulfate–poly(methylene-*co*-guanidine) capsules using a controlled encapsulation process, Enzyme and Microbial Technology, **36** (2005) 118–126.

104. Bučko M., Vikartovská V., Gemeiner P., Lacík I., Kolláriková G., Marison I. W., *Nocardia tartaricans* cells immobilized in sodium alginate–cellulose sulfate–poly(methylene-co-guanidine) capsules: mechanical resistance and operational stability, *Journal of Chemical Technology & Biotechnology*, **81** (2006) 500–504.

105. Brissová M., Lacík I., Powers A. C., Anilkumar A. V., Wang T., Control and measurement of permeability for design of microcapsule cell delivery system, *Journal of Biomedical Materials Research*, **39** (1998) 61–70.

106. Lacík I., Brissová M., Anilkumar A. V., Powers A. C., Wang T., New capsule with tailored properties for the encapsulation of living cells, *Journal of Biomedical Materials Research*, **39** (1998) 52–60.

107. Nurdin N., Canaple L., Bartkowiak A., Desvergne B., Hunkeler D., Capsule permeability via polymer and protein ingress/egress, Journal of Applied Polymer Science, **75** (2000) 1165–1175.

108. Grigorescu G., Rehor A., Hunkeler D., Polyvinylamine hydrochloride-based microcapsules: polymer synthesis, permeability and mechanical properties, *Journal of Microencapsulation*, **19** (2002) 245–259.

109. Sefton M. V., May M. H., Lahooti S., Babensee J. E., Making microencapsulation work: conformal coating, immobilization gels and *in vivo* performance, *Journal of Controlled Release*, **65** (2000) 173–186.

110. Vallbacka J. J., Nobrega J. N., Sefton M. V., Tissue engineering as a platform for controlled release of therapeutic agents: implantation of microencapsulated dopamine producing cells in the brains of rats, *Journal of Controlled Release*, **72** (2001) 93–100.

111. Kovacs-Nolan J., Mine Y., Microencapsulation for the gastric passage and controlled intestinal release of immunoglobulin Y, *Journal of Immunological Methods*, **296** (2005) 199–209.

112. Zhu J. H., Wang X. W., Ng S., Quek C. H., Ho H. T., Lao X. J., *et al.*, Encapsulating live cells with water-soluble chitosan in physiological conditions, *Journal of Biotechnology*, **117** (2005) 355–365.

113. Chia S. M., Wan A. C. A., Quek C. H., Mao H. Q., Xu X., Shen L., *et al.*, Multi-layered microcapsules for cell encapsulation, *Biomaterials*, **23** (2002) 849–856.

114. Yin C, Chia M. S., Chai H. Q., Yu H., Zhuo R. X., Leong K. W., *et al.*, Microcapsules with improved mechanical stability for hepatocyte culture, *Biomaterials*, **24** (2003) 1771–1780.

115. Quek C. H., Li J., Sun T., Chan M. L., Mao H. Q., Gan L. M., *et al.*, Photo-crosslinkable microcapsules formed by polyelectrolyte copolymer and modified collagen for rat hepatocyte encapsulation, *Biomaterials*, **25** (2004) 3531–3540.

116. Mei L. H., Yao S. J., Cultivation and modelling of encapsulated *Saccharomyces cerevisiae* in NaCS-PDMDAAC polyelectrolyte complexes, *Journal of Microencapsulation*, **19** (2002) 397–405.

117. Zhang J., Yao S. J., Ying X. J., Guan Y. X., Lin D. Q., Immobilization of cells by macro-porous NaCS-PDMDAAC capsules and cultivation in shaking flask and bubble bioreactor, *Chinese Journal of Biotechnology*, **21** (2005) 633–637.

118. Zhang X., Wang W., Xie Y., Zhang Y., Wang X., Guo X., *et al.*, Proliferation, viability, and metabolism of human tumor and normal cells cultured in microcapsule, *Applied Biochemistry and Biotechnology*, **134** (2006) 61–76.

119. Zhang Z. W., Dorrell N., Wren B. W., Farthingt M. J., Helicobacter pylori adherence to gastric epithelial cells: a role for non-adhesin virulence genes, *Journal of Medical Microbiology*, **51** (2002) 495–502.

120. Yao S. J., Guan Y. X., Lin D. Q., Mass transfer behavior of solutes in NaCS–PDMDAAC capsules, *Industrial & Engineering Chemistry Research*, **45** (2006) 1811–1816.

121. Orive G., Hernández R. M., Gascón A. R., Calafiore R., Chang T. M., De Vos P., *et al.*, Cell encapsulation: promise and progress, *Nature Medicine*, **9** (2003) 104–107.

122. Opara E., YooJ. J., Saul J. M., Sivanandane S., Atala A., Encapsulated cells for hormone replacement therapy, US9283251 B2, Mar 15 (2016).

123. Bezwada R. S., Hydrolysable linkers and cross-linkers for absorbable polymers, US 8048980 B2, Nov. 1(2011).

124. Hubbell J. A., Pathak C. P., SawhneyA. S., Desai N. P., Hill-West J. L., Photopolymerizable biodegradable hydrogels as tissue contacting materials and controlled-release carriers, , US 5567435 A, Oct. 22 (1996).

125. Cock L. J. De., Koker S. De., Geest B. G. De., Grooten J., Vervaet C., Remon J. P., *et al.*, Polymeric multilayer capsules in drug delivery, *Angewandte Chemie International Edition*, **49** (2010) 6954–6973.

126. He P., Takahashi T., Hoshi T., Anzai J., Suzuki Y., Osa T., Preparation of enzyme multilayers on electrode surface by use of avidin and biotin-labeled enzyme for biosensor applications, *Materials Science and Engineering: C*, **2** (1994) 103–106.

127. Johnston A. P. R., Cortez C., Angelatos A. S., Caruso F., Layer-by-layer engineered capsules and their applications, *Current Opinion in Colloid & Interface Science*, **11** (2006) 203–209.

128. Behravesh E., Sikavitsas V. I., Mikos A. G., Quantification of ligand surface concentration of bulk-modified biomimetic hydrogels, *Biomaterials*, **24** (2003) 4365–4374.

129. Anseth K. S., Metters A. T., Bryant S. J., Martens P. J., Elisseeff J. H., *et al.*, *In situ* forming degradable networks and their application in tissue engineering and drug delivery, *Journal of Controlled Release*, **78** (2002) 199–209.

130. Guillaudeu S. J., Fox M. E., Haidar Y. M., Dy E. E., Szoka F. C., Fréchet J. M. J., PEGylated dendrimers with core functionality for biological applications, *Bioconjugate Chemistry*, **19** (2008) 461–469.

131. Feng X., Taton D., Chaikof E. L., Gnanou Y., Bouquet-type dendrimerlike poly(ethylene oxide)s with a focal aldehyde and peripheral hydroxyls, *Biomacromolecules*, **8** (2007) 2374–2378.

132. Hiemstra C., Zhong Z., Li L., Dijkstra P. J., Feijen J., *In-situ* formation of biodegradable hydrogels by stereocomplexation of PEG-(PLLA)8 and PEG-(PDLA)8 star block copolymers, *Biomacromolecules,* **7** (2006) 2790–2795.

133. Hiemstra C., Zhong Z., Tomme S. R. Van., Steenbergen M. J. Van., Jacobs J. J. L., Otter W. Den., *et al., In vitro* and *in vivo* protein delivery from *in situ* forming poly(ethylene glycol)-poly(lactide) hydrogels, *Journal of Controlled Release,* **119** (2007) 320–327.

134. Holland T. A., Tessmar J. K. V., Tabata Y., Mikos A. G., Transforming growth factor-beta 1 release from oligo(poly(ethylene glycol) fumarate) hydrogels in conditions that model the cartilage wound healing environment, *Journal of Controlled Release,* **94** (2004) 101–114.

135. Leach J. B., Bivens K. A., Collins C. N., Schmidt C. E., Development of photocrosslinkable hyaluronic acid- polyethylene glycol-peptide composite hydrogels for soft tissue engineering, *Journal of Biomedical Materials Research A,* **70** (2004) 74–82.

136. Wieland J. A., Houchin-Ray T. L., Shea L. D., Non-viral vector delivery from PEG-hyaluronic acid hydrogels, *Journal of Controlled Release,* **120** (2007) 233–241.

137. Almany L., Seliktar D., Biosynthetic hydrogel scaffolds made from fibrinogen and polyethylene glycol for 3D cell cultures, *Biomaterials,* **26** (2005) 2467–2477.

138. Bhattarai N., Ramay H. R., Gunn J., Matsen F. A., Zhang M., PEG-grafted chitosan as an injectable thermosensitive hydrogel for sustained protein release, *Journal of Controlled Release,* **103** (2005) 609–624.

139. Yamaguchi N., Chae B., Zhang L., Kiick K. L., Furst E. M., Rheological characterization of polysaccharide-poly(ethylene glycol) star copolymer hydrogels, *Biomacromolecules,* **6** (2005) 1931–1940.

140. Nicodemus G. D., Villanueva I., Bryant S. J., Mechanical stimulation of TMJ condylar chondrocytes encapsulated in PEG hydrogels, *Journal of Biomedical Materials Research A,* **83** (2007) 323–331.

141. Brown C. D., Stayton P. S., Hoffman A. S., Semi-interpenetrating network of poly(ethylene glycol) and poly(D, L-lactide) for the controlled delivery of protein drugs, *Journal of Biomaterials Science, Polymer Edition,* **16** (2005) 189–201.

142. Song S., Faleo G., Yeung R., Kant R., Posselt A. M., Desai T. A., *et al.,* Silicon nanopore membrane (SNM) for islet encapsulation and immunoisolation under convective transport, *Scientific Reports,* **6,** 23679. (2016)

143. Zhan M. J., Li G., Wei Q., Cui H. L., Lin L., Bioactivity of TiO2/Ti composite membrane with different crystral phase, *Advanced Materials Research*, **287–290** (2011) 69–72.

144. El-Ghannam A., Bone reconstruction: from bioceramics to tissue engineering, *Expert Review of Medical Devices*, **2** (2005) 87–101.

145. Sittadjody S., Saul J. M., Joo S., Yoo J. J., Atala A., Opara E. C., Engineered multilayer ovarian tissue that secretes sex steroids and peptide hormones in response to gonadotropins, *Biomaterials*, **34** (2013) 2412–2420.

146. Chang P. L., Hortelano G., Awrey D. E., Tse M., Growth of recombinant fibroblasts in alginate microcapsules, *Biotechnology and Bioengineering*, **43** (1994) 925–933.

147. Salmons B., Saller R. M., Baumann J. G., Günzburg W. H., Construction of retroviral vectors for targeted delivery and expression of therapeutic genes, *Leukemia*, **9(Suppl. 1)** (1995) S53–S60.

148. Salmons B., Groner B., Calberg-Bacq C. M., Ponta H., Production of mouse mammary tumor virus upon transfection of a recombinant proviral DNA into cultured cells, *Virology*, **144** (1985) 101–114.

149. Salmons B., Gunzburg W. H., Targeting of retroviral vectors for gene therapy, *Human Gene Therapy*, **4** (1993) 129–141.

150. Tai I. T., Sun A. M., Microencapsulation of recombinant cells: a new delivery system for gene therapy, *FASEB Journal*, **7** (1993) 1061–1069.

151. Hughes M., Vassilakos A., Andrews D. W., Hortelano G., Belmont J. W., Chang P. L., Delivery of a secretable adenosine deaminase through microcapsules– a novel approach to somatic gene therapy, *Human Gene Therapy*, **5** (1994) 1445–1455.

152. Date I., Shingo T., Ohmoto T., Emerich D. F., Long-term enhanced chromaffin cell survival and behavioral recovery in hemiparkinsonian rats with co-grafted polymer-encapsulated human NGF-secreting cells, *Experimental Neurology*, **147** (1997) 10–17.

153. Löhr M., Hummel F., Faulmann G., Ringel J., Saller R., Hain J*et al.*, Microencapsulated, CYP2B1-transfected cells activating ifosfamide at the site of the tumor: the magic bullets of the 21st century, *Cancer Chemotherapy and Pharmacology*, **49(Suppl. 1)** (2002) S21–S24.

154. Zhang Y., Wang W., Xie Y., Yu W., Teng H., Liu X., *et al.*, *In vivo* culture of encapsulated endostatin-secreting Chinese hamster ovary cells for systemic tumor inhibition, *Human Gene Therapy*, **18** (2007) 474–481.

155. Savelkoul H. F., van Ommen R., Vossen A. C., Breedland E. G., Coffman R. L., van Oudenaren A., Modulation of systemic cytokine levels by implantation of alginate encapsulated cells, *Journal of Immunological Methods*, **170** (1994) 185–196.

156. Gates R. J., Lazarus N. R., Reversal of streptozotocin-induced diabetes in rats by intraperitoneal implantation of encapsulated neonatal rabbit pancreatic tissue, *The Lancet.*, **2** (1977) 1257–1259.

157. Chandy T., Mooradian D. L., Rao G. H., Evaluation of modified alginate-chitosan-polyethylene glycol microcapsules for cell encapsulation, *Artificial Organs*, **23** (1999) 894–903.

158. Desai N. P., Sojomihardjo A., Yao Z., Ron N., Soon-Shiong P., Interpenetrating polymer networks of alginate and polyethylene glycol for encapsulation of islets of Langerhans, *Journal of Microencapsulation*, **17** (2000) 677–690.

159. Lanza R. P., Jackson R., Sullivan A., Ringeling J., McGrath C., Kühtreiber W., *et al.*, Xenotransplantation of cells using biodegradable microcapsules, *Transplantation*, **67** (1999) 1105–1111.

160. Strand B. L., Gåserød O., Kulseng B., Espevik T., Skjåk-Braek G., Alginate-polylysine-alginate microcapsules: effect of size reduction on capsule properties, *Journal of Microencapsulation*, **19** (2002) 615–630.

161. Krol S., del Guerra S., Grupillo M., Diaspro A., Gliozzi A., Marchetti P., Multilayer nanoencapsulation. New approach for immune protection of human pancreatic islets, *Nano Letters*, **6** (2006) 1933–1939.

162. Miyamoto M., Inoue K., Gu Y., Tun T., Cui W., Fujiwara I., *et al.*, Improved large-scale isolation of breeder porcine islets: possibility of harvesting from nonheart-beating donor, *Cell Transplantation*, **7** (1998) 397–402.

163. Goto M., Johansson H., Maeda A., Elgue G., Korsgren O., Nilsson B., Low molecular weight dextran sulfate prevents the instant blood-mediated inflammatory reaction induced by adult porcine islets, *Transplantation*, **77** (2004) 741–747.

164. Petersen P., Lembert N., Zschocke P., Stenglein S., Planck H., Ammon H. P., *et al.*, Hydroxymethylated polysulphone for islet macroencapsulation allows rapid diffusion of insulin but retains PERV, *Transplantation Proceedings*, **34** (2002) 194–195.

165. Lembert N., Wesche J., Petersen P., Zschocke P., Enderle A., Planck H., *et al.*, Macroencapsulation of rat islets without alteration of insulin secretion kinetics, *Experimental and Clinical Endocrinology & Diabetes*, **109** (2001) 116–119.

166. Lembert N., Wesche J., Petersen P., Doser M., Zschocke P., Becker H. D., *et al.*, Encapsulation of islets in rough surface, hydroxymethylated polysulfone capillaries stimulates VEGF release and promotes vascularization after transplantation, *Cell Transplantation*, **14** (2005) 97–108.

167. Sakata N., Encapsulated islets transplantation: Past, present and future, *World Journal of Gastrointestinal Pathophysiology*, **3** (2012) 19–26.

168. Chia S. M., Leong K. W., Li J., Xu X., Zeng K., Er P. N., *et al.*, Hepatocyte encapsulation for enhanced cellular functions, *Tissue Engineering*, **6** (2000) 481–495.

169. Aoki T., Jin Z., Nishino N., Kato H., Shimizu Y., Niiya T., *et al.*, Intrasplenic transplantation of encapsulated hepatocytes decreases mortality and improves liver functions in fulminant hepatic failure from 90% partial hepatectomy in rats, *Transplantation*, **79** (2005) 783–790.

170. Sgroi A., Mai G., Morel P., Baertschiger R. M., Gonelle-Gispert C., Serre-Beinier V., *et al.*, Transplantation of encapsulated hepatocytes during acute liver failure improves survival without stimulating native liver regeneration, *Cell Transplant*, **20** (2011) 1791–1803.

171. Culver K. W., Ram Z., Wallbridge S., Ishii H., Oldfield E. H., Blaese R. M., *In vivo* gene transfer with retroviral vector-producer cells for treatment of experimental brain tumors, *Science*, **256** (1992) 1550–1552.

172. Ram Z., Culver K. W., Walbridge S., Blaese R. M., Oldfield E. H. *In situ* retrovirus-mediated gene transfer for treatment of brain tumors in rats, *Cancer Research*, **53** (1993) 83–88.

173. Burdick J. A., Vunjak-Novakovic G., Engineered Microenvironments for controlled stem cell differentiation, *Tissue Engineering Part A*, **15** (2009) 205–219.

174. Watt F. M., Hogan B. L., Out of Eden: stem cells and their niches, *Science*, **287** (2000) 1427–1430.

175. Chan E.-S., Lim T.-K., Voo W.-P., Pogaku R., Tey B. T., Zhang Z. B., Effect of formulation of alginate beads on their mechanical behavior and stiffness, *Particuology*, **9** (2011) 228–234.

176. Park J., Cho C. H., Parashurama N., Li Y., Berthiaume F., Toner M., *et al.*, Microfabrication-based modulation of embryonic stem cell differentiation, *Lab on a Chip*, **7** (2007) 1018–1028.

177. Chan B. P., Hui T. Y., Wong M. Y., Yip K. H., Chan G. C., Mesenchymal stem cell–encapsulated collagen microspheres for bone tissue engineering, *Tissue Engineering Part C Methods*, **16** (2010) 225–235.

178. Chan B. P., Hui T. Y., Yeung C. W., Li J., Mo I., Chan G. C., Self-assembled collagen–human mesenchymal stem cell microspheres for regenerative medicine, *Biomaterials*, **28** (2007) 4652–4666.

179. Wilson J. L., McDevitt T. C., Stem cell microencapsulation for phenotypic control, bioprocessing, and transplantation, *Biotechnology and Bioengineering*, **110** (2013) 667–682.

180. Batorsky A., Liao J., Lund A. W., Plopper G. E., Stegemann J. P., Encapsulation of adult human mesenchymal stem cells within collagen-agarose microenvironments, *Biotechnology and Bioengineering*, **92** (2005) 492–500.

181. Sakai S., Hashimoto I., Kawakami K., Production of cell-enclosing hollow-core agarose microcapsules via jetting in water-immiscible liquid paraffin and formation of embryoid body-like spherical tissues from mouse ES cells enclosed within these microcapsules, *Biotechnology and Bioengineering*, **99** (2008) 235–243.

182. Paul A., Shum-Tim D., Prakash S., Investigation on PEG integrated alginate–chitosan microcapsules for myocardial therapy using marrow stem cells genetically modified by recombinant baculovirus, *Cardiovascular Engineering and Technology*, **1** (2010) 154–164.

183. West E. R., Shea L. D., Woodruff T. K., Engineering the follicle microenvironment, *Seminars in Reproductive Medicine*, **25** (2007) 287–299.

184. Matsusaki M., Kadowaki K., Nakahara Y., Akashi M., Fabrication of cellular multilayers with nanometer-sized extracellular matrix films, *Angewandte Chemie International Edition*, **46** (2007) 4689–4692.

185. Zhang W., Yang G., Zhang A., Xu L. X., He X., Preferential vitrification of water in small alginate microcapsules significantly augments cell cryopreservation by vitrification, *Biomedical Microdevices*, **12** (2010) 89–96.

186. Choi J. K., Yue T., Huang H., Zhao G., Zhang M., He X., The crucial role of zona pellucida in cryopreservation of oocytes by vitrification, *Cryobiology*, **71** (2015) 350–355.

187. Granicka L. H., Kawiak J. W., Glowacka E., Werynski A., Encapsulation of OKT3 cells in hollow fibers, *ASAIO Journal*, **42** (1996) M863–M866.

188. Jarvis A. P. Jr., Grdina T. A., Sullivan M. F., Cell growth and hemoglobin synthesis in murine erythroleukemic cells propagated in high density microcapsule culture, *In Vitro Cellular & Developmental Biology*, **22** (1986) 589–596.

189. Nilsson K., Scheirer W., Merten O. W., Östberg L., Liehl E., Katinger H. W., *et al.*, Entrapment of animal cells for production of monoclonal antibodies and other biomolecules, *Nature*, **302** (1983) 629–630.

190. Goguen B., Kedersha N., Clonogenic cytotoxicity testing by microdrop encapsulation, *Nature*, **363** (1993) 189–190.

191. Pueyo M. E., Darquy S., Arbet-Engels C., Poitout V., Di Maria S., Gangnerau M. N., *et al.*, A method for obtaining monodispersed cells from isolated porcine islets of Langerhans, *International journal of Artificial Organs*, **18** (1995) 34–38.

192. Guo J. F., Jourdian G. W., MacCallum D. K., Culture and growth characteristics of chondrocytes encapsulated in alginate beads, *Connective Tissue Research*, **19** (1989) 277–297.

193. Loty S., Sautier J. M., Loty C., Boulekbache H., Kokubo T., Forest N., Cartilage formation by fetal rat chondrocytes cultured in alginate beads: aproposed model for investigating tissue–biomaterial interactions, *Journal of Biomedical Materials Research*, **42** (1998) 213–222.

194. Takabatake H., Koide N., Tsuji T., Encapsulated multicellular spheroids of rat hepatocytes produce albumin and urea in a spouted bed circulating culture system, *Artificial Organs*, **15** (1991) 474–480.

195. Hollingshead M. G., Alley M. C., Camalier R. F., Abbott B. J., Mayo J. G., Malspeis L., *et al.*, *In vivo* cultivation of tumor cells in hollow fibers, *Life Science*, **57** (1995) 131–141.

196. Okada N., Fushimi M., Nagata Y., Fukunaga T., Tsutsumi Y., Nakagawa S., *et al.*, Evaluation of angiogenic inhibitors with an *in vivo* quantitative angiogenesis method using agarose microencapsulation and mouse hemoglobin enzyme-linked immunosorbent assay, *Japanese Journal of Cancer Research*, **87** (1996) 952–957.

197. Okada N., Fushimi M., Nagata Y., Fukunaga T., Tsutsumi Y., Nakagawa S., *et al.*, A quantitative *in vivo* method of analyzing human tumor-induced angiogenesis in mice using agarose microencapsulation and hemoglobin enzyme-linked immunosorbent assay, *Japanese Journal of Cancer Research*, **86** (1995) 1182–1188.

198. McMahon J., Schmid S., Weislow O., Stinson S., Camalier R., Gulakowski R., Shoemaker R., Kiser R., Dykes D., Harrison S. Jr., Mayo J., Boyd M., Feasibility of cellular microencapsulation technology for evaluation of anti-human immunodeficiency virus drugs *in vivo*, Journal of the National Cancer Institute, **82** (1990) 1761–1765.

199. Quigley E. M. M., Gut bacteria in health and disease, *Gastroenterology & Hepatology (N. Y)*, **9** (2013) 560–569.
200. Clarke G., Stilling R. M., Kennedy P. J., Stanton C., Cryan J. F., Dinan T. G., Minireview: Gut microbiota: the neglected endocrine organ, *Molecular Endocrinology*, **28** (2014) 1221–1238.
201. Sommer F., Bäckhed F., The gut microbiota — masters of host development and physiology, *Nature Reviews Microbiology*, **11** (2013) 227–238.
202. Guarner F., Malagelada J. R., Gut flora in health and disease, *Lancet.*, **361** (2003) 512–519.
203. O'Hara A. M., Shanahan F., The gut flora as a forgotten organ, *EMBO Reports*, **7** (2006) 688–693.
204. Prakash S., Tomaro-Duchesneau C., Saha S., Cantor A., The gut microbiota and human health with an emphasis on the use of microencapsulated bacterial cells, *Journal of Biomedicine and Biotechnology*, **2011**, 981214. (2011)
205. Lepage P., Leclerc M. C., Joossens M., Mondot S., Blottière H. M., Raes J., *et al.*, A metagenomic insight into our gut's microbiome, *Gut*, **62** (2013) 146–158.
206. Beaugerie L., Petit J.-C., Antibiotic-associated diarrhoea, *Best Practice & Research Clinical Gastroenterology*, **18** (2004) 337–352.
207. Hill C., Guarner F., Reid G., Gibson G. R., Merenstein D. J., Pot B., *et al.*, Expert consensus document. The International Scientific Association for Probiotics and Prebiotics consensus statement on the scope and appropriate use of the term probiotic, *Nature Reviews Gastroenterology & Hepatology*, **11** (2014) 506–514.
208. McFarland L. V., Meta-analysis of probiotics for the prevention of antibiotic associated diarrhea and the treatment of clostridium difficile disease, *American Journal of Gastroenterology*, **101** (2006) 812–822.
209. Borges S., Silva J., Teixeira P., The role of lactobacilli and probiotics in maintaining vaginal health, *Archives of Gynecology and Obstetrics*, **289** (2014) 479–489.
210. Khalesi S., Sun J., Buys N., Jayasinghe R., Effect of probiotics on blood pressure: asystematic review and meta-analysis of randomized, controlled trials, *Hypertension*, **64** (2014) 897–903.
211. Cuello-Garcia C. A., Brożek J. L., Fiocchi A., Pawankar R., Yepes-Nuñez J. J., Terracciano L., *et al.*, Probiotics for the prevention of allergy: A

systematic review and meta-analysis of randomized controlled trials, *Journal of Allergy and Clinical Immunology*, **136** (2015) 952–961.

212. Saez-Lara M. J., Gomez-Llorente C., Plaza-Diaz J., Gil A., The role of probiotic lactic acid bacteria and bifidobacteria in the prevention and treatment of inflammatory bowel disease and other related diseases: asystematic review of randomized human clinical trials, *BioMed Research International*, **2015** (2015) 505878.
213. Rijkers G. T., De Vos W. M., Brummer R. J., Morelli L., Corthier G., Marteau P., Health benefits and health claims of probiotics:bridging science and marketing, *British Journal of Nutrition*, **106** (2011) 1291–1296.
214. Brandtner E. M., Dangerfield J. A., Guenzburg W. H., B. Salmons Salmons. Protection of microbial cells from acidic degradation, WO 2012101167 A1, Aug. 2 (2012).
215. Ding W. K., Shah N. P., Acid, bile, and heat tolerance of free and microencapsulated probiotic bacteria, *Journal of Food Science*, **72** (2007) M446–M450.
216. Krasaekoopt W., Bhandari B., Deeth H., The influence of coating materials on some properties of alginate beads and survivability of microencapsulated probiotic bacteria, *International Dairy Journal*, **14** (2004) 737–743.
217. Lee K. Y., Heo T. R., Survival of *Bifidobacterium longum* immobilized in calcium alginate beads in simulated gastric juices and bile salt solution, *Applied and Environmental Microbiology*, **66** (2000) 869–873.
218. Capela P., Hay T. K. C., Shah N. P., Effect of cryoprotectants, prebiotics and microencapsulation on survival of probiotic organisms in yoghurt and freeze-dried yoghurt, *Food Research International*, **39** (2006) 203–211.
219. Muthukumarasamy P., Holley R. A., Survival of *Escherichia coli* O157 : H7 in dry fermented sausages containing micro-encapsulated probiotic lactic acid bacteria, *Food Microbiology*, **24** (2007) 82–88.
220. Chandramouli V., Kailasapathy K., Peiris P., Jones M., An improved method of microencapsulation and its evaluation to protect *Lactobacillus spp.* in simulated gastric conditions, *Journal of Microbiological Methods*, **56** (2004) 27–35.
221. Mandal S., Hati S., Puniya A. K., Khamrui K., Singh K., Enhancement of survival of alginate-encapsulated *Lactobacillus casei* NCDC 298, *Journal of the Science of Food and Agriculture*, **94** (2014) 1994–2001.

222. Fávaro-Trindade C. S., Grosso C. R., Microencapsulation of *L. acidophilus* (La-05) and *B. lactis* (Bb-12) and evaluation of their survival at the pH values of the stomach and in bile, *Journal of Microencapsulation*, **19** (2002) 485–494.

223. Liserre A. M., Ré M. I., Franco B. D. G. M., Microencapsulation of *Bifidobacterium animalis* subsp. *lactis* in modified alginate-chitosan beads and evaluation of survival in simulated gastrointestinal conditions, *Food Biotechnology*, **21** (2007) 1–16.

224. Lee J. S., Cha D. S., Park H. J., Survival of freeze-dried *Lactobacillus bulgaricus* KFRI 673 in chitosan-coated calcium alginate microparticles, *Journal of Agricultural and Food Chemistry*, **52** (2004) 7300–7305.

225. Annan N. T., Borza A. D., Hansen L. T., Encapsulation in alginate-coated gelatin microspheres improves survival of the probiotic *Bifidobacterium adolescentis* 15703T during exposure to simulated gastro-intestinal conditions, *Food Research International*, **41** (2008) 184–193.

226. Martoni C., Bhathena J., Urbanska A. M., Prakash S., Microencapsulated bile salt hydrolase producing *Lactobacillus reuteri* for oral targeted delivery in the gastrointestinal tract, *Applied Microbiology and Biotechnology*, **81** (2008) 225–233.

227. Kailasapathy K., Survival of free and encapsulated probiotic bacteria and their effect on the sensory properties of yoghurt, *LWT- Food Science and Technology*, **39** (2006) 1221–1227.

228. Sultana K., Godward G., Reynolds N., Arumugaswamy R., Peiris P., Kailasapathy K., Encapsulation of probiotic bacteria with alginate–starch and evaluation of survival in simulated gastrointestinal conditions and in yoghurt, *International Journal of Food Microbiology*, **62** (2000) 47–55.

229. Homayouni A., Azizi A., Ehsani M. R., Yarmand M. S., Razavi S. H., Effect of microencapsulation and resistant starch on the probiotic survival and sensory properties of synbiotic ice cream, *Food Chemistry*, **111** (2008) 50–55.

230. Lian W. C., Hsiao H. C., Chou C. C., Survival of bifidobacteria after spray-drying, *International Journal of Food Microbiology*, **74** (2002) 79–86.

231. Lian W. C., Hsiao H. C., Chou C. C., Viability of microencapsulated bifidobacteria in simulated gastric juice and bile solution, *International Journal of Food Microbiology*, **86** (2003) 293–301.

232. Hsiao H. C., Lian W. C., Chou C. C., Effect of packaging conditions and temperature on viability of microencapsulated bifidobacteria during storage, *Journal of the Science of Food and Agriculture*, **84** (2004) 134–139.

233. Chen M. J., Chen K. N., Kuo Y. T., Optimal thermotolerance of *Bifidobacterium bifidum* in gellan–alginate microparticles, *Biotechnology and Bioengineering*, **98** (2007) 411–419.

234. Hyndman C. L., Groboillot A. F., Poncelet D., Champagne C. P., Neufeld R. J., Microencapsulation of *Lactococcus lactis* within cross-linked gelatin membranes, Journal of Chemical Technology & Biotechnology, **56** (1993) 259–263.

235. Sun W., Griffiths M. W., Survival of bifidobacteria in yogurt and simulated gastric juice following immobilization in gellan-xanthan beads, *International Journal of Food Microbiology*, **61** (2000) 17–25.

236. Mcmaster L. D., Kokott S. A., Slatter P., Micro-encapsulation of *Bifidobacterium lactis* for incorporation into soft foods, *World Journal of Microbiology and Biotechnology*, **21** (2005) 723–728.

237. McMaster L. D., Kokott S. A., Reid S. J., Abratt V. R., Use of traditional African fermented beverages as delivery vehicles for *Bifidobacterium lactis* DSM 10140, *International Journal of Food Microbiology*, **102** (2005) 231–237.

238. Chen H., Ouyang W., Jones M., Metz T., Martoni C., Haque T., *et al.*, Preparation and characterization of novel polymeric microcapsules for live cell encapsulation and therapy, *Cell Biochemistry and Biophysics*, **47** (2007) 159–168.

239. Oliveira A. C., Moretti T. S., Boschini C., Baliero J. C., Freitas O., Favaro-Trindade C. S., Stability of microencapsulated *B. lactis* (BI 01) and *L. acidophilus* (LAC 4) by complex coacervation followed by spray drying, *Journal of Microencapsulation*, **24** (2007) 673–681.

240. Doherty S. B., Gee V. L., Ross R. P., Stanton C., Fitzgerald G. F., Brodkorb A., Development and characterisation of whey protein micro-beads as potential matrices for probiotic protection, *Food Hydrocolloids*, **25** (2011) 1604–1617.

241. Lahtinen S. J., Ouwehand A. C., Salminen S. J., Forssell P., Myllärinen P., Effect of starch- and lipid-based encapsulation on the culturability of two Bifidobacterium longum strains, *Letters in Applied Microbiology*, **44** (2007) 500–505.

242. Picot A., Lacroix C., Encapsulation of bifidobacteria in whey protein-based microcapsules and survival in simulated gastrointestinal

conditions and in yoghurt, *International Dairy Journal,***14** (2004) 505–515.

243. Ainsley Reid A., Vuillemard J. C., Britten M., Arcand Y., Farnworth E., Champagne C. P., Microentrapment of probiotic bacteria in a Ca(2+)-induced whey protein gel and effects on their viability in a dynamic gastro-intestinal model, *Journal of Microencapsulation,* **22** (2005) 603–619.

244. Gbassi G. K., Vandamme T., Probiotic encapsulation technology: from microencapsulation to release into the gut, *Pharmaceutics,* **4** (2012) 149–163.

245. Godward G., Kailasapathy K., Viability and survival of free, encapsulated and co-encapsulated probiotic bacteria in yoghurt, *Milchwissenschaft,* **58** (2003) 396–399.

246. Vivek K. B., Use of encapsulated probiotics in dairy based foods, *International Journal of Food, Agriculture and Veterinary Sciences,* **3** (2013) 188–199.

247. Chen K. N., Chen M. J., Liu J. R., Lin C. W., Chiu H. Y., Optimization of incorporated prebiotics as coating materials for probiotic microencapsulation, *Journal of Food Science,* **70** (2005) M260–M266.

248. Steidler L., Hans W., Schotte L., Neirynck S., Obermeier F., Falk W., *et al.,* Treatment of murine colitis by *Lactococcus lactis* secreting interleukin-10, *Science,* **289** (2000) 1352–1355.

249. Lu L., Walker W. A., Pathologic and physiologic interactions of bacteria with the gastrointestinal epithelium, *American Journal of Clinical Nutrition,* **73** (2001) 1124S–1130S.

250. Prakash S., Chang T. M., *In vitro* and *in vivo* uric acid lowering by artificial cells containing microencapsulated genetically engineered *E. coli* DH5 cells, *International Journal of Artificial Organs,* **23** (2000) 429–435.

Chapter 11

Bacteriophage Encapsulation: Trends and Potential Applications

Jiahui Jeremy Kor[a] and Xian Jun Loh[a,b,c]
[a]*Department of Materials Science and Engineering, National University of Singapore, 9 Engineering Drive 1, Singapore 117576*
[b]*Institute of Materials Research and Engineering (IMRE), 3 Research Link, Singapore*
[c]*Singapore Eye Research Institute, 11 Third Hospital Avenue, Singapore 168751*
lohxj@imre.a-star.edu.sg

11.1 Introduction

A bacteriophage ("bacteria-eater") is a virus that infects and replicates within prokaryotes; single-celled organisms that lack membrane-bound nucleus, mitochondria, or any other membrane-bound organelle, for example bacteria. Bacteriophages are composed of proteins which encase a DNA or RNA genome. This genome is injected into the cytoplasm of bacterium which then allows the bacteriophage to replicate [1]. They are considered to be among

Polymer Capsules
Edited by Ye Liu and Xian Jun Loh

ISBN 978-981-4774-60-4 (Hardcover), 978-0-429-42873-9 (eBook)
www.jennystanford.com

the most common and diverse entities in the biosphere; they are estimated to outnumber bacteria by a factor of ten [2]. There are 13 families of bacteriophage, classified according to their nucleic acid composition and morphology.

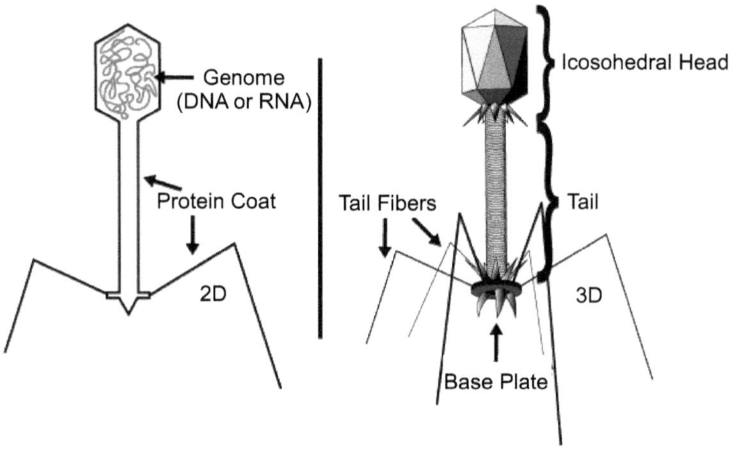

Figure 11.1 Structure of typical tailed bacteriophage. https://commons.wikimedia.org/wiki/File:Tevenphage.png

Phage therapy is the therapeutic application of bacteriophages which are capable of killing bacteria in order to reduce in number or eliminate the target bacteria. Application of phage therapy offers the following advantages over more classical means of treatment:

i. Phages have high target specificity, that is, they only attack their host bacteria, leaving the natural miroflora of the organism unharmed, unlike antibiotics
ii. Phages have the capability of "auto dosing;" they are able to continually infect target bacterial cells as the phage population increases
iii. Phages multiply specifically where host bacteria are located
iv. In the absence of host bacterial cells, the organism's immune system will be able to remove the phages
v. Phages are effective even in low doses
vi. Phage production technology is relatively simple and inexpensive

During infection of the bacterium cell the phage attaches to its surface and injects its genetic material into the cell after which the phage usually takes on a lytic or lysogenic role (Fig. 11.2) [3]. In the former, the phage takes over the machinery of the cell to manufacture phage components. The cell is then lysed to release the produced components which go on to infect other host cells. In contrast, lysogenic phages replicate with the bacterium cell as a unit instead of destroying it. There is no immediate lysing of the host cell, instead the viral genome integrates with the host DNA and replicates along with it. The virus remains dormant until the host cell deteriorates, at which point the reproductive cycle activates and the cell lyses. Lytic phages are therefore more suitable for phage therapy due to exponential growth and rapid reduction of the bacterial cell population.

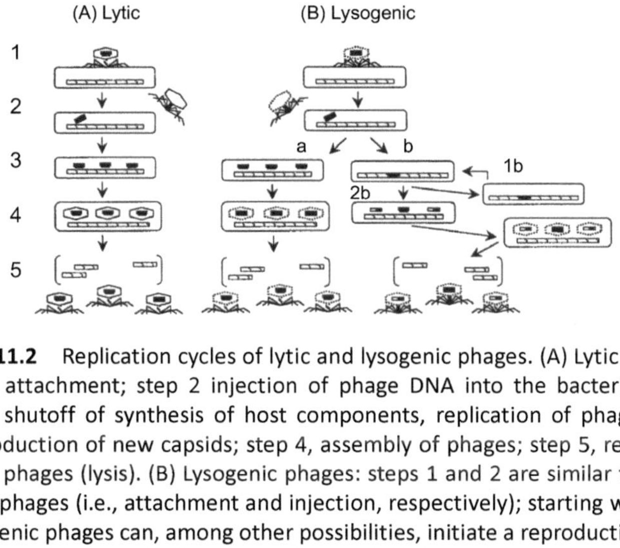

Figure 11.2 Replication cycles of lytic and lysogenic phages. (A) Lytic phages: step 1, attachment; step 2 injection of phage DNA into the bacterial host; step 3, shutoff of synthesis of host components, replication of phage DNA, and production of new capsids; step 4, assembly of phages; step 5, release of mature phages (lysis). (B) Lysogenic phages: steps 1 and 2 are similar to those of lytic phages (i.e., attachment and injection, respectively); starting with step 3, lysogenic phages can, among other possibilities, initiate a reproductive cycle similar to that of lytic phages (a) or integrate their DNA into the host bacterium's chromosome (lysogenization) (b). Lysogenized cells can replicate normally for many generations (1b) or at some point undergo lysogenic induction (2b) spontaneously or because of inducing agents such as radiation or carcinogens, during which time the integrated phage DNA is excised from the bacterial chromosome and may pick up fragments of bacterial DNA.

Bacteriophages attach only to specific receptors on the surface of the host cell, for example lipopolysaccharides and proteins [4]. This specificity gives rise to the narrow activity spectrum of phages as mentioned earlier.

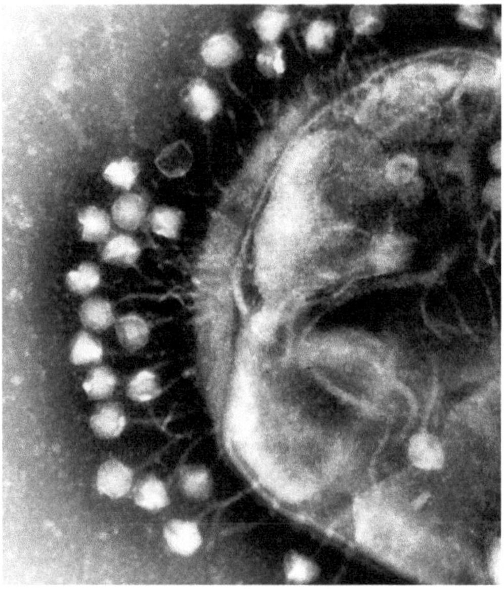

Figure 11.3 Bacterium with multiple adsorbed phages [5].

Since their discovery in the early 20th century, bacteriophages have been used to treat and prevent bacterial diseases such as bubonic plague and cholera, but they were limited in their success and soon disregarded due to the development of conventional antibiotics like penicillin. There were several factors which led to the abandonment of investigating bacteriophages as a means of antibacterial treatment [6]. Firstly, phages have a narrow host range, meaning the patient should only be administered with phage strains that will attack the strain of bacteria that infects their system. However, researchers at the time did not use the correct phage strains in their trails and obtained negative outcomes. Second, bacterial debris such as endotoxins were present in phage preparation due to crude preparation techniques. This was severely harmful to test subjects and negated any positive effects of the bacteriophage. Third, researchers used inadequate techniques to process the bacteriophage during phage preparation. They applied heat and oxidizing agents to remove bacteria; however this had the effect of denaturing the phage coat proteins, and they failed to check for continued viability of the phages. Lastly, prevalent technological limitations meant that researchers could not obtain meaningful results. They were unaware that phages tend to be rapidly cleared from circulation, and also that some phages integrated with their host cell in a lysogenic role.

However, there has recently been renewed interest in phage therapy as a means of healthcare due to the increasing number of multidrug-resistant (MDR) bacterial species, which render conventional antibiotics ineffective. Bacteriophages are able to coevolve; it is therefore unlikely for bacteria to develop a similar general resistance as in the case of drugs which are unchanging chemical compounds. With the benefit of hindsight and learning from the experience of early attempts, modern researchers can come up with solutions to the questions faced by their predecessors. For example, screening should be done before treatment to ensure the bacteriophage will be strongly lytic to the strain of bacteria present in the host. Improved technology and methodology afford better purification of samples. Strains of bacteriophage can be specifically cultivated to be cleared at a slower rate.

These corrections have had observable results. For example, phages were shown to be effective in rescuing rats from fatal systemic infections [7], calves and lambs from fatal diarrhoea [8], and in preventing destruction of skin grafts in burned rabbits [9], and furthermore that a single injection of phage preparation proved to be more effective than multiple doses of antibiotics. Encouraging results were also observed in human subjects: approximately 90% of 550 patients infected with suppurative bacteria were cured when phage preparation was administered orally. More detail regarding the healthcare applications of bacteriophages will be covered in the following section.

The most common method of protecting bacteriophages from adverse environmental conditions is through the use of encapsulation techniques. Encapsulation is defined as the process of entrapment of microorganism cells by means of coating them with appropriate hydrocolloids in order to segregate the cells from the surrounding environment [10]. These cells are retained within a material in order to reduce cell injury and loss, as some applications dictate the need for exposure to harsh conditions and environment.

For example, bacteriophages applied *in vivo* have to be protected from acidic gastrointestinal (GI) conditions, while those used in food preservation technology may have to be processed and stored for long periods of time in dry conditions. An instance whereby *in vivo* protection is required is encapsulated phages administered to livestock, which is currently undergoing research. Early studies show microencapsulated phages fed to pigs led them to shed less *Salmonella typhimurium* as compared to control groups [11].

Therefore, a multitude of industries where bacteriophages are relevant, including the food industry, pharmaceuticals and agriculture, are looking to optimize factors which will afford them the best bacteriophage/encapsulation system in terms of effectiveness, feasibility, and cost.

11.2 Motivations and Potential Applications

The potential effectiveness of phage therapy in combating microbial infections has many widespread applications.

11.2.1 Food Preservation Technology

The food industry is an area where bacterial infection is a pressing concern, as contamination leads to foodborne diseases, food wastage, and negative financial effects. Limiting microbe growth is especially challenging. Despite ensuring acceptable hygiene standards and complying with safety guidelines, bacteria will still be found in food and food handling surfaces [12]. There is a chance of infection along the entire food supply chain. Microbes also have the ability to attach to living or inert surfaces and form microbial communities (biofilms), developing resistances to anti-microbial agents [13]. From 2000 to 2008, 9.4 million foodborne illnesses occurred every year in the US alone, 39% of which were caused by bacterial infection. Contaminated food also led to 55,691 hospitalizations and 1351 deaths annually [14]. Microbiological spoilage also leads to food products being discarded, causing an estimated $161.1 billion loss in 2010 [15]. Clearly, food preservation techniques have to be continually improved to meet safety standards as well as consumer demand. Because of the huge impact food has on our health and economy, coupled with the increasingly negative consumer attitude toward chemical preservatives in food products, interest has shifted toward more natural anti-microbial solutions [16]. US regulatory agencies Food and Drug Administration (FDA) and the Department of Agriculture (DA) have approved several phage-based preparations which will be used to tackle frequent foodborne pathogens like *Salmonella* and *Escherichia coli* (*E. coli*).

Apart from food storage, application of phages may benefit the entire food chain, from growth and cultivation, to processing, to final packaging of the product (Fig. 11.4).

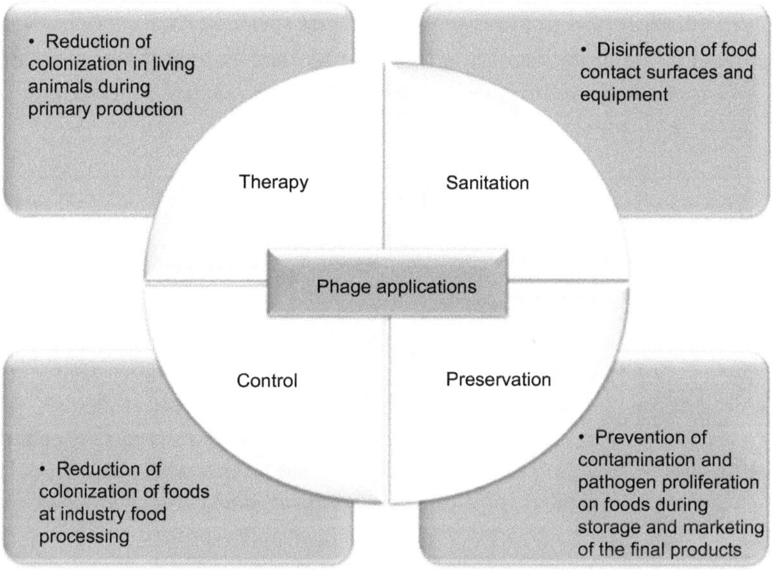

Figure 11.4 Improvements in food safety via application of phages in the food chain. Reprinted from Ref. [16], Copyright 2012, Sanna M. Sillankorva et al.

These preparations are available either as additives to animal feed or are applied to live animals prior to slaughter (phage therapy), or to decontaminate food products for human consumption (biocontrol) [17]. These aspects are related and complementary; improvements in phage therapy will usually benefit biocontrol. For example, using phage therapy to reduce *Campylobacter* loads in poultry intestines by 2 log leads to a reduction in the incidence of campylobacteriosis caused by the consumption of chicken by a factor of thirty [18]. Biocontrol is done by applying phages onto meat, fresh produce, or mixed into raw milk. Biosanition can be carried out by using phages to combat microbial colonies that grow on processing equipment, especially hard-to-clean surfaces. However, this may not be as feasible due to the large variety of bacteria that grow in such places. Biopreservation shows more promise, as phages have been shown to lyse host cells even at temperatures as low as 1°C, meaning that they can be applied in both refrigerated and non-refrigerated foods [19].

Below is a summary of the main foodborne pathogens and the current progress made thus far towards treating them, using phage therapy [16]. Research in employing phages against such infections mainly focuses on both preharvest (phage therapy) and postharvest (biocontrol) interventions.

Bacteria	Mode of infection	Treatment
Escherichia coli	Survives well under intestinal conditions, commonly transmitted by undercooked contaminated meat	Biotech company Intralytix released a coliphage product called EcoShield in 2011. It targets *E. coli* O157:H7 in red meat and has been proven to eliminate 95%–100% of bacteria (postharvest). Finalyse is another product which is sprayed on cattle prior to slaughter (preharvest).
Campylobacter	Found in the intestines of poultry and cattle. Can also be spread by drinking contaminated water.	Phage therapy (preharvest) has shown to reduce cecal content of the bacteria [20]; however phage-resistance strains have been reported [21]. Phages which target *Campylobacter* may trigger evolution of phage-resistant phenotypes. Limited results have been achieved in postharvest strategies.

Bacteria	Mode of infection	Treatment
Salmonella	Colonizes the gastrointestinal tract of livestock. Commonly transmitted by contaminated eggs and meat.	Laboratory tests show significant decrease in bacterial population in major tissues after phage therapy. BacWash and BIOTECTOR S1 are 2 existing products to control salmonella in livestock.
Listeria	Able to persist in food storage conditions (low pH, oxygen and temperatures) [22]. Commonly transmitted in minimally processed foods for example ready-to-eat meals.	With the most common sources of contamination in mind, research has been focused on postharvest strategies. LISTEX P100 is an example of a phage-based product currently being commercialized to prevent listeria contamination.

11.2.2 Healthcare

Aside from applications in the food industry, the unique capability of phages to target bacteria to a high degree of specificity, coupled with its rapid and exponential rate of reproduction, also implies potential in fighting infectious disease. Bacteriophages have thus found numerous potential uses in healthcare, namely as a possible replacement or supplement for antibiotics and as prophylaxis and treatment of various bacterial infection in humans, such as pulmonary infection.

Our reliance on antibiotics to treat bacterial infections has resulted in the increase in antibiotic-resistant bacterial strains, severely limiting the effectiveness of current treatment methods. Microbial resistance is most distinct in diarrheal disease, respiratory tract infections, and hospital-acquired infections. Infections caused by such pathogens are usually unaffected by conventional treatment, therefore prolonging infection times, increasing risk and exposure to the public [23]. For example, the CDC reported more than a decade ago that "95% of hospital staphylococcal (staph) infections failed to respond to first-line antibiotics" [24]. Current estimates place the number of hospital patients infected with methicillin-resistant *S. aureus* at 80,000 every year. The difficulty in treating staph infections has also led to community-acquired MRSA becoming an increasingly serious threat. Another deleterious side effect of this reliance is the overuse or misuse of antibiotics. 50% of antibiotic prescriptions are estimated to be unnecessary as they are being used to treat coughs, colds, or other viral infections, which are not responsive to antibiotic treatment. Educational and systemic reviews have to be undertaken to ensure rational use and prescription, so that this issue does not repeat itself with alternative forms of treatment. Before implementing phage treatment, it is essential to first identify the bacterial species that is causing the infection. This allows doctors to then choose the most effective phage for treatment. Phages should exhibit strongly lytic behaviour to prevent bacteria from developing phage resistance [25]. One recent study which demonstrated the efficacy of phage therapy was carried out in 2015, wherein phages were used to treat phenotypically diverse, antibiotic-resistant *Pseudomonas aeruginosa* in cystic fibrosis patients. Phages were reported to have up to 98% effectiveness against 84 different strains of the bacteria [26]. Earlier in 2009, private company Biocontrol Ltd. also reported successful application of bacteriophages, which required less dosage yet showed longer lasting effects as well as higher recovery rates than antibiotics in treating chronic otitis in 24 volunteers [27]. Therapy trials from Eliava Institute showed that a combination of bacteriophages and antibiotics proved more effective in treating a wide range of infections than monotherapy alone, showcasing bacteriophage as a potential supplement to conventional therapeutic methods [28].

As a prophylactic, phages have been proven to achieve some success, where they received a higher degree of attention in European countries like Poland and Russia. Below is a selection of five significant studies with positive results.

Infection	Infecting (etiologic) agents	Phage treatment results
Lung and pleural infections	*Staphylococcus*	223 patients diagnosed with lung and pleural infections underwent phage therapy, and the results compared against another group of 117 cases in which antibiotics were used. In the phage-treated group, 82% fully recovered, as compared to 64% of the patients treated with antibiotics [29].
Bacterial dysentery and salmonellosis	*Shigella* and *Salmonella*	Treatment of salmonellosis was carried using phages and a combination of antibiotics and phages. Cases in which antibiotics alone was ineffective responded well to treatment by a combination of the two [30].
Inflammatory urologic diseases	*Staphylococcus, E. coli*, and *Proteus*	Chronic and acute urogenital inflammation in 46 patients was treated by adapted phages. Phage treatment was reported to show marked clinical improvements in 92% of the patients and 84% rate of bacteriological clearance [31].
Postoperative wound infections in cancer patients	*Staphylococcus* and *Pseudomonas*	2 groups of cancer patients afflicted with postsurgical wound infections were subjected to different treatment: phages and antibiotics. The phage-treated group showed an 82% success rate while antibiotic treatment was successful in 61% of the cases [32].
Peritonitis, osteomyelitis, lung abscesses, and postsurgical wound infections	*Staphylococcus, Streptococcus*, and *Proteus*	Phage treatment was administered in 236 patients having antibiotic-resistant infections, either subcutaneously or through surgical drains. Phage therapy eliminated infections in 92% of the patients [33].

As mentioned earlier, drug-resistance is especially pronounced in bacteria which cause respiratory tract infections. One specific example is the multidrug-resistant *Mycobacterium tuberculosis*, which causes approximately 400,000 cases annually [34]. Matinkhoo *et al.* reported that low temperature spray drying (elaborated on in the following section) was an efficient method of phage preparation, preserving their viability as well as achieving high dispersiblity and a respirable particle size. The phage powder was then delivered using a dry powder inhaler, and *in vitro* lung dosage of phages delivered by a single actuation was found to reach minimum effective *in vivo* levels (10^7–10^8 plaque-forming units) [35].

Despite these and other promising results, Kutateladze *et al.* maintain a pessimistic outlook, noting that due to the high cost and duration of trials complying with EU and US guidelines, coupled with the lack of interest from major pharmaceutical firms, it is unlikely for phage therapy to gain widespread application in the near future.

11.3 Biomaterials Involved in Encapsulation

A biomaterial in this context refers to both organic and inorganic macromolecules made up of covalently bonded monomers. Their structure and conformation convey the biomaterial specific functions, such as the ability to form gels. In selecting the appropriate material for bacteriophage encapsulation, several factors have to be considered [36]:

i. Physical and chemical properties of the biomaterial
ii. Toxicity testing results of the material
iii. Processability
iv. Sterilizability

Among the most commonly used biomaterials in encapsulation processes include chitosan, alginate, and cellulose.

Chitosan is a linear polysaccharide produced by the deacetylation of chitin, a compound found in the exoskeleton of crustaceans. It can be used in electrospinning techniques (described below) to encapsulate phages [37]. More commonly, it is used as a coating for alginate gel beads to improve their stability [38]. One drawback of using chitosan is its insolubility at pH levels above 5.4; this prevents complete release of its loaded agent in acidic environments [39].

Alginate is another naturally occurring polymer which can be extracted from algae and certain forms of bacteria. It is popular as an encapsulation material as it provides good resistance against acidity [40]. Similar to chitosan, it can also be used in electrospinning techniques, encapsulating the core as a gel matrix. Alginate capsules provide certain benefits given their ability to deliver proteins with high efficiency. It also provides good control over the amount of protein released, and enhances protein stability. However, these properties are not yet fully optimized; therefore improving protection against acidic degradation and increasing maximum molecule load are still existing areas of interest [41].

Lastly, cellulose, the most abundant natural polymer, is also an appealing choice as an encapsulation material due to its high biocompatibility and good physical properties. Furthermore, cellulose and modified cellulose-based materials are readily available. They are affordable and do not harm the environment [42]. A variety of therapeutic agents, substances, and bacteriophages have been successfully incorporated into cellulose and employed in food and medical applications [43–45].

11.3.1 Encapsulation Techniques

An encapsulated phage may have a spherical or irregular shape. This is dependent on the active agent and the encapsulation technique employed; most industrial applications demand a regular, spherical shape and uniformly small particles. It typically consists of a core (the phage or active agent) surrounded by an external coating or shell (Fig. 11.5). This coating serves to protect the active agent from the environment. It is important to make the distinction between encapsulation and immobilization as these terms are sometimes used interchangeably: the former refers to the formation of a shell or coating around the core, whereas the latter is based on trapping the core within a matrix.

The products of the encapsulation process are usually termed "microcapsules" as they tend to be in the magnitude of microns in size. Further categorization can be done (Fig. 11.6): They are referred to as nanocapsules if their size is less than 1 micron, and macrocapsules if larger than 1000 microns [47].

Figure 11.5 (Mononuclear) encapsulation in core–shell capsules. Reprinted from Ref. [46], Copyright 2015, with permission from Elsevier.

10^{-10}	10^{-9}	10^{-8}	10^{-7}	10^{-6}	10^{-5}	10^{-4}	10^{-3}
100 pm	1 nm			1 μm			1 mm
Atoms	Small size moleeculess	Proteins	Viruses	Prokaryotes		Eukaryotes	
→	nanoparticles →			microparticles →			macroparticles →

Figure 11.6 Size classification of capsules in relation to other biomolecules.

Encapsulation serves the following purposes:

i. Protects microorganisms and sensitive compounds
ii. Improves physical and technical properties
iii. Small capsule size helps to reduce mass transfer limitations
iv. Protects against oxidation
v. Provides controlled release of active agent
vi. Safer handling of toxic materials
vii. Masks undesired properties such as taste and odor
viii. Semipermeable membrane facilitates diffusion of nutrients and metabolites

There are various approaches to naming and classifying encapsulation techniques. The table in Fig. 11.7 shows, with the most common techniques mentioned in bold.

Chemical processes	Physico-chemical processes	Physico-mechanical
▶ **Suspension, dispersion, emulsion polymerization**	▶ Coacervation and phase separation	▶ **Spray drying and freezing**
▶ Polycondensation	▶ Sol-gel encapsulation (ionotropic gelation)	▶ Vacuum encapsulation
▶ *In situ* polymerization	▶ Supercritical CO_2-assisted microencapsulation	▶ (Co-) **extrusion dripping**
	▶ Layer-by-layer assembly	▶ Fluid-bed coating
		▶ Emulsion stirring
		▶ Ultrasonic atomizer
		▶ Centrifugation
		▶ Pan coating
		▶ Electrospun nanofibres

Figure 11.7 Different techniques used for encapsulation.

11.3.2 Emulsification

Emulsification consists of mixing the active agent and a polymer within a large volume of oil. Homogenization is maintained by stirring the mixture continuously; this is important to obtain water-in-oil (w/o) emulsion systems of the desired shape and size. Once the emulsion is broken, the capsules can be recovered by settling [36].

As a proof-of-concept, Balcão *et al.* designed an effective multiple emulsion (w/o/w) system to encapsulate phages with a broad lytic spectrum able to infect *Salmonella* and *E. coli* [48]. W/o/w emulsions are small water droplets within larger oil droplets, which are themselves dispersed in a continuous aqueous phase. The multiple emulsion system has the advantage of being able to carry both polar and non-polar molecules, and offers better control over the release of therapeutic molecules [49]. To prepare the multiple emulsion system, an oily phase was prepared and maintained at 40°C. An aqueous phase containing the phages was prepared separately and brought to the same temperature. These were mixed thoroughly in a thermostated bath, forming the w/o emulsion. Finally the external aqueous phase was added to the emulsion and homogenized until the mixture returned to room temperature. In the study, the development and optimization of the encasing nanovesicles were characterized by determining their size profile using DLS, surface morphology using CRYO-SEM and thermal properties via DSC. Their viability was also assessed *in vitro* using various bacterial strains. Optimized emulsion systems showed no phase separation and the nanovesicles had an average diameter of 114 nm. The Zeta potential was −13 mV and they retained their stability for up to 3 months, after which particle aggregation was observed which led to increase in particle size. It was also discovered that increasing the molecular weight of the lipids used and decreasing the dielectric constant indicates higher hydrophobicity. This led to greater impregnation of the interface and thus the formation of more stable nanoemulsions. Finally they propose that an aerosol method of delivery may be effective since the bacteriophages are applied directly to the site of infection, decreasing the time required for the active agents to take effect. Additionally, the lipid nanovesicles provide protection for the bacteriophages against immune response.

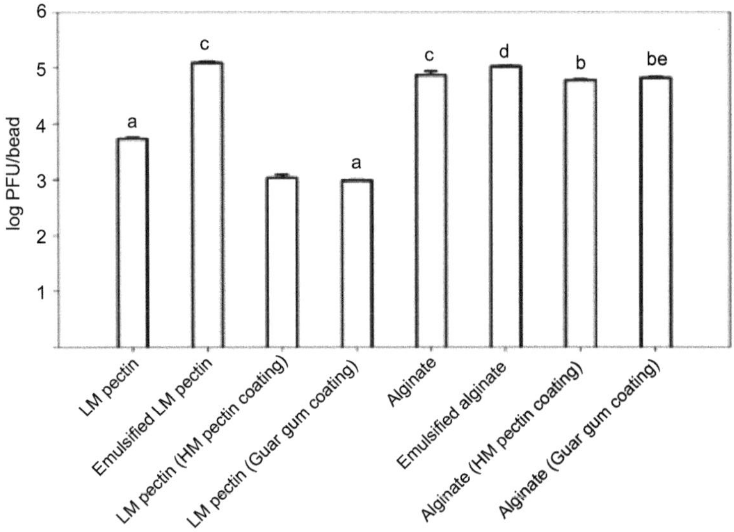

Figure 11.8 Encapsulation efficiency. In general, alginate-based matrices showed higher encapsulation efficiencies than pectin-based beads.

Emulsification can also be used to protect phages from a harsh acidic environment. Dini *et al.* [41] developed biopolymeric matrices (base polymers alginate and pectin) for application in phage therapy via oral administration, to reduce *E. coli* infestation in bovines. These encapsulation materials are termed "smart" molecules due to their resistance to the acidic environmental conditions of the stomach but responsiveness to alkali conditions in the intestines. Thus the encapsulated system can reach the target site without swelling and release the active agents upon arrival. Optimal results were judged by encapsulation efficiency as well as protection of phages under extreme acidic conditions. The best combination was achieved by the emulsification of pectin with oleic acid. This gave maximum encapsulation efficiency (1.2 x 10^6 PFU per bead) and the highest number of surviving, viable phages [10^3] after 30 minutes of exposure to pH 1.6 conditions. The authors noted that given the low toxicity of the emulsion system components, low material costs and simple manufacturing processes, the proposed formulation presents a potential alternative to controlling bacterial infection in the food supply chain.

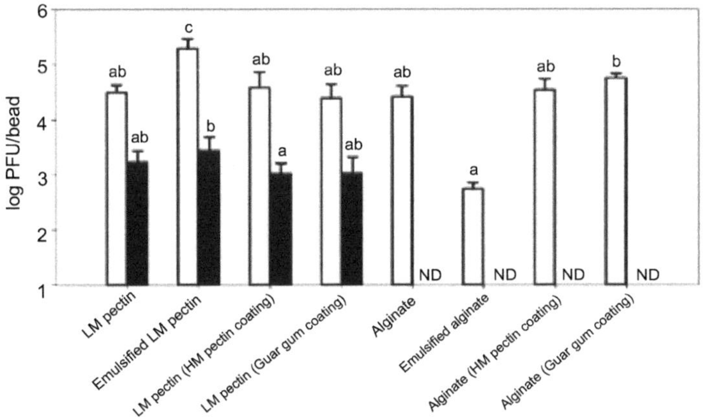

Figure 11.9 Protective effects. Emulsified pectin displayed the best protection rates against acid (pH 1.6) while emulsified alginate offered lowest protection.

11.3.3 Extrusion

The extrusion process is one of the oldest and most common encapsulation techniques. It is a simple process which minimizes cell injury and maintains a relatively high level of cell viability [10]. Other advantages include high biocompatibility and flexibility. Furthermore, no additional raw materials are needed, unlike the emulsion process which requires the addition of oil and emulsifiers. The extrusion process itself consists of first preparing a hydrocolloid solution, mixing in the microorganism, and extruding the cell suspension through a syringe needle. The droplet is allowed to fall into the hardening solution (setting bath). The hardening solution typically consists of multivalent cations (usually calcium in the form of calcium chloride). After dripping, polymers in the mixture surround the added cells and form three-dimensional lattices by crosslinkages of calcium ions. The majority of generated beads are 2–3 mm in diameter. This parameter is strongly influenced by factors such as the diameter of the extruder orifice, distance between orifice and setting bath, and nature of polymer used, as bead diameter is inversely related to the concentration and viscosity of the encapsulation solution.

Tang *et al.* [50] reported successful findings in encapsulating Felix 01, a *Salmonella* targeting bacteriophage, using the extrusion method. They found that alginate microspheres protected the phage in simulated gastric fluid (SGF), but hindered its release rate in simulated intestinal fluid (SIF). This was confirmed by *in vivo* tests, where a large proportion of phages were recovered after passing through the GI tract. A new encapsulation solution was implemented, which combined alginate with whey protein. Whey protein was hydrated in deionized water for 1 h to ensure complete hydration of the proteins. This solution was adjusted to pH 8.0 and heated for 30 minutes at 80 °C to denature the proteins. Sodium alginate was dissolved in 50 mM Tris–HCl buffer, forming the alginate stock solution. The whey protein and alginate stock solutions were used to prepare the encapsulation mixtures. Various ratios and concentrations of alginate and whey protein were mixed and tested for its suitability in producing phage microspheres. The phage solution was mixed with encapsulation solution and extruded through a 500 μm nozzle into a 0.1 M $CaCl_2$ solution while stirring with a magnetic bar. The resulting microspheres were allowed to harden in the $CaCl_2$ solution for 30 minutes, rinsed with distilled water, filtered, and sealed at 4°C. The new formulation was proven to provide significant protection for phages in SGF (pH 2.5). Free phages were completed deactivated after 60 seconds of incubation at low pH. Even viable counts of alginate-encapsulated phages were undetectable after 60 minutes of incubation. In contrast, the new formulation led to only a 2 log unit reduction from 10^{11}–10^{9} pfu/mL after 2 h of incubation. The release profiles also showed that phages were released much faster and with less time delay. For example, alginate microspheres showed a 2 h time delay before phages were released, whereas alginate–whey protein capsules had already released 63% of loaded phages within 2 h. These marked improvements in survivability and release rates are good indications that this technique may be used in oral delivery of phages.

Ma *et al.* [51] also conducted a similar study into improving the encapsulation formulation for oral delivery of *Staphylococcus aureus* phage K. Microsphere fabrication was similar to the technique described above. They also found that free phage K was destroyed after 1 minute in SGF. Alginate microspheres were found to offer slightly better protection here, with a viable count reduction of 2.4 log units after 60 minutes, and complete deactivation after 90. The encapsulation formulation was modified by adding 1% calcium carbonate, significantly improving the stability of phage K, which experienced a count reduction of just 0.17 log units after 2 h of exposure. This was attributed to the buffering capacity of carbonate ions which neutralized the attacking H^+ ions. A lag time of at least 1 h was also observed before release of phage K occurred. This was due to the swelling stage of the dissolution process of the capsule.

However, one distinct disadvantage of extrusion is the difficulty involved in scaling up this process. It is unfeasible for large scale production due to the slow formation of the microbeads, which are generally 2–5 mm larger than those obtained from the emulsion process. Modifications to existing processes and equipment are needed to overcome long fabrication times and to decrease particle size.

11.3.4 Spraydrying

Spraydrying has been widely recommended for use in encapsulation processes in the food industry due to its effectiveness and low economic cost. The process involves the dispersion of the core phages into a polymer solution forming an emulsion or dispersion. The mixture is then homogenized. Atomization of the mixture is then done in a hot drying chamber which removes water from the mixture. This causes the shell polymer to solidify onto the core due to rapid removal of water from the emulsion which leads to the formation of matrix type microcapsules [52]. The resultant dry product also makes spray drying a good option for use in preparing pharmaceutical ingredients for delivery via dry powder inhalers (DPIs) [53]. One limitation of this technique is that the high temperatures required for the drying process prohibits certain microorganisms from spray drying. Since bacteriophages are sensitive to thermal stress [54], lower drying temperatures are needed to reduce phage inactivation. For instance, by lowering the air outlet temperature to 40–45°C, it was found to reduce titre loss to less than 1 log pfu/mL.

Vandenheuvel *et al.* [55] investigated the effects of drying *Pseudomonas* podovirus LUZ19 under different temperatures and excipients to provide bulk mass and protect phage structure. The phage suspension was mixed with a sugar solution and atomized through a bi-fluid nozzle at a constant feed rate with drying inlet air temperature at 85°C and again at 100°C. They found that the lower drying temperature in the presence of trehalose was an effective measure in preventing phage loss; phage titre only decreased by 0.02 log units. The particle size range was also suitable for delivery with DPIs. They concluded that on an industrial scale, spray drying is a feasible method for preparing phages into dry powder suitable for pulmonary delivery.

Pasternack *et al.* [56] also describe spray drying as a possible means of storage preparation for their newly developed bacteriophage, ECML-4 which displays lytic behavior against *E. coli*. The dried bacteriophage can then be reconstituted and used to control *E. coli* growth in food products, manufacturing equipment and animal environments via spraying or misting an aqueous medium containing the dispersed phage or direct application as a freeze-dried powder.

11.3.5 Electrospun Nanofibers

Electrospinning is a fiber-production method which produces nanofibers of diameters in the range of tens of nanometers. A high voltage is applied to the liquid which imparts charges; the electrostatic repulsion within the liquid, due to the charge, then acts against the surface tension of the liquid. The liquid is then stretched which causes a stream to emerge from the surface, forming a charged liquid jet. The jet is elongated and thinned by a spiral motion caused by the repulsive forces and deposited on a grounded collector (Fig. 11.10). This spinning action causes the jet to decrease in diameter, thus obtaining fibers on the nanometer scale.

This method is then used to encapsulate bacteriophages, due to ease of production, good stability and high viability of phages after encapsulation [58].

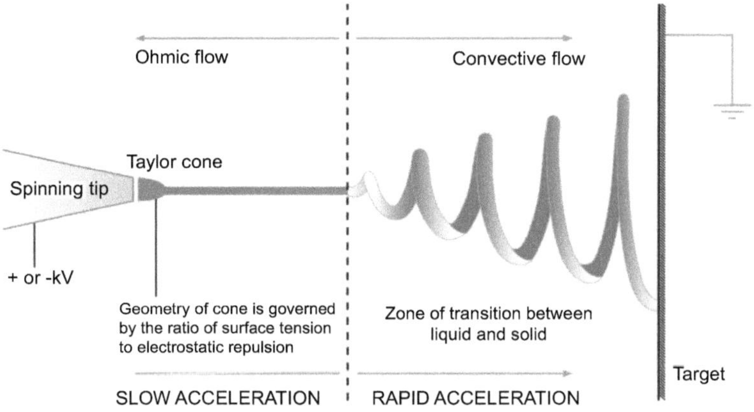

Figure 11.10 Formation of fiber via electrospinning. Reproduced from Ref. [57], image courtesy: Joanna Gatford, The New Zealand Institute for Plant and Food Research Ltd.

One study done by Salalha *et al.* [59] investigated the use of electrospinning to encapsulate living material such as bacteria and bacteriophages T4, T7, and λ. They focused on infectivity of the material before and after encapsulation in the nanofiber to study the viability of this technique for use in the preparation of phages. The biological material was suspended in a mixture of 14% poly(vinyl alcohol) in water, with the applied electric field set at 1.1 $kVcm^{-1}$. The suspension was spun from a hypodermic needle with an inner diameter of 0.5 mm toward a grounded collector plate with a collection time of 1 h. This method resulted in fibers of uniform thickness ranging from 250–400 nm, with the biological material aligned longitudinally and distributed along the fiber. Infectivity of the phages was measured by the ability of the encapsulated material to form plaque on host bacteria. The rate of survivability is low but usable, with less than 10% of the phages being viable after electrospinning. This was attributed to the high sensitivity of the studied phages to shear forces. In addition, it was found that encapsulated phages may be safely stored at low temperatures for months. The authors conclude that electrospinning holds potential as an encapsulation method as the pressure within the fiber and the electrostatic field is survivable by a large range of microorganisms. Their work is also noteworthy as no previous study has successfully encapsulated cells using this method.

Similar results were obtained by Korehei [60], who also measured a drastic loss in T4 phage activity after electrospinning. He attributed this loss to the rapid evaporation of water during the electrospinning process, which led to a drastic change in the osmotic environment of the polymer solution and spun fibers. This resulted in the dehydration and deactivation of the encapsulated phages. He hypothesized that the survivability of the phages can be enhanced by pre-encapsulating with a protective alginate capsule via an emulsion step before the electrospinning stage. This resulted in a reduced but still significant drop in viability. He then advocated the use of coaxial electrospinning, which results in a composite fiber. The two component solutions are drawn using twin capillaries. The organic material is spun into the core of the fiber, which is surrounded by a protective polymer shell. This shell reduces solvent evaporation, protecting the phages from dehydration and degradation. It also has the added benefit of providing more control of release rates and preventing a burst release effect. Coaxial electrospinning maintained full viability of the encapsulated phages. Much longer storage times were also observed. The encapsulated bacteriophage exhibited no loss in viability after 1 month of storage at 4 °C. Korehei concluded that these results show that coaxial electrospun phages are a viable option for use in bacterial control, especially in the storage and packaging of perishable foods.

Korehei then went on to investigate the effect of using different polymer blends on the release profile of the loaded phages [58]. It was found that using a pure poly(ethylene oxide) shell layer resulted in an almost 100% release of the T4 phage within 30 minutes. The water soluble PEO enabled the rapid leaching of T4 bacteriophage, thereby being an unsuitable option for long-term application. He then applied a PEO/CDA (cellulose diacetate) blend, a molecule noted for its ability to sustain its molecular integrity for long periods before undergoing hydrolytic degradation. The blended shell was effective in suppressing the burst release effect of the T4 phage. Increasing the CDA/PEO ratio also increased the release duration of the phages. This was due to the following two reasons: A high CDA content increases the hydrophobicity of the electrospun fibers, which resulted in fiber swelling being the preferred release mechanism. This controlled diffusional release is more stable than the fiber dissolution mechanism as a result of using soluble PEO.

Secondly, the addition of CDA increased the structural integrity of the fiber and restricted fiber expansion during the post-treatment period. Therefore CDA offered improved control over the release of T4 phage. Increasing the molecular weight of PEO also had a similar effect of reducing release rates, as the fibers surrounding the phages increase in both diameter and viscosity.

Besides coaxial electrospinning, Lee and Belcher [61] proposed an alternative mechanism to improve the viability of encapsulated phages. They accomplished this by blending the virus M13 with polyvinyl pyrrolidone, a highly water soluble polymer. They found that this method protected the virus from the toxic effects of the solvent hexaflouropropanol, thereby increasing the processability, structural integrity, and infectibility of the M13 virus.

11.4 Conclusion

Phage therapy and its encapsulation techniques have been shown in studies to be effective against a wide range of bacteria. They have made some progress in the food industry, where a few phage products have been approved by the FDA and USDA. The next step should be to anticipate any potential hazards or limitations with the applications of these products. For example, currently proposed delivery systems of phages include dipping or spraying directly onto food produce and equipment. However, these delivery methods carry certain drawbacks. Phages may be deactivated by other wash fluids or cleaning agents present in the system. Phages may also be diluted, and phage medium containing nutrients may encourage the growth of phage-resistant bacteria. This may occur by the modification of phage receptors on the bacterial cell wall. However, this issue may be countered by regular disinfection using established protocols or by using a phage cocktail, since it is unlikely that the bacteria will be resistant to all strains of phages.

The two main objectives of encapsulating a phage are to improve its viability during application, and to extend its storage lifespan. In food industry applications some of the most common hazards to phage survival are extremes in pH, heat, and salinity. However, susceptibility to these hazards varies among phage families. Viruses with weaker structures or long tails are more susceptible to damage by stress, which can destroy the infectibility of the phage. Naked phages are easily damaged by high temperatures. Other factors which might interfere with the function of phages are the chemical composition and hydration of the application environment. The method of administration also plays a role; some examples were described above. Typical techniques currently explored to protect phages from such hazards are described above, namely emulsification, extrusion, spray drying, and electrospinning. The main criteria to consider when choosing an option are the stability, survivability, and specific requirements of the encapsulated phages. Ideally, a universal encapsulation technique should be developed. Encapsulation methods are being explored and developed in many research fields. Hence, results from studies of encapsulation techniques are more protected, particularly systems used for oral delivery.

Another aspect of phage therapy to consider is the storage time of the phages. For economic reasons, they should be able to be stored for a reasonable amount of time without a severe drop of viability. Current estimates show that phage samples lose about 1% activity per day, even when stored under optimal conditions. Generally, lower temperatures are more suitable for long-term storage of phages. However, none of the tested storage conditions are able to guarantee indefinite stability for all phages, due to differences in sensitivity to temperature and storage media. Therefore, while phages have been proven effective against a wide range of bacterial infections, it will be a while before the technology required in encapsulation as well as storage processes reach a level which will allow the widespread usage of phages in the food industry, and even longer before we see phage usage in clinical applications.

There are many convincing arguments in favor of the pursuit of phage therapy. The first type of classification that can be made regarding these reasons are factual arguments. These have been listed throughout this literature review, from the impact of multidrug-resistant bacteria on the food and healthcare industry, to the potential health and economic benefits presented by phage therapy and encapsulation techniques. There are currently almost no new antibiotics being developed to deal with these challenges. Despite these facts, as well as the growing interest in phage therapy, four main obstacles to its growth have been identified:

1. Clinical studies done in the past are not considered valid proof of the safety and effectiveness of phage therapy
2. There is an insufficient legal framework to protect the interests and property rights of companies who conduct research on phage therapy
3. Current regulations concerning drug development are incompatible with phages. For example, produced drugs are required to have identical chemical compositions, whereas phages are designed to be customisable to each patient
4. It is too early to determine if over-dependence on phages will lead to the same negative outcomes as the excessive and improper use of antibiotics

The second classification that can be used is the moral argument in favor of phage therapy [62]. The first moral argument is the stance that a patient should have the right to access phage therapy if there is the possibility of a cure, even though it has not yet obtained official approval, or if its effects are not yet fully known. This is a concept of "ethically justified medical therapy." This view states that it is morally wrong to completely obstruct access to a potential cure for the patient, and that phage therapy should be reviewed on a case-by-case basis before a decision is made. There are six factors that play a role in influencing these decisions. First of all, there should be just cause in subjecting the patient to a risky and potentially harmful procedure, for instance if his life or limb is being threatened by bacterial infection. Secondly, the agents making the decision should have morally good intentions. For example, the practitioner should balance the interests of his patients fairly with the interests of his employer and act accordingly. Third, there should be a reasonable chance of success by employing the phage therapy. This can be determined by the extent of study and number of existing case studies which provide evidence of the efficacy of the phage under consideration. Next, phage therapy should only be employed when all other forms of available therapy have already been exhausted. That is, the non-approved phage therapy is being used only as a last resort. Also, there should be certainty that any potential health benefits of phage therapy outweigh the risks presented to the patient. It is important to try and minimize adverse health effects. Lastly, the patient's right to autonomy should be respected and taken into consideration. His decision should be taken into account together with the other five factors before a final decision is made.

The second moral argument is that pharmaceutical companies should play a more active role in the development of phage therapy. The principle of beneficence states that pharmaceutical companies, as primary drivers of phage research, bear the brunt of the responsibility for providing this good for society. Of course, public interest is not the only priority of these companies—economic and commercial factors have to be considered as well. One proposed solution to this problem is to align the interests of the firm with that of society. For example, firms can be made aware of the importance of maintaining high moral standards (e.g., public image) or legal and financial incentives can be given for the relevant research.

References

1. McGrath S., van Sinderen D., *Bacteriophage: Genetics and Molecular Biology,* Caister Academic Press (2007), pp. 1–32.
2. Dabrowska K., Switała-Jelen K., Opolski A., Weber-Dabrowska B., Gorski A., Bacteriophage penetration in vertebrates, *Journal of Applied Microbiology*, 98 (2005) 7–13.
3. Sulakvelidze A., Alavidze Z., Morris J. G. Jr., Bacteriophage therapy, *Antimicrobial Agents and Chemotherapy*, 45 (2001) 649–659.
4. Gabashvili I. S., Khan S. A., Hayes S. J., Serwer P., Polymorphism of bacteriophage T7, *Journal of Molecular Biology*, 273 (1997) 658–667.
5. Smith H. L., Trevino R. T., Bacteriophage infection dynamics: multiple host binding sites, *Mathematical Modelling of Natural Phenomena*, 4 (2009) 109–134.
6. Carlton R. M., Phage therapy: past history and future prospects, *Archivum Immunologiae et Therapiae Experimentalis*, 47 (1999) 267–274.
7. Smith H. W., Huggins M. B., Successful treatment of experimental *Escherichia coli* infections in mice using phage: its general superiority over antibiotics, *Journal of General Microbiology*, 128 (1982) 307–318.
8. Smith H. W., Huggins M. B., Effectiveness of phages in treating experimental Escherichia coli diarrhoea in calves, piglets and lambs, *Journal of General Microbiology*, 129 (1983) 2659–2675.

9. Soothill J. S., Treatment of experimental infections of mice with bacteriophages. *Journal of Medical Microbiology*, 37 (1992) 258–261.
10. Mortazavian A., Razavi S. H., Ehsani M. R., Sohrabvandi S., Principles and methods of microencapsulation of probiotic microorganisms, *Iranian Journal of Biotechnology*, 5 (2007) 1–18.
11. Wall S. K., Zhang J., Rostagno M. H., Ebner P. D., Phage therapy to reduce preprocessing Salmonella infections in market-weight swine, *Applied and Environmental Microbiology*, 76 (2010) 48–53.
12. Holah J. T., Taylor J. H., Dawson D. J., Hall K. E., Biocide use in the food industry and the disinfectant resistance of persistent strains of Listeria monocytogenes and Escherichia coli, *Journal of Applied Microbiology*, 92(Suppl) (2002) 111S–120S.
13.. Lewis K., Multidrug tolerance of biofilms and persister cells, *Current Topics in Microbiology and Immunology*, 322 (2008) 107–131.
14. Scallan E., Hoekstra R. M., Angulo F. J., Tauxe R. V., Widdowson M. A., Roy S. L., *et al.*, Foodborne illness acquired in the United States—major pathogens, *Emerging Infectious Diseases*, 17 (2011) 7–15.
15. Buzby J. C., Hodan, F. Wells, Hyman J., The estimated amount, value, and calories of postharvest food losses at the retail and consumer levels in the United States, *Economic Information Bulletin*, EIB-121 (2014) 39.
16. Sillankorva S. M., Oliveira H., Azeredo J., Bacteriophages and their role in food safety, *International Journal of Microbiology*, 2012 (2012) 863945.
17. Greer G. G., Bacteriophage control of foodborne bacteria, *Journal of Food Protection*, 68 (2005) 1102–1111.
18. Rosenquist H., Nielsen N. L., Sommer H. M., Norrung B., Christensen B. B., Quantitative risk assessment of human campylobacteriosis associated with thermophilic Campylobacter species in chickens, *International Journal of Food Microbiology*, 83 (2003) 87–103.
19. Bigwood T., Hudson J. A., Billington C., Carey-Smith G. V., Heinemann J. A., Phage inactivation of foodborne pathogens on cooked and raw meat, *Food Microbiology*, 25 (2008) 400–406.
20. Carvalho C. M., Gannon B. W., Halfhide D. E., Santos S. B., Hayes C. M., Roe J. M., *et al.*, The in vivo efficacy of two administration routes of a phage cocktail to reduce numbers of Campylobacter coli and Campylobacter jejuni in chickens, *BMC Microbiology*, 10 (2010) 232.
21. Scott A. E., Timms A. R., Connerton P. L., Loc Carrillo C., Adzfa Radzum K., Connerton I. F., Genome dynamics of Campylobacter jejuni in

response to bacteriophage predation, *PLoS Pathogens,* 3 (2007) e119.

22. Seeliger H. P. R., Jones D., Genus Listeria Pirie 1940, In *Bergey's Manual of Systematic Bacteriology*, Vol. 2, Sneath P. H. A., Mair N. S., Sharpe M. E., Holt J. G. (Eds.), Williams & Wilkins, Baltimore (1986) pp. 1235–1245.
23. Kutateladze M., Adamia R., Bacteriophages as potential new therapeutics to replace or supplement antibiotics, *Trends in Biotechnology*, 28 (2010) 591–595.
24. CDC., Overview of Healthcare-associated MRSA. (Prevention, C. f. D. C. a. ed., U.S. Department of Health and Human Services, (2008).
25. Goodridge L. D., Designing phage therapeutics, *Current Pharmaceutical Biotechnology*, 11 (2010) 15–27.
26. Sahota J. S., Smith C. M., Radhakrishnan P., Winstanley C., Goderdzishvili M., Chanishvili N., *et al.*, Bacteriophage delivery by nebulization and efficacy against phenotypically diverse *Pseudomonas aeruginosa* from cystic fibrosis patients, *Journal of Aerosol Medicine and Pulmonary Drug Delivery*, 28 (2015) 353–360.
27. Wright A., Hawkins C. H., Anggard E. E., Harper D. R., A controlled clinical trial of a therapeutic bacteriophage preparation in chronic otitis due to antibiotic-resistant *Pseudomonas aeruginosa*; a preliminary report of efficacy, *Clinical Otolaryngology*, 34 (2009) 349–357.
28. Chanishvili N., Phage therapy—history from Twort and d'Herelle through Soviet experience to current approaches, *Advances in Virus Research*, 83 (2012) 3–40.
29. Meladze G. D., Mebuke M. G., Chkhetiia N., Kiknadze N., Koguashvili G. G., Efficacy of staphylococcal bacteriophage in the treatment of purulent lung and pleural diseases, *Grudnaia khirurgiia*, 1 (1982) 53–56.
30. Miliutina L. N., Vorotyntseva N. V., Current strategy and tactics of etiotropic therapy of acute intestinal infections in children, *Antibiot Khimioter*, 38 (1993) 46–53.
31. Perepanova T. S., Darbeeva O. S., Kotliarova G. A., Kondrat'eva E. M., Maiskaia L. M., Malysheva V. F., *et al.*, The efficacy of bacteriophage preparations in treating inflammatory urologic diseases, *Urologiia i Nefrologiia*, (1995) 14–17.

32. Kochetkova V. A., Mamontov A. S., Moskovtseva R. L., Erastova E. I., Trofimov E. I., Popov M. I., *et al.*, Phagotherapy of postoperative suppurative-inflammatory complications in patients with neoplasms, *Sovetskaia Meditsina*, (1989) 23–26

33. Sakandelidze V. M., Meipariani A. N., Use of combined phages in suppurative-inflammatory diseases, *Zhurnal Mikrobiologii, Epidemiologii, I Immunobiologii*, 51 (1974) 135–136.

34. Prasad R., Multidrug and extensively drug-resistant TB (M/XDR-TB): problems and solutions, *The Indian Journal of Tuberculosis*, 57 (2010) 180–191.

35. Matinkhoo S., Lynch K. H., Dennis J. J., Finlay W. H., Vehring R., Spray-dried respirable powders containing bacteriophages for the treatment of pulmonary infections, *Journal of Pharmaceutical Sciences*, 100 (2011) 5197–5205.

36. Gbassi G. K., Vandamme T., Probiotic encapsulation technology: from microencapsulation to release into the gut, *Pharmaceutics*, 4 (2012) 149–163.

37. Kuhn J. C., Zussman E., Encapsulation of bacteria and viruses in electrospun fibers. US9096845B2 , Aug. 4 (2015).

38. Anal A. K., Singh H., Recent advances in microencapsulation of probiotics for industrial applications and targeted delivery, *Trends in Food Science & Technology*, 18 (2007) 240–251.

39. Huguet M. L., Neufeld R. J., Dellacherie E., Calcium-alginate beads coated with polycationic polymers: Comparison of chitosan and DEAE-dextran, *Process Biochemistry*, 31 (1996) 347–353.

40. Gbassi G. K., Vandamme T., Ennahar S., Marchioni E., Microencapsulation of *Lactobacillus plantarum* spp in an alginate matrix coated with whey proteins, *International Journal of Food Microbiology*, 129 (2009) 103–105.

41. Dini C., Islan G. A., de Urraza P. J., Castro G. R., Novel biopolymer matrices for microencapsulation of phages: enhanced protection against acidity and protease activity, *Macromolecular Bioscience*, 12 (2012) 1200–1208.

42. Pelton R., Bioactive paper provides a low-cost platform for diagnostics, *Trends in Analytical Chemistry*, 28 (2009) 925–942.

43. Sawayama S., Rao K. K., Hall D. O., Immobilization of *Rhodobacter capsulatus* on cellulose beads and water treatment using a photobioreactor, *Journal of Fermentation and Bioengineering*, 86 (1998) 517–520.

44. Zhu M., Du M., Cordray J., Ahn D. U., Control of *Listeria monocytogenes* contamination in ready-to-eat meat products, *Comprehensive Reviews in Food Science and Food Safety*, 4 (2005) 34–42.
45. Murthy K., Engelhardt R., Encapsulated bacteriophage formulation. WO2006047871A1 , May 11 (2006).
46. Choińska-Pulit A., Mituła P., Śliwka P., Łaba W., Skaradzińska A., Bacteriophage encapsulation: Trends and potential applications, *Trends in Food Science & Technology*, 45 (2015) 212–221.
47. Jyothi N. V., Prasanna P. M., Sakarkar S. N., Prabha K. S., Ramaiah P. S., Srawan G. Y., Microencapsulation techniques, factors influencing encapsulation efficiency, *Journal of Microencapsulation*, 27 (2010) 187–197.
48. Balcao V. M., Azevedo A. F., Castro L. M., Costa C. I., Santos S., Matos C. M., *et al.*, Design of a lipid nanovesicle system encapsulating bacteriophages integrated in a multiple emulsion formulation: a proof-of-concept *NSTI-Nanotechnology*, 3 (2010) 459–462.
49. Davis S. S., Walker I. M., Multiple emulsions as targetable delivery systems, *Methods in Enzymology*, 149 (1987) 51–64.
50. Tang Z., Huang X., Baxi S., Chambers J. R., Sabour P. M., Wang Q., Whey protein improves survival and release characteristics of bacteriophage Felix O1 encapsulated in alginate microspheres, *Food Research International*, 52 (2013) 460–466.
51. Ma Y., Pacan J. C., Wang Q., Sabour P. M., Huang X., Xu Y., Enhanced alginate microspheres as means of oral delivery of bacteriophage for reducing *Staphylococcus aureus* intestinal carriage, *Food Hydrocolloids*, 26 (2012) 434–440.
52. Kailasapathy K., Microencapsulation of probiotic bacteria: technology and potential applications, *Current Issues in Intestinal Microbiology*, 3 (2002) 39–48.
53. Cabral-Marques H., Almeida R., Optimisation of spray-drying process variables for dry powder inhalation (DPI) formulations of corticosteroid/cyclodextrin inclusion complexes, *European Journal of Pharmaceutics and Biopharmaceutics*, 73 (2009) 121–129.
54. Müller-Merbach M., Rauscher T., Hinrichs J., Inactivation of bacteriophages by thermal and high-pressure treatment, *International Dairy Journal*, 15 (2005) 777–784.
55. Vandenheuvel D., Singh A., Vandersteegen K., Klumpp J., Lavigne R., Van den Mooter G., Feasibility of spray drying bacteriophages into respirable powders to combat pulmonary bacterial infections,

European Journal of Pharmaceutics and Biopharmaceutics, 84 (2013) 578–582.

56. Pasternack G., Sulakvelidze A., Novel *E. coli* 0157:H7 bacteriophage and uses thereof. US7625741B2, Dec. 1 (2009).
57. Electrospinning. https://en.wikipedia.org/wiki/Electrospinning#/media/File:Electrospinning_Diagram.jpg
58. Korehei R., Kadla J. F., Encapsulation of T4 bacteriophage in electrospun poly(ethylene oxide)/cellulose diacetate fibers, *Carbohydrate Polymers*, 100 (2014) 150–157.
59. Salalha W., Kuhn J., Dror Y., Zussman E., Encapsulation of bacteria and viruses in electrospun nanofibres, *Nanotechnology*, 17 (2006) 4675.
60. Korehei R., Kadla J., Incorporation of T4 bacteriophage in electrospun fibres, *Journal of Applied Microbiology*, 114 (2013) 1425–1434.
61. Lee, S.-W., Belcher A. M., Virus-Based Fabrication of Micro- and Nanofibers Using Electrospinning. *Nano Letters*, 4 (2004) 387–390.
62. Verbeken G., Huys I., Pirnay J. P., Jennes S., Chanishvili N., Scheres J., Gorski A., De Vos D., Ceulemans C., Taking bacteriophage therapy seriously: a moral argument. *BioMed Research International*, 2014 (2014) 621316.

Index